Practical Electronic Design for Experimenters

Practical Electronic Design for Experimenters

Louis E. Frenzel, Jr.

New York Chicago San Francisco Athens London Madrid
Mexico City Milan New Delhi Singapore Sydney Toronto

Library of Congress Cataloging-in-Publication Data

Names: Frenzel, Louis E., Jr., 1938- author.
Title: Practical electronic design for experimenters / Louis E. Frenzel, Jr.
Description: New York : McGraw-Hill Education, [2020] | Includes index. | Summary: "This book covers the basics of electronics but the focus is on the overall design of the product. The author argues that design is really a product-level consideration and not something that happens at the circuit level. Therefore, the book provides the reader with information about ready-made circuits, off-the-shelf components, and other readily available parts. Then the focus shifts to the fundamentals of design, resources, power supplies, digital designs, embedded controllers, wireless applications, prototyping, as well as testing, troubleshooting, and debugging. The applications of strong design training can be fun and far-reaching, from autonomous robots to smart clothing"--Provided by publisher.
Identifiers: LCCN 2019047087 | ISBN 9781260456158 (paperback) | ISBN 9781260456165 (ebook)
Subjects: LCSH: Electronic circuit design--Amateurs' manuals. | Electronic apparatus and appliances--Design and construction--Amateurs' manuals.
Classification: LCC TK7867 .F7533 2020 | DDC 621.3815--dc23
LC record available at https://lccn.loc.gov/2019047087

McGraw-Hill Education books are available at special quantity discounts to use as premiums and sales promotions, or for use in corporate training programs. To contact a representative, please visit the Contact Us page at www.mhprofessional.com.

Practical Electronic Design for Experimenters

1 2 3 4 5 6 7 8 9 LHS 24 23 22 21 20

ISBN 978-1-260-45615-8
MHID 1-260-45615-3

The pages within this book were printed on acid-free paper.

Sponsoring Editor
Lara Zoble

Editorial Supervisor
Donna M. Martone

Acquisitions Coordinator
Elizabeth Houde

Project Manager
Sarika Gupta,
Cenveo® Publisher Services

Copy Editor
Kevin Campbell

Proofreader
Cenveo Publisher Services

Indexer
ARC Indexing

Production Supervisor
Pamela A. Pelton

Composition
Cenveo Publisher Services

Art Director, Cover
Jeff Weeks

Contents

Introduction

This book is for you experimenters and makers who want to design your own electronic circuits and equipment. There are not too many books like this. Most books tell you how electronic devices work and provide some projects to learn from. But now you have in your hands a book that is actually going to show you how to design your own electronic circuits and equipment. It is written in a way so that any of you who have a background in electronic fundamentals can create a circuit or device to do something you want to do. You don't have to be an engineer to design things.

With the knowledge and procedures in this book, you can create products for resale, implement scientific projects that need special equipment, or produce circuits for your own DIY (do-it-yourself) idea. The book relies upon the availability of popular integrated circuits and the many finished modules and subassemblies. Using existing products and legacy circuits eliminates most of the difficult circuit design. In many cases, you can piece together existing circuits and modules to make a device with minimal electronic design. However, some basic circuit design is usually necessary and hopefully, this book will help with that.

The design approach in this book focuses on making a working device using standard parts and circuits. The recommendations in each chapter suggest that you use chips and circuits that have been used before. Why reinvent the wheel? The result is lesser design time and greater success at lower cost. Your design may not always be "leading-edge" but it will do the job.

You Are the Target Audience

When writing this book, I had the following people in mind:

- Hobbyists, experimenters, DIYers, and makers who want to create their own equipment.
- New engineers—graduates who are well versed in math, physics, and electronic fundamentals but have not yet learned to apply that knowledge to creating products.
- Technicians who are knowledgeable in electronics but have not designed.
- Scientists like physicists, chemists, geologists, and other users of electronic equipment who often need custom noncommercial equipment but can learn to design their own.
- Students who can supplement their theoretical studies with practical design knowledge. Students in an introductory college design course or taking a design capstone course or culminating design project course where the theory is applied to a specific circuit or device.

It is likely that you are part of one those groups.

Book Rationale

Where does one generally go to learn electronic design? At colleges or universities offering a BSEE degree, of course. These institutions teach all the science, math, theory, components, and circuits. Some courses actually teach related design. Much of the design taught in college is how to design integrated circuits using special software created for that purpose. Yet, many colleges and universities rarely address modern practical product design. There is a need to learn how to translate theory into practice by creating useful end products. This book addresses that void. It is unique and serves a need. Not everyone can go to college, but that does not mean you cannot learn design. In fact, in the real world BSEE graduates go out to jobs and that is where they really learn the design process. On the job training (OJT) is where you get actual design experience. Now you can get a taste of that with this book.

Functions of This Book

- Illustrates a practical, almost "cook book" approach that you can use to create new devices or design equipment to solve a unique problem not met by available existing products.
- Shows you how to create your own devices from scratch.
- Introduces basic systems design processes to define the product.
- Shows how standard off the shelf (OTS) products can be used to create the desired end product.
- Emphasizes that large segment of electronic design today is actually at the product level rather than the component and circuit level.
- Illustrates that you do not necessarily need a college degree to design some types of electronic products.

Prerequisites

The book assumes that you have some minimal level of knowledge or experience in electronics. While formal college-level electronic education is preferred, any training or instruction in the fundamentals from the military service, company classes, or by personal self-learning will probably be adequate. At a minimum you should be familiar with these topics:

- Ohm's law
- Kirchhoff's laws
- Resistors and capacitors in series and parallel
- How transistors operate (BJT and MOSFET)
- Basic digital logic

This book reviews some of this material and uses the "teaching moment" that explains selected necessary basic theory along with the design processes.

Math

You should know up front that this book does use some mathematics. After all, design is the process of calculating electronic values to implement a specific circuit. It is a necessary part of design. That may be bad news for some of you who hate math. Get over it. The good news is that most of the math is pretty simple. For example, many calculations are just the process of plugging numerical values into a given formula and grinding out the math. Other math is basically just algebra. You may have to rearrange a formula to solve for a different variable but it rarely gets more complex than that. Get yourself a good scientific calculator, use the calculator that is in your smartphone, or tap the calculator in the Windows operating system. It's not that hard.

Book Features

- A first design book for the inexperienced maker and experimenter.
- Chapters covering the most common types of circuits and equipment.
- Provides the knowledge to immediately create new devices.
- Describes well-known circuits and short cuts that always work.
- Related theory, basic principles, or background covered briefly as needed.
- Provides design projects that will help you apply and test your design ability.
- Recommends standard available parts.
- Includes design examples.
- Provides a collection of popular circuits that always work, which you can use as building blocks for new designs.
- Math level: Mainly algebra, some elementary trigonometry, and basic logarithms. No calculus.

Design Projects

Included at the end of each chapter are several Design Projects. These are provided to help you apply the design procedures. The project may be just a demonstration or a major design assignment. Be sure to do these as they provide the practice you need to become competent in design. Simulate, build, and test your design. Typical solutions are given in Appendix B to further illustrate proper techniques as well as the kinds of decisions that you may need to make. Be sure to read all of the Design Projects and their solutions in Appendix B. These solutions give you a significant amount of additional design tips, approaches, and processes.

What This Book Does Not Cover

- Leading-edge circuit design. Once you learn and practice the basic design procedures given in this book, you can then move on to more sophisticated and complex designs.
- Integrated circuit design. This is usually done with expensive electronic design automation (EDA) software. It also does not cover the current semiconductor processes and chip-making techniques.
- Mechanical design and packaging. This includes printed circuit board (PCB) design and manufacturing. Electronic packaging is a whole different field of expertise where you need to know about chemicals, metals, plastics, and other related technologies.
- PCB design. Another mechanical function that is mostly handled by software these days. A world of its own.
- Programming. Software and programming are mentioned in the chapter on microcontrollers, but no programming languages or techniques are taught. Hopefully you know some programming but if not, don't worry. The amount of software coverage included here is minimal.

Book Content

This is basically a hardware book. Its approach encourages hands-on experimentation by building things. The book is also a bit "retro." The book includes many older circuits and techniques. Why? Mainly they are still available, affordable, proven to work, and easy to design with. Your goal is to design some useful device so why try to devise a complex high-tech circuitry device when cheap simple circuits and processes work?

The main focus is analog or linear circuits but an extensive digital chapter is included. Introductory chapters on PLDs/FPGAs and microcontroller design are provided as a starting point for your future work with these subjects.

What's in It for You?

You will be spending time and money working with this book. Why should you do that? Here are a few benefits to consider.

- You get to satisfy your interest in electronics by working with hands-on projects.
- You will develop your natural human desire to create things and solve problems.
- You will be able to design and build practical and useful electronic devices.
- You will learn more electronics. Design is a great teacher. It makes the theory come alive in the circuit or device you are designing. And you will never find a better way to really understand electronics until you have to design actual circuits and equipment.
- You may even improve your knowledge and skills to the extent that they could be useful in your job if you work in the electronics industry.
- Have fun with your hobby.

Three Pieces of Advice

Failure Is an Option

First, do not be afraid to fail. Experiment. If in doubt, try it out. If that does not work, try something else. Failure is a common occurrence in design. Examples are a circuit that does not work at all and one that works but does not meet the specifications. Each failure is just a learning process. Be patient. You will eventually figure out something that works. Failure is just part of the overall learning experience that design provides.

Enjoy the Process

I can tell you right now that there is probably nothing more satisfying than to design something, build it and see it works successfully. There is true delight in that accomplishment. Have the fun and feel the reward of achievement.

Invest in Yourself

Finally, let me say this to you experimenters and makers. Plan to invest in some test equipment, prototyping hardware, software, and components. Without good test and measurement capability, you cannot actually evaluate what you are designing. Full prototype construction is recommended despite the availability of excellent simulation software. You never really know for sure how a product performs until you actually build and test one.

How to Use This Book

Start by rereading this Introduction again. Definitely read Chapter 1 first to get the big picture about electronic design. Then go on and read Chapters 2 through 4 and do what they say. Specifically, put together a basic book library (Appendix A), so you will have some ready references if you need them. Next, set up your workbench. That includes acquiring the necessary tools, test equipment, and breadboarding equipment. Track down a circuit simulator software like Multisim and get it installed on your PC.

As a first project, I suggest you next go to Chapter 5. It has multiple circuits that you will use again and again in other designs. Complete the Design Projects given. Simulate them and/or breadboard them and run the physical tests. Get some experience in breadboarding and testing. You will come to appreciate how time consuming all this is.

Now you can go on to the chapters on specific designs. If you do not have a good laboratory power supply, you may want to go to Chapter 6 next and build your own power supply.

You are on your own after that. You can go to any other chapter as it fits your needs.

Again, I urge you to build and test the Design Projects given in each chapter. It will give you practice in breadboarding and/or using the simulation software. Some possible design solutions are given in Appendix B. Finally, as you go through the book, you will discover a product or circuit that interests you. Start the design and follow through.

Now, go design something.

CHAPTER 1

Introduction to Electronic Design

Product design is the process of creating an electronic circuit, device, or piece of equipment. It may be a new commercial product for sale to generate new revenue and profit. Or it could be a highly specialized device needed as part of a scientific research effort. Then again it may be the brainchild of a hobbyist or experimenter for entertainment or learning purposes. The design process varies widely from engineer to engineer or from company to company. Yet despite the differences, the processes have common elements or essential steps. This chapter attempts to identify these common and necessary steps and to generate a cookbook design method that you can use to create a product.

Defining Design

The formal definition of design is to conceive and plan from your own mind some idea, process, or object—to create something using one's intelligence and experience by defining look, function, and operation. The result is often original and may be patentable. While that definition applies to this book, there is another definition that is more applicable. That is, design is developing an electronic product, circuit, or device for some useful purpose. That design can include existing circuits, components, and techniques. Design is combining standard, well-known circuits, parts, and methods to solve a problem or produce some useful new device. Using proven circuits, parts, and methods will improve the chances of success, reduce costs, and save design time.

Design Perspective

Designing electronic circuits has evolved over the years from designing circuits with discrete components to designing in two other major ways. The most sophisticated and original circuit design today is done by engineers in the semiconductor companies. These engineers use computer-aided design software to facilitate both the circuit design and the manufacturing of semiconductors. This is where the real innovative designs come from. You still get to design with individual transistors and capacitors, but at the software level. This is probably the highest-level design because it requires significant theoretical knowledge, experience, and natural creativity. Engineers who do this probably have advanced degrees as well as plenty of experience.

The second form of electronic design is what some call the connect-the-pins approach. What that means is that the engineer designs products by selecting appropriate integrated circuits (ICs) and then connecting their pins to produce the final product. Some say this is not a very creative process because it can be done without a whole lot of theory and experience. This is probably the most common form of electronic design where engineers are developing products, not chips. If you know how the chips work, you can probably do this type of design. The most challenging parts of the design are printed circuit board (PCB) design and writing the software or firmware for the ubiquitous microcontroller that is usually part of most designs.

This book covers design much like the connect-the-pins approach. It offers an almost cookbook-like method for conceiving of a product and making it. With literally thousands of different types of ICs out there, you need to be creative to put them together in one of the almost infinite number of ways possible to accomplish your design goal. Best of all, you really do not have to be a graduate engineer to do it. But you do need to meet the prerequisites discussed elsewhere.

Two key points to consider are:

- Know your chips. Get familiar with the available ICs, dig out the details of those of interest, and get relevant data sheets, app notes, etc. Keep track of new chip introductions by monitoring the semiconductor company Web sites and keeping up with industry magazines and Web sites.
- Become software literate. Learn to code in a popular language, and become proficient in writing programs for micros. The future is firmware.

The remainder of this book will take you down that path.

Get a Design Notebook

Before listing the design steps, you should acquire a notebook that you will use to document your design. It will contain statements of purpose, goals, features, benefits, specifications, test results, identified problems, and other defining data. The notebook will also be used to contain your calculations, draw your block diagrams and schematics, and record test and measurement data. The design notebook can be anything you are comfortable with. A standard-size spiral bound school notebook is a good choice. They are available in most big box stores like Walmart and Target, office supply stores, grocery stores, and pharmacies. Special engineering notebooks containing grid paper are great and useful but also expensive and not really necessary.

Do not skip this first step. You must keep notes and record details so that you know what works and what does not. You want to maintain all facts and figures, schematics, calculations, test results, and debugging notes in one place so you can reference them later if needed. And be sure to get into the habit of putting the date on each page.

Many companies absolutely require engineers to maintain a design notebook to document the progress and retain the experience and knowledge they acquire during the design. It also documents the activities in case the outcome is a patentable circuit, process, or product. Writing everything down will take some getting used to, and it may aggravate you at first. Eventually you will discover how useful the notebook is since we all tend to forget. You must document everything. This is especially true in writing software code. Chances are your design will include a microcontroller for which you will write some programs. Documenting this process is critical. If you or someone else needs to revise or fix the software, you will appreciate any explanations or other details you find there. Documenting can be aggravating, but get over it—before long you will grow to appreciate the record-keeping process.

Get a Calculator

When you design, you will be making calculations. Most of the calculations are simple formulas to solve or at worst, some basic algebra. You may need to rearrange a formula to solve for a different variable, for example. For these calculations you need a scientific calculator. The calculator should include scientific notation, trigonometry functions, logarithms, and other engineering calculation functions. Individual calculators are available, but you do not need anything fancy. Programmable calculators are nice but are not needed for this book. You can also use the calculator built into your Windows operating system or your smartphone. I still use an old Texas Instruments calculator I have had for decades. Use whatever works best for you.

A Standard Design Approach

The design approach introduced here is covered in these basic stages:

- Definition
- Detail design
- Simulation (optional)
- Prototype
- Testing
- Packaging
- Use

Definition Stage

Figure 1.1 shows a flowchart of the key steps in the definition stage of the design.

1. Name the product. Give it a name that tells others what it is or does.
2. Describe the product by writing out a short paragraph. State why it is needed, by whom, and what it is supposed to do. Be as detailed as possible, but keep it short if you can. This description is mainly for you, but it could be used by your supervisor, a marketing person, or a customer.
3. List the main features of the product. What will it do? What are some important characteristics that allow it to solve a problem or perform a function not available elsewhere? What do the sales and marketing people want? If this is a product for sale,

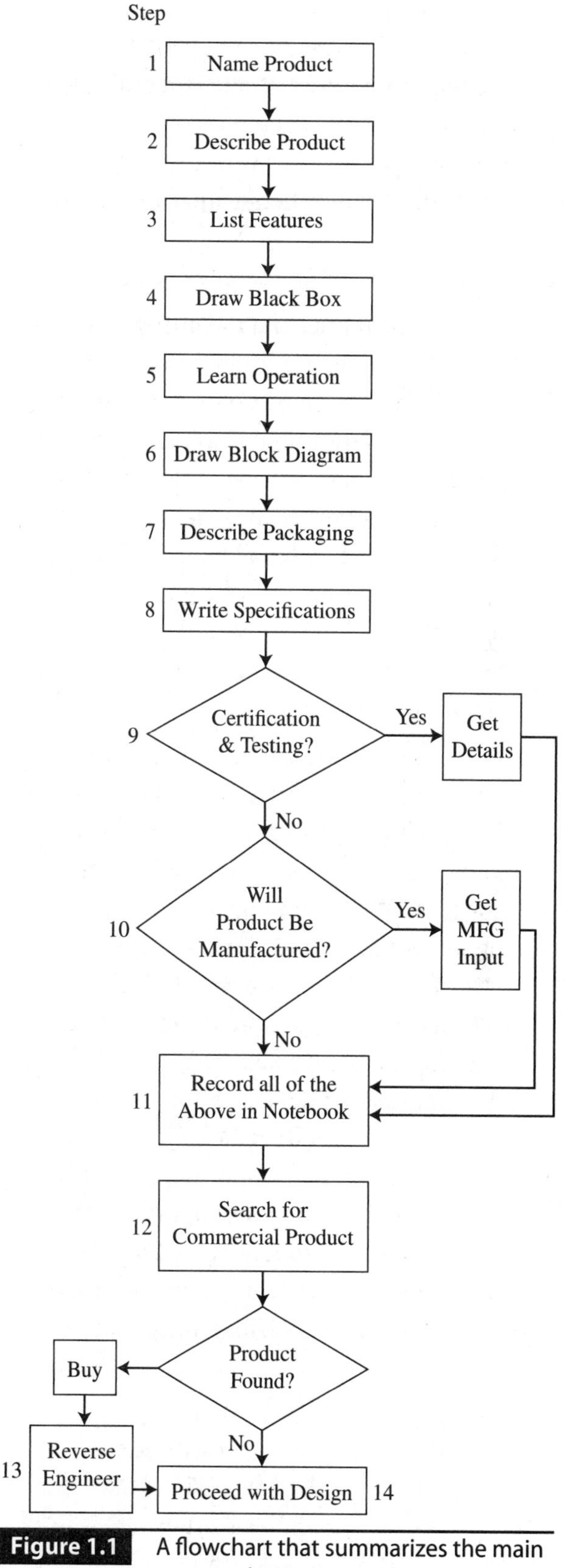

Figure 1.1 A flowchart that summarizes the main steps in designing an electronic circuit or product.

what features should be included to make the product attractive to buyers? Who are the competitors, if any? Identify their products and do a comparison. If necessary, negotiate the features with the sales and marketing team. Then finalize the features list.

4. Treating the product as a "black box," identify the inputs and the outputs. See Fig. 1.2. An electronic circuit or product can be viewed as a box of signal processing and manipulation that responds to one or more inputs and then generates one or more outputs. Describe each signal in as much detail as possible. Also identify where the inputs come from and where the outputs go. Be sure to include the ac and/or dc power requirements. Again, be as detailed as you can at this early stage.

5. Learn how the device or circuit works. This is the time for detailed research. You can't design something unless you know how it works. You must learn the concepts, theory, and operational details of something before you can design it. Use any available textbooks to identify circuits and configurations. At this point you should also do an internet search to determine some details about your project. Search on the product name you assigned, and use alternative descriptions to gather as much information as you can. Search on circuit names, component part numbers, or whatever other detail you may have or need. Look for block diagrams, schematic diagrams, how-it-works descriptions, and the like. If competitive products are available, acquire them if the budget allows. If you want to design something but don't know where to start, it is probably because you do not know how it functions. As an example, at one time I wanted to design a metal detector. How hard could it be? But I was stumped. I did not even know what circuits it contained or how they work together to locate metal objects. So go back and reread this step.

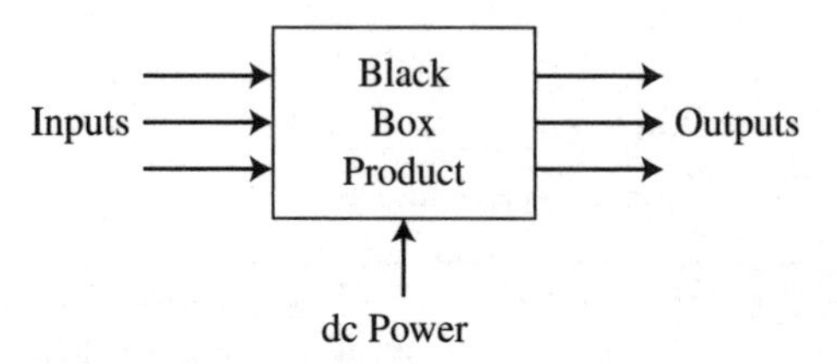

Figure 1.2 The "black box" concept that applies to most electronic circuits or products.

6. Draw a block diagram of the device. Knowing the theory of operation and the inputs and outputs, you can probably generate a first attempt to see what circuits and modules or other subassemblies you may need. State the purpose of each stage. Do the best you can at this point, and keep in mind it will change as you learn more. For you real beginners, Table 1.1 summarizes all the most common circuits engineers use.

7. Describe the most desirable physical packaging. What will the end product look like mechanically? Will it use a printed circuit board (PCB)? This book does not delve into the physical packaging of a design, but you still need to determine it.

8. Write out a set of specifications. Knowing the inputs, outputs, and power requirements, you should be able to list key specifications. Some of these are frequency range, input and output voltage levels, specific signal shapes and conditions, estimated power supply voltages, data rate, interface requirements, and current limits. You also need to add any special environmental conditions, such as temperature range, humidity, vibration/shock, and electromagnetic interference (EMI) considerations. Add physical specifications like desired size, weight, and

Table 1.1 The most commonly used electronic circuit building blocks and what each does

Circuit	Function	Available IC or module? Other
Amplifier	Takes a small input voltage or power and boosts it up by some gain factor to a larger output.	Yes, usually. All types available.
Analog-to-digital converter (ADC)	Samples an analog signal and produces a digital output.	Yes, many.
Attenuator	A circuit that reduces the amplitude of voltage or power by some loss factor usually expressed in decibels (dB).	There may be. A resistive voltage divider or network is more common.
Buffer	A circuit, usually an amplifier, that isolates one circuit from another to prevent or minimize loading that affects output voltage, power, or frequency.	Maybe. Could be analog or digital.
Clock	An oscillator that generates accurate rectangular pulses used for timing and operating an MCU or other digital circuits.	Yes.
Comparator	Takes two input signals, one usually a fixed reference voltage and the other a varying amplitude signal, and provides an output signal that indicates which input is equal to, less than, or more than the other.	Yes.
Counter	A circuit that keeps track of the number of binary input pulses that are applied to its input. The output is in binary format.	Yes, multiple.
Driver	A circuit that operates some other device like a motor, relay, LED, or servo.	Yes, usually. Could be just a transistor or an IC.
Decoder	A digital circuit that looks at multiple lines of binary signals and detects one or more separate output conditions, each recognizing a unique code.	Yes.
Detector	A circuit that indicates the presence of a signal. Another name for a demodulator.	Maybe.
Digital-to-analog converter (DAC)	A circuit that translates its binary input into the analog equivalent.	Yes, many.
Divider	A digital circuit that produces an output that is lower in frequency than the input. Also, an analog circuit that splits an input into two or more equal outputs.	Some. An often-designed circuit to fit the application. A digital counter makes a good frequency divider.
Filter	A circuit that allows some frequencies to pass and others to be stopped or at least greatly attenuated.	Yes. Some filters can be designed with discrete components to fit the application.
Follower	A high input impedance–low output impedance circuit that provides some isolation between amplifier stages as well as some power gain. See Buffer.	No. Usually a circuit designed for the application.
Frequency synthesizer	A signal source like an oscillator that generates a sine wave or rectangular wave.	Yes. Multiple kinds. Two major versions like phased-locked loop (PLL) and direct digital synthesis (DDS).

(Continued)

Table 1.1 The most commonly used electronic circuit building blocks and what each does (*Continued*)

Circuit	Function	Available IC or module? Other
Function generator	A signal source that usually generates sine, square, and triangular waves for testing.	Yes.
Gate	Basically, just a switch that blocks a signal until another enabling signal is applied. May also perform some digital (boolean) logic function.	Yes, both analog and digital types.
MCU	Microcontroller unit. An embedded microcontroller or single chip computer that is the basis of most other products today.	Yes, many. Something for every application.
Memory	A storage circuit for binary words or data.	Yes, Many kinds.
Mixer	Two types, linear and RF. Linear mixers combine multiple analog inputs into a composite signal, as in an audio mixer that adds multiple musical instruments and two or more microphones. The RF mixer serves as an up converter or down converter for translating signals to a higher or lower frequency, as in radio transmitters or receivers.	A few.
Modulator	A circuit that varies the amplitude, frequency, or phase of a higher frequency carrier signal for the purpose of transmitting information by wireless.	Yes, not many.
Multiplexer	A circuit with two or more inputs and a single output and a means of selecting any one of these to appear at the output.	Yes, both digital and analog versions are available.
Oscillator	A circuit that generates a signal, analog or digital, at a specific frequency or over some variable frequency range.	Yes, multiple types. RC, LC, crystal.
Rectifier	A circuit that converts ac into pulsating dc that is usually smoothed into a continuous voltage by a capacitor.	No. Usually made with discrete diodes.
Regulator	A circuit that maintains a fixed output voltage despite changes in other operational factors like input voltage or output load.	Yes, many types for all occasions.
Register	A circuit made up of flip flops that can store a binary value or manipulate it.	Yes, multiple types.
Voltage-controlled oscillator (VCO)	A signal source whose output frequency can be varied by applying a dc control voltage.	A few.
Voltage divider	A circuit made primarily with resistors that produces an output that is lower than the input.	A circuit that must usually be designed. Capacitor and inductor dividers can be made but are not common.

power consumption. Also consider ease of use, maintenance, and potential repair. The end product should be simple to operate with minimal training or instruction.

9. Consider required testing and certification. If you are designing a product for resale, you may need to meet some required set of standards mandated by law. Examples are ac-powered devices that may have to be tested by the Underwriters Laboratories (UL) or the Federal Communications Commission (FCC).

10. Will the product be manufactured? If this is a one-off product, skip this step. If the product will be made in volume, be sure to involve the manufacturing people in evaluating the design and getting their input regarding steps to make the device from initial PCB construction through final testing and packaging.
11. Record all of this information in your notebook.
12. Next, you should look to see if what you defined is already a product available for sale. Maybe you won't have to design it if you can purchase a ready-made version. Do an extensive internet search. Use different product names or descriptions to be sure you will locate something similar. If you find something similar, acquire as much information as you can, and compare its features and specs to your definition. Buy the product if you can afford it.
13. Reverse engineer the product. Take it apart, being careful not to damage anything. Do the following:
    a. Take photos along the way.
    b. Identify all of the major subassemblies and larger components, and document any wiring between these sections.
    c. Identify the power source like the ac line or batteries and the related power supply.
    d. If PCBs are involved, remove them, but record any interconnections by way of connectors or wiring.
    e. Develop the schematic diagram from the PC board. Identify how the copper traces on the PCB connect the various components. Your initial schematic diagram will be messy and crude, but you can redraw it later in a more useful way.
    f. Identify the individual components. Read the resistor color codes, read capacitor values, any read numbers or part numbers on the ICs, and transistors. Record all this on the schematic diagram.
    g. Redraw the schematic diagram and part numbers and values. NOTE: In many products, the labels on the ICs will be omitted to prevent someone from identifying the part and copying the circuit. If that is the case, maybe later you can deduce what it is.
    h. Given your copied design, consider whether you could duplicate this item. If you can, you can adopt the design for your own version. If you do not believe that you can duplicate it, just reassemble the product and use it. Then move on to another project.
14. If no commercial product turns up to buy, press on with the design.

Detail Design Stage

This is where you fill in the boxes in your block diagram with specific circuits, modules, or other units. Think of the various circuits available and how you can use them as building blocks. From your searches you should have identified the circuits or ICs you want to use. Identify specific circuits where you can. Search for specific ICs that do what you need. You should be able to determine that you need an amplifier, a filter, a digital counter, LCD display, or whatever. You may do some rough partitioning at this point as you identify different parts of the design. For example, you may have an analog signal or linear segment, a digital segment, and a power supply segment. Then, for the first time, try to draw a schematic diagram of the design.

Next you will choose components to match your circuit specifications. You will be selecting ICs, diodes, transistors, capacitors, resistors, potentiometers, transformers, and a mix of other parts. You should have catalogs on hand from the major distributors if you can get them. Otherwise, go online to the major distributor sites to select your parts. You can also do additional internet searches to find what you need and to get additional information from data sheets, application notes, and other sources.

Simulation Stage

You can also call this the verification stage. This is an optional process where you validate that your circuits will work. You can use circuit simulation software to build the circuits and product on the computer before making an actual prototype. More details are given in Chap. 3. Simulation is a good learning experience, but it does take time to learn the software and the simulation process. You could go directly to a hardware prototype for testing. But I recommend you give simulation a try.

Prototype Stage

Now you start building your prototype. Build each circuit one at a time and make each work alone. Guidelines for prototyping and breadboarding are given in Chap. 3. Once you define each circuit's function, you can begin connecting circuits together to form the final product.

Testing Stage

Testing is the final stage. This is covered in Chap. 4. You will test your device to see that it implements all of the desired features. You will also test to see that it meets the specifications you assigned earlier. You can expect to do some troubleshooting at this time to fix problems, fine-tune the design to meet specifications, or correct errors. Chapter 15 covers troubleshooting. Occasionally you will, as they say, "have to go back to the drawing board."

Packaging Stage

At his point your product is finished and it works. And by now you should have thought about how the product should be packaged. What is its housing? How are the circuits wired? No doubt a PCB is required. Packaging is a mechanical design process beyond the scope of this book. Yet it is important, especially if you plan to market the device as a commercial product.

Use Stage

Manufacture, sell, or use the product.

One final thing. As indicated earlier, this product design process does not include considerations for high-volume manufacturing. The design process for manufacturing is similar, but serious consideration is given to cost of manufacturing, ease and speed of manufacturing, special testing or alignment procedures, and parts availability.

Design examples using the process described here are given in the design chapters to come. Here is a summary.

- Chapter 5 Common Circuit Design Techniques. Basic circuits and concepts you will use in most designs.
- Chapter 6 Power Supply Design. Battery and ac to dc supplies.
- Chapter 7 Amplifier Design. Mostly op amps, but some discrete designs.
- Chapter 8 Signal Source Design. Oscillators, clocks, synthesizers.
- Chapter 9 Filter Design. RC, LC, active, and modules.
- Chapter 10 Electromechanical Design. Switches, relays, motors, servos.
- Chapter 11 Digital Design. Discrete IC logic.
- Chapter 12 Programmable Logic Devices.
- Chapter 13 Designing with Microcontrollers. Interfacing and I/O and programming.

Design Doctrine Dozen

The rules for design in this book are based on the premise that you are trying to design something that will work reliably, have a reasonable cost, and take less time to create. Your goal should be to create a product that works, solves a problem, or fills a need. Here are the guidelines for design as recommended in this book.

1. Keep it simple, stupid (KISS). Simple designs are always best. They are less complex, less expensive, and take less time to create. And they are generally more reliable.
2. Do not reinvent the wheel. Use existing circuits and designs. Borrow liberally from magazine articles, books, manufacturer's data sheets, application notes, and online sources. Why spend extra time experimenting with new approaches when there is probably already a design you can access and use? Seek out and maintain a library of standard circuits that work. I keep a file folder for different types of useful circuits (amplifiers, oscillators, logic circuits, etc.) When I come across a magazine article, internet printout, or data sheet, I file it for future use. Put together these existing designs in different combinations as needed. Modify these circuits as needed to create your design. There should be no embarrassment in using the designs of others if you can. Most basic electronic functions have been discovered and implemented. These are mostly in the public domain, and you are free to use them. Take advantage. It is OK to be creative and design some things from scratch, but just remember it takes more time, and you may need to redesign it multiple times to get what you want. While not everything has been invented yet, it is difficult to devise a design that is totally original. Most common operations have already been developed many times in a variety of forms.
3. Old designs are just fine. Old circuits and components are not bad. If it works and solves the problem, use it.
4. Cheaper is usually better.
5. It does not have to be leading edge. You will not be designing your own ICs. Use existing chips when you can.
6. Use manufacturers' reference designs. Many semiconductor manufacturers have already designed what you may need. A reference design is a predesigned device using the manufacturer's ICs. It is usually a prewired PCB with connectors and in some cases software—everything you need to get started without having to design it yourself. These evaluation boards are recommended because they save time and money.
7. Use manufacturers' design tools. Design tools are software that semiconductor manufacturers develop to help engineers design selected circuits. The software that is typically available online simply walks you through the design process and leads to a design for you. Of course, the tools will typically lead you to the company's ICs or other devices to implement it.
8. Use free or low-cost design software from the internet. Circuit simulation software is available from multiple sources. Feel free to use it to develop your design. However, always build and test a real prototype to be sure it works.
9. Use existing well-known ICs, transistors, and other components. There are multiple sources, and the cost is low. For commercial designs, some companies require that there be one or more secondary sources for ICs, transistors, and other parts. Then if one manufacturer discontinues the part, you will

still be able to get it elsewhere. Chapter 14 gives you some recommendations.

10. A microcontroller design is not always the best approach. Most products today are based upon a central embedded controller. These microcontroller units (MCUs), or micros as I refer to them here, are flexible and cheap, but they require software and programming in addition to the electronic interface design. Sometimes a simple hardware design is the fastest and cheapest solution.
11. Focus your budget on good test equipment and prototyping equipment. You cannot really design without testing and measuring equipment. So, plan to set aside a budget for a good multimeter, oscilloscope, and breadboarding gear.
12. Learn and have fun. Experiment. Screw up. Fail. Learn what works and what does not. Then eventually achieve success.

WARNING!

In designing commercial products, some circuits and methods may really be new and patentable. These circuits or methods become valuable intellectual property (IP) to your employer. Such IP may give some company competitive benefits. Or the company could license the design to generate royalty income. This is especially true of IC designs, but it could apply to some other arrangement. Just be sure to document everything in your notebook in case it comes up.

Types of Design

When designing any electronic circuit or product, you will discover that there are lots of ways to do it. My own view of this is that there are three basic design approaches. Here is a summary of each.

Textbook Design

You could also call this the traditional approach. This is the process of designing a circuit or product by using standard textbook theory and procedures. There are multiple textbooks to help you do this. The procedures are well known and generally proven. They are taught in college. This approach does not use cookbook recipes but offers the theory with examples, then tests you with end-of-chapter problems. The theory is given, but its interpretation and its implementation are left to you. This book generally uses this approach, but it is supplemented with a bit of experience that yields some step-by-step procedures that save time.

Empirical Design

This approach is design by experimentation. You can also call it the cut-and-try method. You essentially start with something you know, then observe the result. If it is not what your goal is, you experiment. You change or add something, observe the outcome, then change again if the end result does not turn out as you want. You go back and learn some more. You keep on learning, testing, experimenting until you get what you want. It sounds crazy, but it works, especially for those with some experience in the subject. After a while you get to know what works and what does not.

Intuitive Design

This is an approach that is based upon years of acquiring knowledge and experience that in turn give you the intuition to create something new. Your design is based upon your intuition without supporting facts. You go with what you know and believe to be true. Or as they say, you go with your gut.

After years of design experience, I have come to believe that a person inherits some of each approach. You start with the textbook approach, learn more as you experiment with the empirical approach, then finally with sufficient knowledge and experience you go with the intuitive approach.

Prerequisites for Design

This book will give you a basic process for designing. Along the way, it will also review some of the related electronic fundamentals. This book assumes that you already have a general working knowledge of electronics. Ideally you will have had some formal training or education in electronics or relevant experience. Self-taught is OK, too. At a bare minimum, you should know Ohm's law, how transistors work, some basic digital logic, and which end of a soldering iron gets hot.

To help in this regard, you should acquire some books to use for reference if you need to learn some fundamentals or refresh your knowledge. Appendix A is a list of books I recommend. They cover the fundamentals and provide some additional information on design. Build a library of such books. Some are expensive, but remember you can always find used ones on Amazon or other internet sources. You can never have too much information. If you are designing, you must seek out and acquire as much reference material from multiple sources as you can.

A Design Example

Here is how you might approach a project using the steps described earlier. This is the process I went through on one project.

The Definition Stage

1. Metal detector.
2. The metal detector should detect buried metal items to a depth of 10 to 20 cm. It will be used to see what treasures are buried in beaches, back yards, and other patches of ground that might provide potential targets.
3. Main features:
    a. Identify metal at a depth of 10 to 20 cm.
    b. Portable.
    c. Lightweight.
    d. Battery operated.
    e. Speaker and/or headphone output.
    f. Relatively easy to build as a first or early hobby project.
4. The black box concept is relatively simple:
    a. Input, search coil
    b. Output, speaker or headphones
5. Searching on the topic of metal detectors produces a ton of information. A detailed search using the terms "How do metal detectors work" and "metal detector circuits" produces a considerable amount of detail. Summary:
    a. There are four basic types of metal detectors: very low frequency (VLF), pulse, heterodyne, and variable tone.
    b. The first two (VLF and pulse) require some sophisticated circuitry that may be difficult to understand without extensive research and experience. These two types are also the most expensive. The third type uses a heterodyne method that mixes two signals together to get a difference frequency in the audio range. The simplest detector is the variable tone type.
    c. One type uses an audio oscillator whose tone changes if metal is near the search coil. The second type uses two oscillators, one at a fixed frequency and another whose frequency can be varied by the presence of nearby metal. The two oscillator frequencies are mixed together. The result is a frequency in the audio range that represents the difference between the two oscillator frequencies. Tone changes signal the presence of metal.
6. I was able to draw a crude block diagram of both methods. See Fig. 1.3.

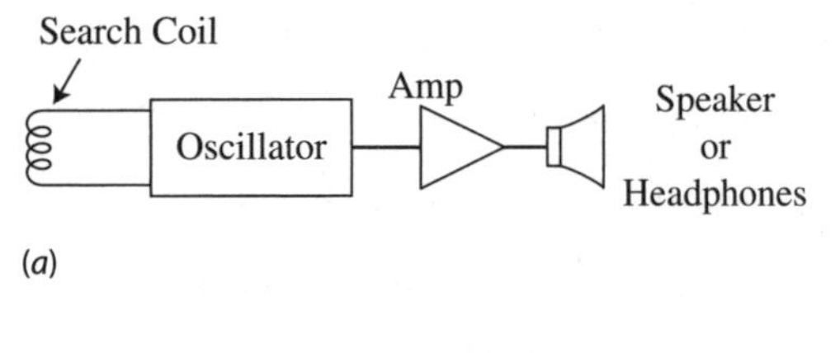

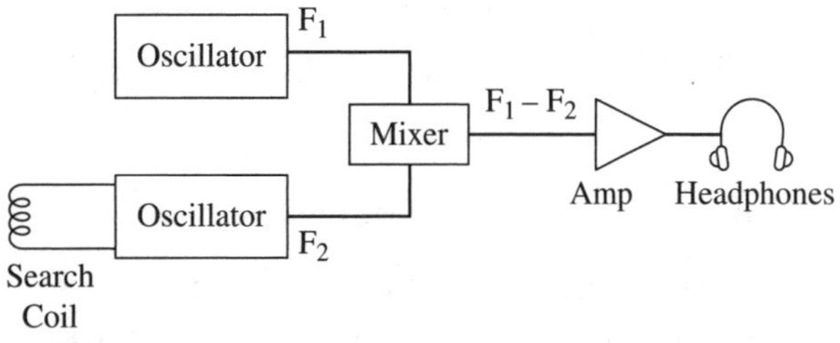

Figure 1.3 First black box diagrams for the project. The upper drawing shows the single oscillator approach (*a*) and the lower drawing showing the two oscillator design (*b*).

7. Physical packaging is not critical since you are making the product for your own use. It should have an enclosure to hold a battery and the circuitry, a search coil that can reach the ground, and a place to plug in some headphones.
8. No specifications other than those features described in step 3 were written since they are not applicable to noncommercial devices.
9. No testing or certification is required for personal products.
10. The product will not be manufactured in quantity.
11. I did write down all the details in my notebook.
12. A search for a commercial product did occur. There are many. From low-cost hobby models for less than $100 to military-grade units for finding mines costing thousands of dollars. No commercial product was purchased.
13. No reverse engineering took place.
14. The design process will go on.

The Design Stage

The design stage started with expanding and improving upon the block diagram produced earlier. The heterodyne type of detector was chosen. Two oscillators are used, so a search for oscillators started. Search data obtained earlier showed several oscillator possibilities. The difficult part was deciding on a frequency of operation. Apparently a wide range is used. I decided upon something in the 50 to 200 kHz range.

One of the oscillators uses a large coil of wire as the search head but also as the inductance in a tuned circuit for one of the oscillators. I searched for as much detail as possible here and more or less summarized the current form and size of circuits I discovered. A popular inductance value as a target is 10 mH. I suspect some experimentation will be needed here.

Next I needed a mixer that would produce the difference frequency between the two oscillators. The designers of this circuit use an XOR gate. I'm not sure how that works, but it apparently does.

I also needed an audio amplifier to operate the headphones.

A second method of detection that I call variable tone uses an oscillator whose frequency is changed by being close to a metal object. It is a simple circuit, and it seems cheap and easy to try. I suspect that the heterodyne method is much better.

Figure 1.4 shows the simple circuit. A 555 timer IC is used as an oscillator with a tuned RLC circuit setting the frequency. This frequency is in the audio range so you can hear it. With the values shown in the figure, the frequency computes to 1073 Hz.

The big design obstacle is the inductance that not only sets the frequency but also serves as the search coil. The desired inductance is 10 mH. The search coil is many turns of copper wire whose diameter is in the 4- to 10-inch-diameter range. Approximately 140 to 150 turns of wire

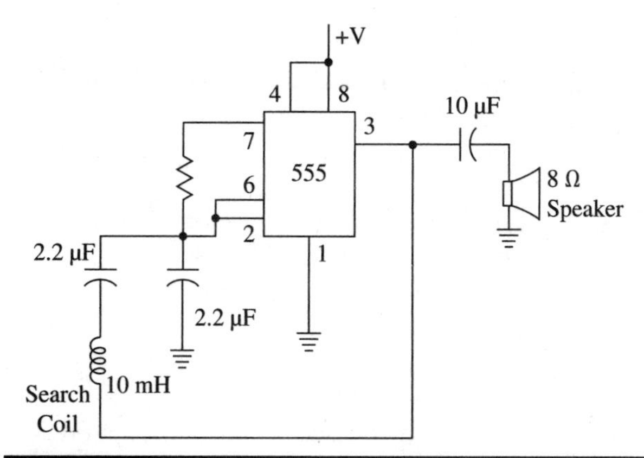

Figure 1.4 A 555 timer IC as an oscillator whose frequency is set by the 10 mH coil and the 2.2 μF capacitor.

should give you an inductance value close to what you need. You will just have to experiment with this to make it work. I ended up with 140 turns of #22 wire on a 6-inch-diameter form. I never did know the exact inductance, but it was in the correct range.

Normally you will just hear the audio tone. If the coil is passed over metal, it affects the inductance of the coil, usually increasing its value and thereby lowering the audio tone.

A design using the heterodyne method was hard to come by. I found two references illustrating this approach. One of them uses two complementary metal oxide semiconductor (CMOS) oscillators operating in the 160 kHz range. (Low Power Metal Detector, from Encyclopedia of Electronic Circuits, Rudolf F. Graf, Tab Books, 1985.)

The circuit uses a CMOS XOR IC to create the two oscillators, a mixer and an audio amplifier. Another similar design used the same CD4030 IC, which is no longer available. This design (Simple Metal Detector Circuit with CD4030, HYPERLINK "http://www.next.gr" www.next.gr) operated at a frequency of 300 kHz. It used a standard 330 μH inductor for the fixed frequency oscillator.

Again, the main design problem is getting the search coil right. An inductance value of

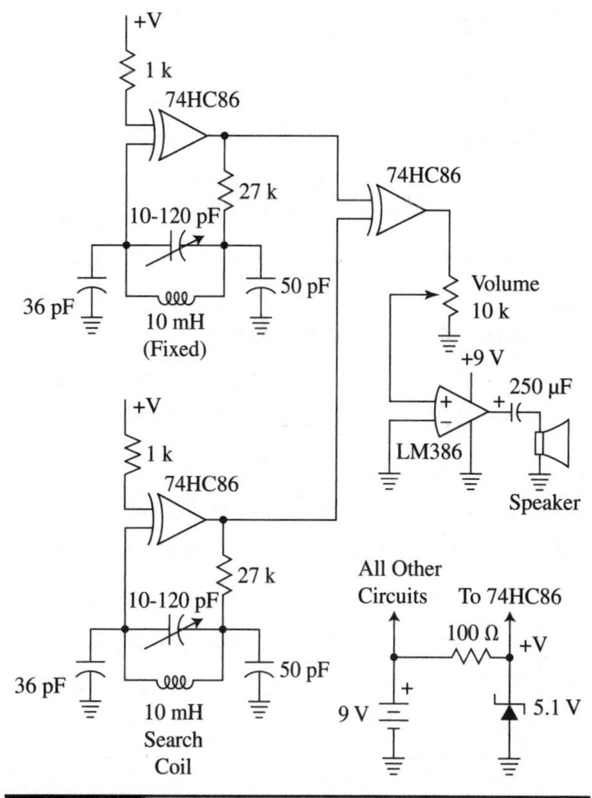

Figure 1.5 A final circuit diagram of the metal detector.

10 mH is the target, with about 140 turns of wire six inches in diameter. I also changed the audio amplifier. The final circuit is shown in Fig. 1.5. The biggest hassle is the coil, and it required a great deal of testing to zero in on an audio frequency. That frequency is not particularly critical, but it must be something you can hear.

If the 300-kHz design is used, the search coil can be much smaller. The target inductance is 330 μH. A suggested search coil size is 40 turns about 7.5 cm in diameter.

The main circuit modifications were the replacement of the no-longer-available 4030 IC with a 74HC86. I selected the design using 160 kHz oscillators. One of the oscillators called for a variable inductor that I could not find. I used a fixed 10-mH inductor and added a 100 pF variable capacitor so the frequency could be varied over a narrow range.

I used the popular LM386 amplifier for the speaker or headphones.

After the design, I skipped the simulation and went right to the prototype. I spent many hours tinkering with it, which is typical of some designs. The oscilloscope was useful in determining the frequency of the oscillators. I also tested the circuit on several pieces of metal such as a coil, a ring, and a small piece of aluminum I had. It does work; as the metal is sensed a tone change occurs to signal the find. The goal of achieving a detecting range of 10 cm or more seems remote, however. More testing is needed.

As for power, I used a 9-V battery. All the circuitry seems to be OK with that except the 74HC86 IC. I used a 5.1-V zener diode to reduce the voltage for the IC to 5 V.

I won't bore you with the packaging other than to say I left the circuit wired on breadboarding socket and put the whole thing in a box.

The metal detector did work after much tinkering. Determining the actual value of the coil inductance was an issue. I solved it using a measurement method described in Chap. 4. Inductance increases with the number of turns of wire. If measurement determines that the inductance is too low, add more turns of wire and measure again. If your inductance is too high, reduce the number of turns of wire. Repeat until you get close to your target value. Once I got close to 10 mH, the circuit worked. The device does detect metal, but it did not live up to its goal of picking up metal objects 10 to 20 cm below ground level. This is typical of many projects. More experimenting is needed.

An Alternative Beginning

If you are new to electronics, and the metal detector example seems too much of a challenge, here are several alternatives.

- Build a kit. You do not have to design anything. You will simply build the kit as detailed in the instructions. It will get you familiar with the components, and you will learn some soldering and other related skills. A small robot kit is a good choice. Some kit sources are given in Chap. 2.
- Build an existing design. Locate a circuit for a product or device you are interested in and just copy and build it. Again, there is no design involved. However, you will become comfortable with breadboarding and prototyping and some testing and troubleshooting. Look online and in magazines for circuits to copy and build. Each of the design chapters contain some useful circuits you can easily build and use in future projects.
- Initiate a so-called system design. A system design is a higher-level approach than circuit design. It involves putting together multiple existing parts and pieces, including software, to create a working system. Some examples are: radio/controlled airplane or drone, audio system, TV streaming, video doorbell, home security system, or something PC controlled. No circuit design is involved, but you will work at the block diagram level, following signal flow between items, and you will make multiple devices all work together.

The idea here is to start small and easy to ensure some success that will give you the confidence to go on. Practice on larger and larger projects until you feel ready to try some design. A small robot kit today, a Battlebot tomorrow.

CHAPTER 2

Design Resources

As you design, you will need a lot of additional information. And there is a boatload of it out there for your use. You just need to know what it is and where to get it. The internet will be the major source of it, but there are other sources like books, catalogs, and magazines. You will certainly find much of the needed information on your own with internet searches. This chapter summarizes some of these sources and shows how to get them.

Books

It is always good to keep a few good reference books on hand to use as needed to understand a component or circuit, check on related theory, or solve a problem. You may already have a library of books you use. Some recommended suggestions for books that are useful when designing are given in App. A.

If you can't find what you need, search the internet for other books. You can search on the publisher's Web sites or do a general search. Some publishers to consider follow:

- Elsevier/Newnes
- Maker Media
- McGraw-Hill
- Pearson/Prentice Hall
- Wiley

Amazon is a good place to search. They are the largest bookseller in the world these days. Barnes and Noble also has a good site to search.

Some free online ebooks on electronics are available. These are mainly addressed to college instructors and professors, and they cover the fundamentals. It is good to have some theory books on hand when you need to look something up. Try at these sites:

Open Educational Resources (OER)
OERCommons.org or Merlot.org

http://www.freebookcentre.net/Electronics/Engineering-Books-Online.html

https://ebookpdf.com/basic-electronics-books

https://easyengineering.net/power-electronics-books/

https://www.allaboutcircuits.com/textbook/

http://www2/mvcc.edu/users/faculty/jfiore/freebooks.html

Sources of Catalogs, Components, and Equipment

You will be buying parts and equipment as you design. There are multiple sources for what you need. It is extremely helpful to have a listing of parts and equipment available and where to get it. The following is a list of parts sources you can use as reference in selecting parts and equipment. Go to their Web sites to select parts. Some of these sources still offer print catalogs that are helpful to have on hand. Most distributors have discontinued print catalogs because of their size and cost to print and update. Everything is online, so you should not have a problem finding the information you need.

Generic Sources of Parts

- All Electronics
- Digi-Key Electronics
- Fry's Electronics
- Jameco Electronics
- Marlin P. Jones & Assoc. Inc.
- Mouser Electronics
- Newark
- RadioShack

Sources for Microcomputer Hardware and Parts

- Adafruit
- Digilent
- Elektor
- Elenco
- MAKE
- Pololu
- SparkFun

Some Kit Sources

- Adafruit
- All Electronics
- Ebay (huge listings of many kits)
- Electronics Hub (multiple kit suggestions)
- Elektor
- Elenco
- Jameco
- Knight Kit Electronics (list of over 60 kit sources)
- littleBits
- MAKE
- SparkFun

Magazines

Magazines will be the source of many design projects, ideas, circuits, and related product ads. Here are a few to consider.

The following magazines are still available in print form. If you are serious, subscribe to keep a steady flow of information coming your way.

- *Circuit Cellar*—Their major focus is on embedded controllers and related topics.
- *Elektor*—A magazine and Web site based in the Netherlands. A bit expensive, but very useful.
- *Make*—Covers all aspects of do-it-yourself (DIY) projects, such as mechanical, woodworking, welding, and 3-D printing. It has many good electronic projects as well.
- *Nuts & Volts*—Long-time publication for hobbyists and experimenters. Good projects.
- *QST*—A publication of the American Radio Relay League (ARRL) for amateur radio operators—You need to join ARRL to get it, but it offers excellent technical articles and electronic, test, and radio projects.
- *Electronic Design*—Written mainly for engineers, this magazine has good articles on the latest technologies and online resources.

Here are a few magazines that still publish online but not in print that are worth subscribing to.

- *ECN*
- *EDN*
- *EE Times*

Go to their Web sites and sign up.

Data Sheets

A data sheet is a publication issued by a company describing in detail the ratings, specifications, and features of a specific transistor, integrated circuit, capacitor, or other device. You will need this information to use the device correctly. For example, if you are implementing a circuit using a 2N7000 metal oxide semiconductor field effect transistor (MOSFET), it is handy to know its maximum voltage, current ratings, and other specifications.

The internet is probably the fastest way to get a data sheet. Simply search on the part number, and typically multiple sources of data sheets will show up. Select one and print it out.

Applications Notes

These are articles published by manufacturers that show how to use their products. A high percentage of these are from semiconductor manufacturers. They regularly publish app notes that explain the operation of the product (a new chip, etc.) and give examples of its use, often including design information. Test equipment manufacturers also publish a massive amount of information on test instruments, measurement procedures, and advanced technology topics. Go to the company Web site and search for app notes, white papers, documentation, and manuals, and see what comes up.

The following are some selected semiconductor manufacturers to search on:

- Altera
- AMD
- Analog Devices
- Cypress Semiconductor
- Infineon
- Intel
- Intersil
- Linear Technologies (Analog devices)
- Maxim Integrated
- Microchip Technology
- NXP
- ON Semiconductor
- Qualcomm
- Rohm
- Silicon Laboratories
- ST Microelectronics
- Texas Instruments
- Xilinx

Some test equipment companies to search follow:

- Anritsu
- Fluke
- Keysight
- National Instruments
- Rhode & Schwarz
- Tektronix

Useful Web Sites

There are many Web sites devoted to electronics. Your internet searches will turn up some of them. Some good ones to look at follow:

AllAboutCircuits.com

Changpuak.ch

Circuitstoday.com

Designfact.com

Electroschematics.com

Electro-Tech-Online.com

www.electronic-notes.com

Elektor.com

element14.com

fritzing.org

Hackster.io

Hobby-electronics.info

Homemade-circuits.com

Instructables.com

Next.gr

Predictable designs.com

Radioelectronics.com

These are the only Web sites I have discovered so far. I keep finding new ones as I search for information. You should do the same. And don't forget the Web sites of the magazines mentioned earlier. All are useful.

Educational Sources

As you design, you are going to want to learn new subjects and procedures. All of the resources given earlier are learning sources, but there are some Web sites that are designed for more formal instruction. For example, there are many educational videos on YouTube and other sites. The way to find these is to do your usual internet search but include the terms *tutorial*, *learning*, *instruction*, or *education* along with the subject.

For formal college-level online courses go to the following Web sites:

coursera.org

edx.org

Udemy.com.

They all cover electronics, software, and programming along with many other related topics. I have used these multiple times and recommend them if you just need the formal instruction. They work.

Back to Basics

If you need to review electronic fundamentals before you do a deep dive into this book, I recommended the video course titled *Understanding Modern Electronics* by Richard Wolfson. It is published by The Great Courses, Chantilly, VA. It features 24 video lectures and a reference book. It is a good review for those of you who have learned it before. It also makes a good first course if you are just beginning.

A good source of math instruction if you are a bit rusty or just never learned it is Khan Academy. It offers online instruction in all levels of math from grade school level to university graduate level: algebra, trig, calculus, statistics, and others. They also offer some online courses in physics and electrical engineering you may wish to explore. And it's free.

YouTube videos are another good source of the basics. There is so much out there it is hard to describe. Just search on what you need to know. Recently, a friend of mine needed to repair or replace the headliner in his 1997 Dodge vehicle. A search turned up five videos on that specific topic. All amateurish but collectively useful. I suspect most of you have already discovered this resource. Something to try when all else fails.

Searches

I am sure you already do searches for information and answers to your questions, whatever the subject. There are lots of electronic and design materials out there. You can search on topics like the following:

- Part number
- Type of circuit
- Kits
- How to design a……
- How to test a…..

Anyway, you know what to do. Print out what you find so you can have a hard copy with you on the work bench.

CHAPTER 3

Simulation and Prototyping

After you design something, you will need to build a prototype of it. Prototyping is the process of creating a real physical version of your circuit or device so that you can see if it works. It is the critical part of the design process. There are also instances when you will discover a circuit you like and may be able to use. Your first step of the design would be to build the circuit and evaluate it based upon your needs. Then you can modify it to fit your specifications. In any case, you must build a prototype. It is the only real way to validate its operation. Some common prototyping techniques are described in this chapter.

There is also another way to build a prototype. Use software simulation. Circuit simulation software lets you build your design on your computer and test it. It is not a necessary step in design, but it is a handy tool to use before you commit to a full hardware prototype.

Many of you may already be using the techniques covered here. In that case, you can just skip this chapter and move on.

Circuit Simulation

Circuit simulation is the process of building a mathematical model of your circuit on a PC. Special software does this for you. The most common approach today is that the software lets you enter your circuit as a schematic diagram. Using predesignated component symbols and interconnecting them, the software builds a mathematical model of the circuit internally. Then you can run the software model to see if your circuit works. Most simulators also have virtual instruments (scope, meters, etc.) to use in making measurements. Then, depending on the outcome of the simulation, you can make changes until you get the circuit to perform as you wish. Figure 3.1 gives a glimpse of National Instruments' Multisim.

There are literally dozens of such simulators. Many are free and others cost a small fortune. The following list shows just a few of the simulators I have discovered and used. Like any software, simulators will take up a good deal of your time while learning how to use them. My recommendation is to research the following links, download a free trial version of one, and give it a try. Start with some simple circuits to get the feel for it, then move on to the more complex circuits as you gain some experience and competence with the program. Some simulators I have discovered follow.

- Circuitlab
- LTSpice (Linear Technologies/Analog Devices)
- Micro-Cap
- Multisim (National Instruments)
- TINA (Texas Instruments)

Some others to check out are

- 123D Circuits
- CircuitLogix
- Circuit Simulator

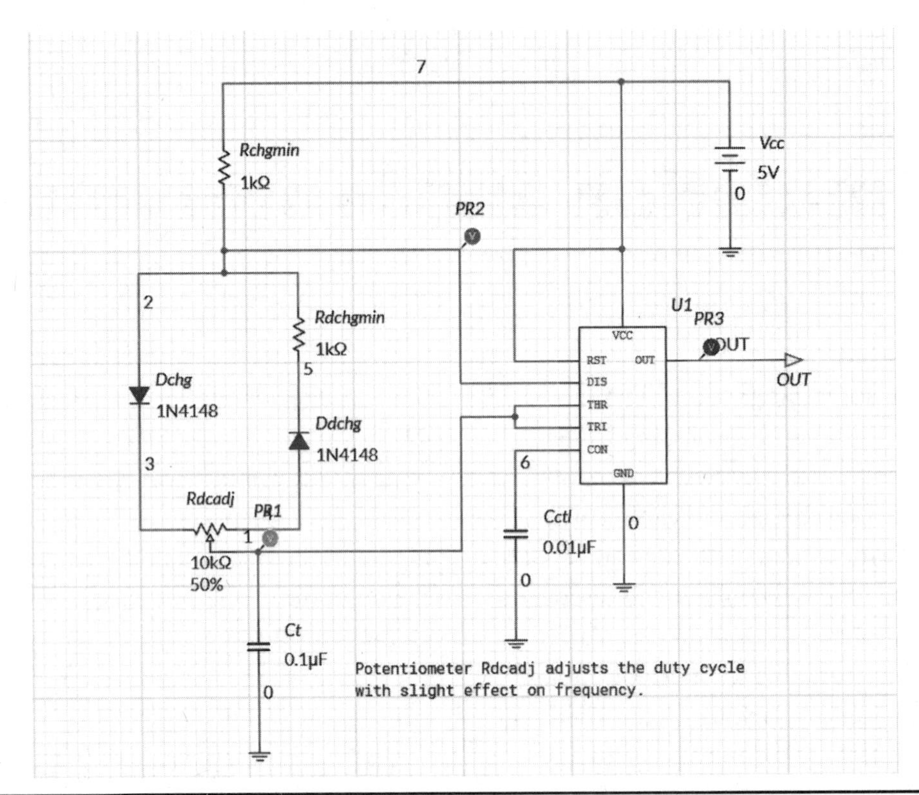

Figure 3.1 A glimpse of the screen of National Instruments' Multisim showing a simulated circuit. (Circuit courtesy National Instruments' Website www.multisim.com)

- DcAcLab
- DoCircuits
- EasyCircuit
- EasyEDA
- Gecko Simulations
- Micro-Cap
- NgSpice
- PartSim
- Proteus
- PSPICE
- Simulator.io

Some are free and others require a payment of some sort.

The main features to look for in a simulator are

- Schematic entry. The software provides component symbols that you place on the screen. Then you interconnect them with lines following the schematic diagram of your design.
- Virtual instruments. These are test instruments simulated by the software that let you monitor the voltages at any point in the circuit. These are mainly a digital multimeter (DMM) and an oscilloscope. A function generator may also be available.
- Bode plotter. This feature lets you plot the frequency response curves of filters, amplifiers, and other circuits.
- Digital logic. Get a simulator that also simulates digital logic circuits.
- Standard parts. All simulators have generic software models that define the characteristics for each component. Other simulators also have models of actual commercial parts like op amps or logic gates

and flip flops. The real parts simulations will produce a more accurate simulation.

- Tech support. Choose a simulator that has sufficient documentation you can learn from. A desirable feature is online chat or telephone help to answer your questions and provide additional information.

One of the oldest and most widely used simulator is Multisim by National Instruments (NI). It has been widely expanded and updated over the years. The main problem with Multisim is its very high price. Only professional engineers or their companies can afford it. Multisim is also used in colleges and universities.

To make Multisim more available, NI has established MultisimLive, a free version for students and experimenters. Its functionality is somewhat limited, but it is definitely useable. I encourage you to try this oneGo to www.multisim.com. I highly recommend that you access all the various introductions on this site, including a step-by-step example of how to build and test a circuit. Also look at all of the various circuits that have been simulated. Spend some time on this site, and you will come away with the concept of simulation.

If you get a simulator, please keep in mind that you will need to put in some time figuring out how to use it. Some love it, others hate it. You have to try it for yourself to see where you stand. I must admit that it is a good test of your design before you spend money on parts and time on construction of the prototype. This prequalification usually minimizes experimentation, testing, and troubleshooting time.

Recommendation

The whole circuit simulation process is complex and difficult to describe in a book chapter. I urge you to get a simulator and try it out. You must experience it for yourself and learn how to use it.

In addition, use the simulator for some of the suggested circuits, then test the software using the design projects in each chapter to verify the outcome. Once you find out if the circuit works, you can build one and test it. The more you practice with the simulator, the better you will get at it.

Breadboarding

Breadboarding is the name given to the construction of prototypes for evaluation. There are multiple ways to do it. It involves connecting components together to build a circuit or device. These include breadboarding sockets, perf board, PC board blanks, and a few others.

Breadboarding Sockets

The most popular way to do this today is to use breadboarding sockets like the one shown in Fig. 3.2. The "rat's nest" of components are several prototype circuits. These solderless connectors accept wire, component leads, IC pins, and special jumper wires to interconnect everything. They are also referred to by some as protoboards.

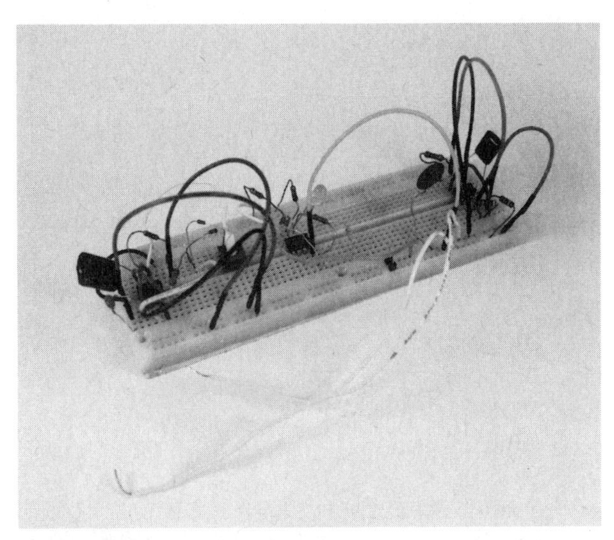

Figure 3.2 Several circuit prototypes on the breadboard. Messy as usual, but they all work.

Figure 3.3 shows a more detailed layout. For each five-hole group, a pressure connector on the backside of the socket electrically connects up to five leads or wires together. The long string of holes running along the sides of the boards are buses or power rails used to connect the circuits to the power supply voltage and ground.

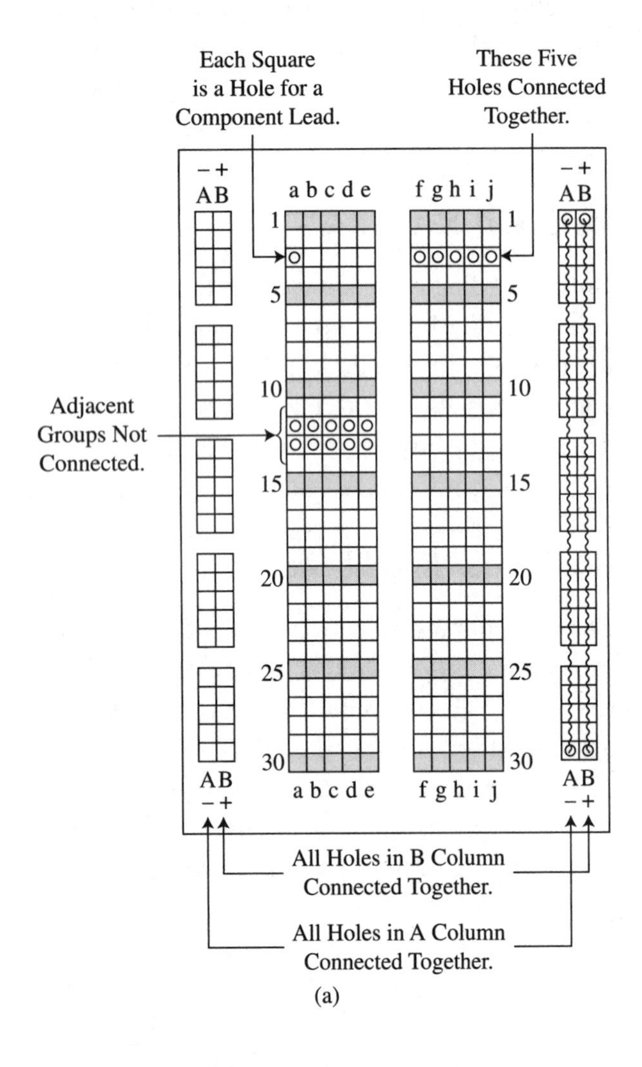

(a)

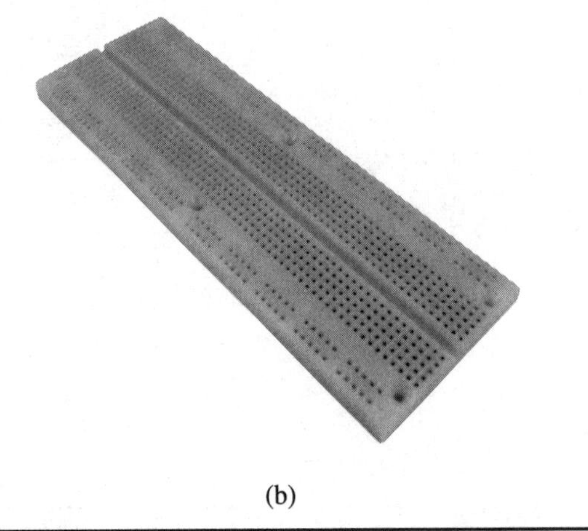

(b)

Figure 3.3 (a) A common breadboard layout. (b) The most popular breadboard size.

These breadboards provide a fast and easy way to put a circuit together with standard components with leads. Newer surface-mount components do not have wire leads, so they cannot be used with these sockets directly, although some special adapters are available to allow surface-mount parts to be used.

These prototyping sockets are available in multiple sizes to fit your project. The most common size is the 840-terminal board that measures 6.5 in long and 2.25 in wide.
I recommend that you have several of different sizes on hand at all times. If you do a lot of experimentation, it is handy to use some of the smaller boards and just build one circuit on it. Don't tear it down as you may need that circuit again in another project, and you won't have to build it, test it, or troubleshoot it again.

Figure 3.4*a* shows a schematic diagram of an op amp test circuit. Figure 3.4*b* shows the IC pin numbering. The typical wiring details of an IC op amp with resistors connected to a +V and −V power supplies.

There are no particular rules to translate your schematic diagram into actual wiring. The only common format is to connect the positive lead of the power supply to the + bus and the negative lead of the power supply to the − bus. This lets circuits on any part of the socket tap into power and ground where needed.

The ICs straddle the center line. A good beginning point is to plug in any ICs first then connect the dc power and ground wires. Then add the other wiring. The preferred wire is insulated #22 gauge solid copper. Buy a roll of it. Special wiring jumpers for these sockets are also available. These are precut to several lengths and have connectors on each end to make

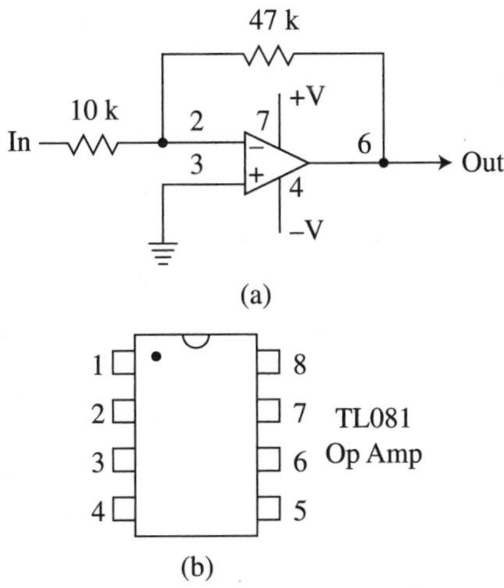

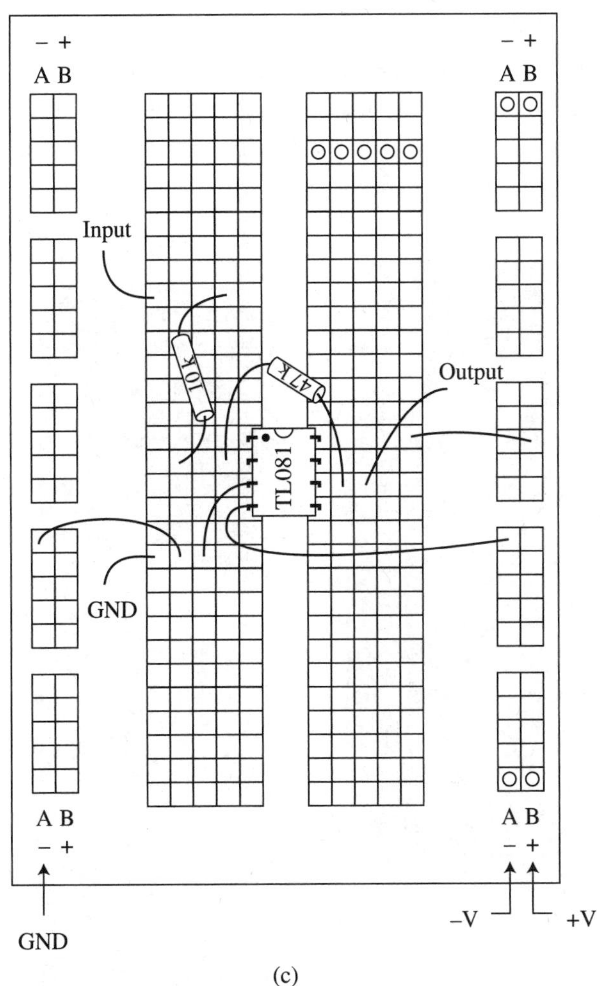

Figure 3.4 (a) Schematic diagram an op amp inverter circuit. (b) C pin numbering. (c) Wiring details.

connection to the socket easy and reliable. The extra cost is worth it, but #22 gauge wire also works just fine.

When building circuits, care should be taken to be sure of the correct orientation of each part. Be sure you have a data sheet for any IC to ensure the correct pin orientation. All ICs have some orientation feature that will identify pin 1. Other critical components are transistors, diodes, LEDs, and electrolytic capacitors. Obviously, the wrong connections can cause damage to a component. Polarized capacitors like aluminum electrolytics and tantalum types will be damaged if inserted backwards. A reverse connected tantalum capacitor, for example, will actually explode when power is applied. So be careful. If in doubt, get a data sheet to verify the connection points.

Just So You Know

Building circuits on breadboarding sockets is OK for dc and low-frequency projects. However, if you are making higher frequency circuits, you could introduce too much stray, distributed, or parasitic inductance and capacitance and create a circuit that does not perform well. It could oscillate, have a limited frequency response, do weird things, or just not work. At higher frequencies, say those beyond about 1 MHz, the component leads and wires become significant inductors. The capacitance between component leads or the capacitance between the connectors in the breadboarding socket may create a circuit that you did not intend. If you suspect this is happening, then there are several things you can do.

First, shorten all component leads as much as is practical, and shorten all wires. You may even want to try another layout on the socket, putting the components closer together. If that does not help, you may need to use some other prototyping method as discussed elsewhere in this chapter.

Perf Board

Perf board is a sheet of insulating material with perforations or holes punched into it. (See Fig. 3.5*a*.) A common material is designated FR4, an insulating base used to make many PCBs. Hole spacing is usually 0.1 in to match up with common IC lead spacing. The idea is to put the component leads through the holes, bend the leads over, and make the circuit interconnections with solder on the back of the board. You can use the component leads themselves or use wires as necessary to make the connections. Connections should be soldered. For higher frequency circuits, keep the components close together and the leads short for best results. Figure 3.5*b* shows an example.

This method of breadboarding is a little neater and closer to what a printed circuit board (PCB) layout would look like. It is a bit more difficult to build than using a prototyping socket. And it is also a bit more difficult to troubleshoot and to make changes. The upside is that it should perform better and look more like a finished project.

There are several kinds of perf board that provide an even more finished look to a prototype. One type has copper wiring patterns on one side that you can use to make the interconnections. (See Fig. 3.6.) These boards come in different sizes and have different solder pad arrangements. You still put the component leads through the holes, but you solder the leads to the copper pads. You have to use the pads that are available to make the links between the different parts. In some cases, you may have to add a wire here or there to make the connections. The result is a very organized and neat package. For one time designs this could be your finished project.

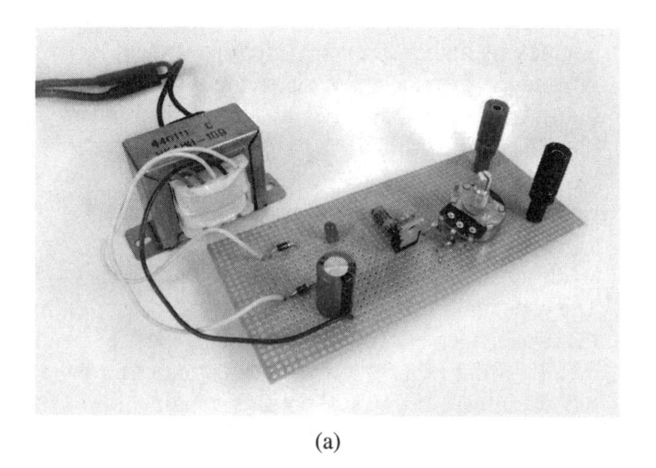

(a)

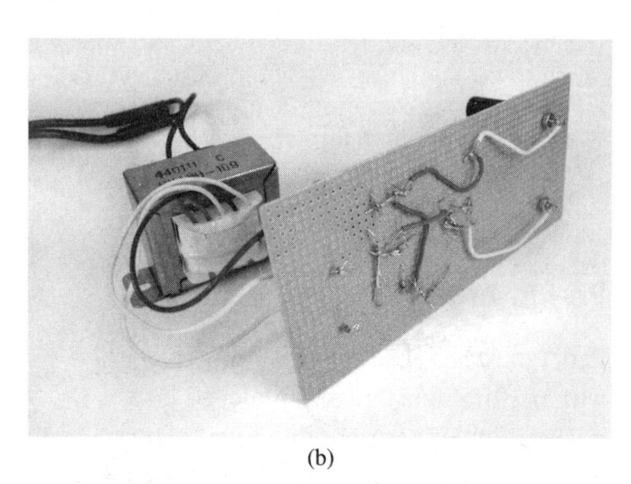

(b)

Figure 3.5 Typical perf board construction (a) and the bottom wiring (b).

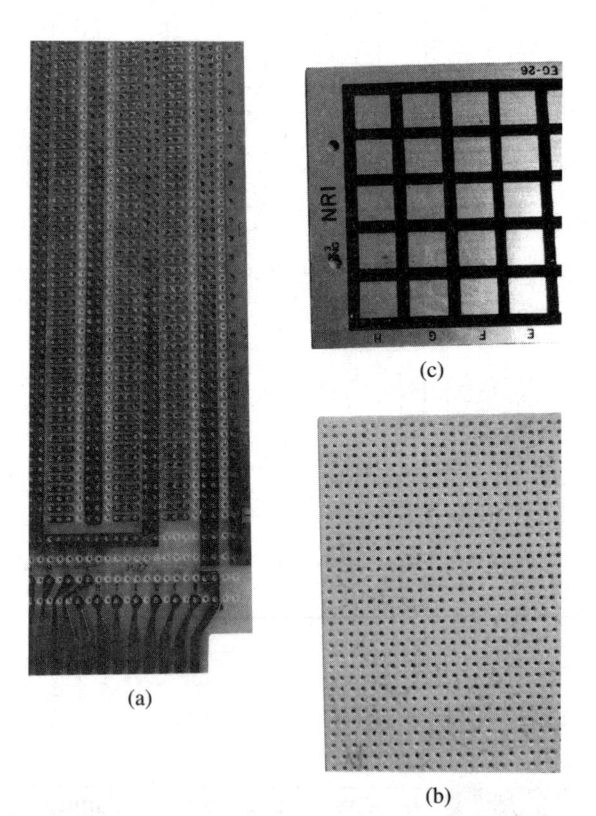

(a) (b) (c)

Figure 3.6 (a) on the left is a larger board with power buses designed to hold many ICs, (b) in the lower right a generic all-purpose board for small circuits, and (c) in the upper right is a board with surface-only pads perfect for dead-bug prototyping.

Dead Bug Method

One alternative way to build prototypes is to use the so-called dead bug method. It uses an unetched PCB with full copper cladding on one side. This becomes your base and common ground for the circuit. If your design uses an IC, you lay it leads-up with the top of the IC on the copper. You will then bend the IC ground lead over and solder it to the copper. If the circuit calls for other pins to be grounded, bend them over and solder them to the copper. The other IC pins you will solder to other components. Resistors, capacitors, and diodes are simply soldered to the IC pins and/or each other. The component leads are sturdy enough to support one another in free space above the copper base. The result is a self-supporting messy circuit. Wires are added to the circuit as needed for external power and signals.

I do not recommend this method. It is hard to follow or trace on a schematic diagram, and it is difficult to solder all the parts together. I do admit that the entire circuit is out in the open and that makes it easy to connect to test equipment and other external devices. Some designers swear by this approach. It is the best way to test high-frequency circuits. Apparently, this method is good up to 1 GHz if you keep the leads short to minimize stray inductance and capacitance.

Some who use this technique have found a way to separate segments of the copper so that signal leads and other nonground connections can be made. You can actually score the copper with a knife until you create an insulating space between copper sections. You can also cut copper strips with your knife and peel the copper off. Other not-so-elegant mechanical methods can be used to create separate islands for multiple connections, such as using a Dremel tool.

If you want to try it out, get a blank piece of PCB, such as FR4 with full copper cladding on one side. Then build a simple circuit yourself. Make your decision from there.

Working with Surface-Mount Components

In general experimenting and prototyping, you will use ICs in the standard dual in-line packages (DIP). They plug right into breadboarding sockets and are larger and easier to handle than the surface-mount devices (SMD) that are used in most new electronic equipment. Most of the older parts still available today are in a DIP. Use these whenever you can as they are less expensive and work well with breadboarding sockets.

There may be times when you want to use a particular IC, but it is only available in a surface-mount package. These are difficult to work with but there are helpful prototyping accessories you can buy. An example is a small PCB to which you can solder the IC. This small board has pins that plug into a breadboarding socket. These adapter boards are made by SchmartBOARD and Capital Advanced Technologies. Keep them in mind if you ever need one. Most of the newer ICs come only in surface-mount packages. You will definitely have to deal with them at some time.

Components

To test the circuits and products you plan to design, you will need the components to build them. You will choose these parts yourself according to the design. This book suggests and recommends many of the parts. See Chap. 14. In some of the projects, specific parts like ICs will be given. You will need to buy these parts as needed from the sources listed in Chap. 2.

The best preparation is to have on hand some of the most common parts you will always need like resistors and capacitors. One good approach is to buy available sets or kits of resistors and capacitors that include multiple pieces of all the most common values. It is maddening and off-putting not to have a specific value of a resistor or capacitor when you are ready to build the circuit. These kits are relatively inexpensive and a good value. You can also get kits of transistors, diodes, and ICs if you want to go all out. Check with the distributors to see what is available. With most common parts on hand, your orders will only be for the special parts required.

One Good Prototyping Practice

When building a multicircuit project it is desirable to keep each circuit separate from the others. Put each circuit on a separate small breadboarding socket, perf board, or dead bug board. In many projects, you may be using one of the small microcomputer boards like the Arduino. Keep it separate. If you are using a reference design kit that some IC manufacturers sell, keep it separate, too. Then you can wire all the individual boards and pieces together to make the final product.

This separation-of-circuits method makes design, testing, and troubleshooting faster and easier. Design and test each circuit separately without the clutter of multiple circuits all on one breadboarding socket or perf board. Later, when you have proved that everything works, you can always repackage it.

A Workbench

It is difficult to work on electronic projects without the proper space. You cannot always commandeer the kitchen or dining tables for your designs. If possible, try to set up a dedicated bench where you can do your design and prototyping. Use the basement, garage, or a corner of a bedroom. A small table is all you usually need. You can set up your test equipment and other stuff and leave it there between work sessions.

Summarizing

A general recommendation is to use the flexible and convenient breadboarding sockets for your initial prototyping efforts as you can build the circuit fast and make changes and corrections quickly. You should keep several of these sockets in different sizes on hand. Also get some perf boards to build your finished circuits. There are instances where you may need a socket for a new project but none are available. That means you may have to tear down an existing prototype to free up the socket. This is usually a big mistake. You never know when you are going to need that circuit again, so it is a good idea to keep previously used circuits for later applications.

As you perfect the circuit, you can move from the breadboarding socket to a perf board for a more finished appearance. As a next step, take a look at what is available in a catalog or online and search for these sockets and boards to see what may fit your projects.

A Word About Tools

If you have done any electronics work before, you may already have the necessary tools for prototyping, kit building, or repair. If not, here is a short list of the minimum tool inventory needed for most projects.

- Needle-nose pliers
- Standard pliers
- Side cutters
- Wire stripper
- Sharp knife
- Several sizes of both standard and Philips screwdrivers

- Soldering iron and holder, 15 to 35 W. Small tip to solder tiny PCB and SMDs
- Solder removal tool or solder wick for desoldering
- A good light for your bench
- A magnifying glass

Some tools that fall into the maybe category are a small electric drill and a PCB holder. A PCB holder has clips to hold a board or other object steady while you solder it. It is like a third hand. Some even come with a magnifying glass to help you see the small details like part numbers and color codes on ICs, transistors, or capacitors.

Your Projects

Each chapter in this book will give you some design projects for practice. Design and build the circuits recommended then build and test them. After that, you should be prepared to conceive your own projects. Maybe you already have something in mind. If so, go for it. Otherwise, you may be looking for a new design project to tackle. A good starting place is the magazines. Hopefully you have already subscribed to *Circuit Cellar*, *Elektor*, *MAKE*, *Nuts & Volts*, or *QST*. These publications always have interesting projects you can duplicate or otherwise give you some ideas to explore.

CHAPTER 4

Testing and Measuring

An essential part of designing is the actual testing of your circuits and end products. That means measuring inputs and outputs as well as other internal voltages and signals. For that reason, you must have some suitable test equipment. You cannot work without it. Most of you probably already have a multimeter to measure voltage, current, and resistance. A decent digital multimeter (DMM) is relatively inexpensive and endlessly useful. But you will need other equipment, specifically power supplies, a signal generator, and an oscilloscope. This equipment should be your first investment if you have decided to get serious about designing your own gear. This chapter summarizes what you need and some options available.

Multimeters

Your first investment, if you have not already made one, is to get a good DMM. Prices run from less than $10 to more than $300. All of them measure dc and ac voltage and current and resistance. Many also test diodes and transistors and continuity. Others also include a capacitor testing capability. Still others include a frequency counter. The bulk of the DMMs use a 3½ digit display (e.g., 1999). By all means, get a good one because you will probably use all of these functions.

How to Use a Multimeter

You should already know how to use a multimeter. If not, here is a review of how to measure voltage, current, and resistance. If you have the instruction manual for your meter, read it now.

Voltage

You connect the meter directly to the voltage source. Mostly you will be measuring dc voltages with positive and negative connections. Be sure you have the test leads, usually red (+) and black (−), plugged into the correct jacks on the meter. Red goes to V and black goes to COM for common or GND for ground.

Estimate the voltage level you will be measuring. Then set the range switch to the next highest value on the switch. If you set the range too low, the voltage may be higher and damage the meter or at least give you an error reading. If you don't know the voltage, guess and use one of the higher voltage ranges.

Connect the test leads to the voltage source. It could be the output of a power supply, a battery, or the voltage drop across a resistor. In any case, just connect the meter in parallel with the voltage source, red to + and black to −. Then readjust the range switch to get closer to the actual value. The more digits of readout you get on the display, the more accurate the measurement. If you get a negative reading, the voltage reading is correct, but you have the test probes reversed.

Current

To measure current, you must connect the meter in series with the line carrying the current. First, the red test lead is usually connected to a different jack on the meter. It is labeled A or 10A for ampere or 10 amperes. The black lead stays in the COM jack. Next, set the meter to measure current. Check your meter instructions for this. Then estimate the amount of current you suspect and set the range switch to next highest range.

Now, connect the meter in series with the current path. Figure 4.1 shows how to measure the current being drawn from a power supply. Just picture the electrons flowing out of the power supply; through the meter or into the black lead; through the meter and out of the red lead through the load; and back to the + output of the power supply. If you get a negative reading, you have the leads reversed, but the reading is correct.

Don't screw this up. If you connect the leads across or in parallel with the voltage source, you will damage the meter. Some meters have an internal replaceable fuse in case you accidentally make the wrong connection.

Resistance

Put the red lead into the V jack. It may also be labeled with an ohms symbol (Ω). Estimate the value of the resistance and set the range

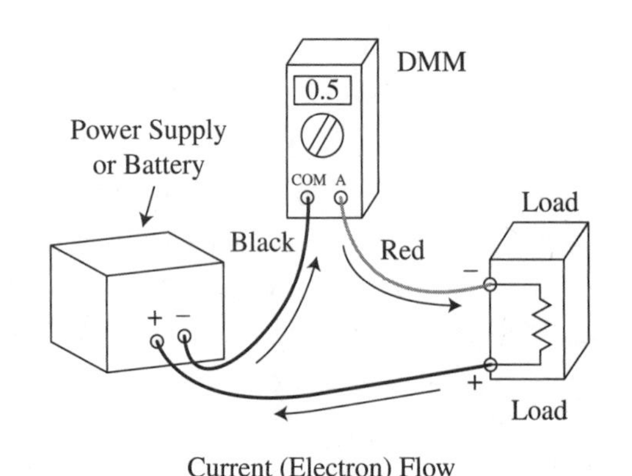

Figure 4.1 DMM connections for measuring current.

switch to a range that should accommodate the measurement. When you do this, most meters will register the digit 1 on the LCD readout. If the leads are not connected, the meter is measuring an open circuit or infinite resistance.

Now connect the test leads to the part whose resistance you wish to measure. This part must not be connected into a circuit. Nor should there be any voltage applied to it. Take the part out of the circuit or at least disconnect one lead. Any voltage applied to the test leads will damage the meter or give you a false reading. If you are measuring a resistor, adjust the range switch to give you the most digits and accurate value.

The multimeter can also be used to measure the continuity of a wire or other connection. To see if a wire is good, connect the test leads to each end of the wire and set the meter range to its lowest range, R × 1. If the wire is good, you will get a zero reading or close to it. A wire has very little resistance unless it is very long or thin or both.

Power Supply

You will need a source of dc power for your circuits. A variable supply is the most desirable as it allows you to set the voltage to any desired level, usually in the 1.5- to 24-V range. Common IC voltage levels are 3.3, 5, 9, and 12 V. A desired feature is dual supplies—one for positive voltages and another for negative voltages. Some ICs like op amps need a dual supply. A good bench power supply will set you back a hundred dollars or more, but you do have other options.

Battery Supplies

AA cells can be combined to create a very useful supply at very low cost. Each AA cell

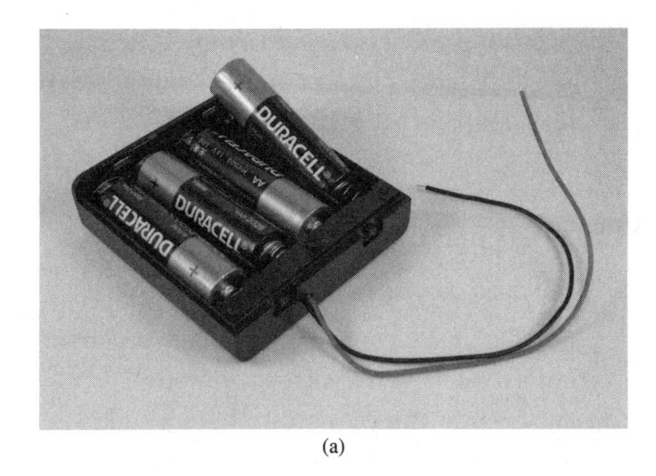

(a)

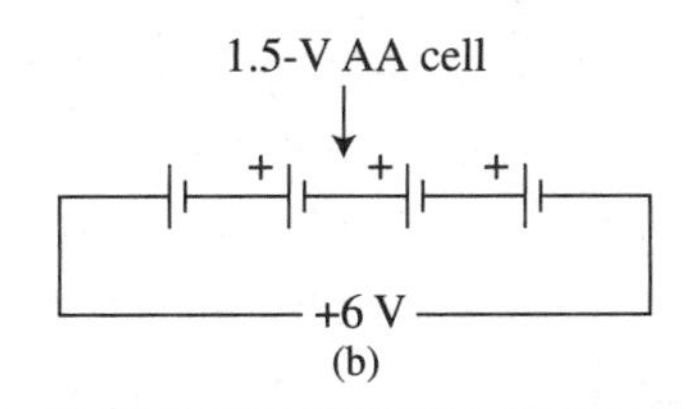

(b)

Figure 4.2 (*a*) Popular AA cells in a holder, (*b*) how the cells are connected electrically to form a 6-V power supply.

can deliver 1.5 V with a current capability of up to 100 mA or so. Putting two or more cells in series can produce a supply of 3, 4.5, 6, 7.5, 9, or 12 V. Figure 4.2*a* shows four AA cells in a holder that can supply 6 V. Figure 4.2*b* shows the connections.

For some circuits, a common 9-V battery can be used. You will also need the connector for this battery. The leads on the connector are usually fine-stranded wire that will not plug into a breadboarding socket. You can solder some short #22 solid wires to these leads so they will plug in easily as illustrated in Fig. 4.3.

If you need some voltage other than 9 V, you can use a zener diode or IC regulator connected to the battery output. A 5-V supply is the most common need. Figure 4.4 shows two ways to get 5 V. The first circuit uses a zener diode to provide 5.1 V.

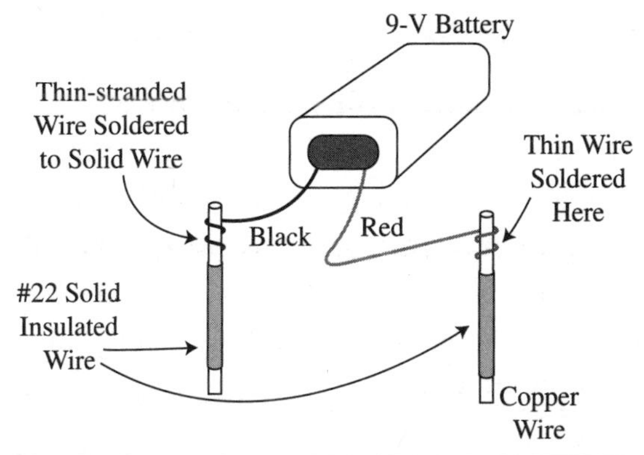

Figure 4.3 A 9-V battery with clip and wiring leads shown connecting heavier wires to the small battery clip wires to make them suitable to plug into a breadboarding socket.

This works for loads that are only a few milliamperes. The second source comes from a 9-V battery and a 7805 integrated circuit regulator. The regulator can handle up to 1 A loads, but the 9-V battery cannot. Limit the load current to 50 mA or so. More details are given in Chap. 6 on power supplies like this.

Battery supplies are a cheap and simple and will last quite a while. It is a way to get started quickly. You can build two of them if you need both positive and negative supplies. Just remember to turn them off or unplug the battery when you are not using it.

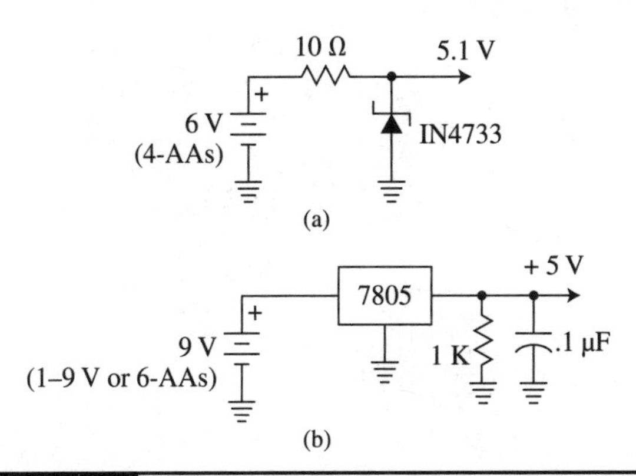

Figure 4.4 Deriving 5 V from a battery supply. (*a*) zener diode and (*b*) IC regulator.

Virtual Instrument Power Supplies

Virtual instruments (VIs) are test instruments that use a computer for a display and software to provide the measurement capability similar to a multimeter or oscilloscope. More details are provided later in this chapter. Most include dual power supplies to provide dc voltages to prototype circuits.

Special Power Supply

You could just make a supply for each project. A good way to do this is to get a "wall wart" or wall adapter. These are cheap and are available for most of the common voltages. See Chap. 6 for further alternatives.

Kit Power Supply

You may be able to find a kit supply that will serve your needs. An example is the JE215. I have used the JE215 for years as a prototyping supply. It has dual supplies with an adjustable output voltage from about 1.5 to 30 V. See Fig. 4.5. It is available from Jameco. Others are available.

Power supply Chap. 6 gives complete details on how to design a supply yourself and provides several examples that you can copy.

Oscilloscopes

Your biggest expense will be an oscilloscope (scope). Don't say you won't need one because you surely will if you design and build anything. Once you get used to it, you will realize its importance. So just bite the bullet and commit to spending some money on a scope. I started years ago by buying a used unit for $100. It was a Tektronix 2235 dual trace unit with a 5-in CRT screen. It held up well but finally died, and fixing it would have cost more than it was worth. So I acquired a new Tektronix TBS 2000 series 100 MHz scope with a color LCD screen. See Fig. 4.6. I recommend this scope or one like it. Charge it on a credit card and pay it off monthly. It is definitely worth it. It will last many years.

Your lower cost options are used scopes, new foreign-made units, deals available online, and hamfest (conference) sales. If those choices do not work for you, your best bet is to get a VI.

Figure 4.5 A popular kit power supply from Jameco supplies both + and – voltages variable over a 1.5- to 30-V range.

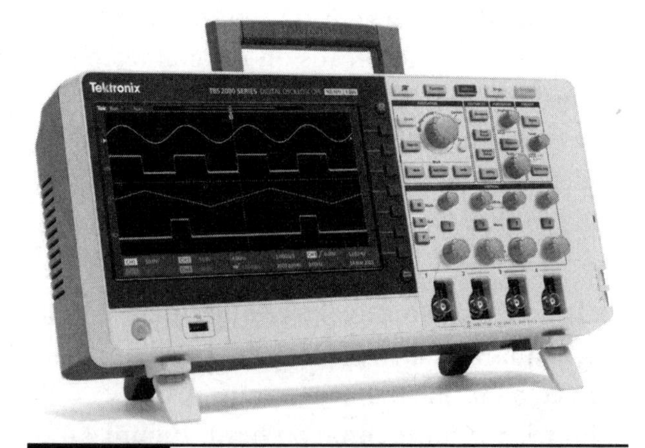

Figure 4.6 The Tektronix TBS 2000 series modern digital oscilloscope, a necessity for design and troubleshooting. Courtesy Tektronix.

The Most Critical Scope Specifications

When selecting a scope, you will want one with the best specifications you can afford.

First, get a dual-channel unit. This means two inputs so that you can display two signals at the same time on the screen. It is extremely handy to be able to see both the input and output of a circuit simultaneously. It helps to diagnose problems and to assess the circuit operation. Other useful features are a Bode plotter and a spectrum analyzer.

Second, scope vertical amplifier bandwidth is perhaps the most critical specification. This spec dictates the upper limit on the frequency of signals you can display. A scope with a 100-MHz bandwidth means you can see a maximum of a 100-MHz sine wave. For rectangular waves like those from digital circuits, it means you may only be able to see undistorted signals up to about 20 MHz. According to the Fourier theory, nonsinusoidal waves are made up of a fundamental frequency sine wave plus odd and even harmonics. Beyond the upper bandwidth limit of the scope, the amplifier acts like a low-pass filter and attenuates signals beyond that bandwidth cutoff.

As a rule of thumb, if you can see up to at least the fifth harmonic, a rectangular wave will look pretty normal. Therefore, try to estimate the upper frequency limit of your projects and get a scope with a bandwidth that is five times that value—if you can afford it.

For example, 1 GHz scope will let you see rectangular waves up to about 1 GHz/5 = 1000 MHz/5 = 200 MHz.

Function Generators

A function generator is a signal generator that can supply sine, square, and triangular wave signals for testing. The frequency range is typically in the <1 Hz to 3 MHz or more. A good bench top unit will cost you several hundred dollars. You may be able to get by without one by building some signal sources yourself as needed for testing. Sine wave oscillators and astable multivibrators for rectangular waves are easy to make. Details will be given later in Chap. 8. Also, most VIs include a function generator. Some units contain an arbitrary waveform generator (AWG or ARB). This is a signal generator that allows you to program the exact characteristics of the signal you need for testing.

Virtual Instruments

Virtual instruments (VIs) are test instruments that use a PC or laptop for a display and its processor and software to provide the measurement features of a multimeter, oscilloscope, and function generator. When measuring voltage or signals, the VI digitizes the input with an analog-to-digital converter (ADC) and sends the binary file to a memory. The memory passes the binary values to a processor and software that interprets the readings and provides an appropriate output on the video monitor. These instruments have become practical since fast ADCs are finally available to provide sufficient sampling speed to give the scope a wide enough bandwidth. Available VIs include stand-alone scopes or a complete set of instruments. Here are a few examples.

PC Oscilloscopes

Pico Technology offers a complete line of PC-based oscilloscopes called PicoScopes. Most are dual or four channels, with vertical bandwidths from 10 MHz to 25 GHz. Prices run from $79 to well over $7000. They are powered over the USB cable that connects them to the computer. Another PC scope maker is Velleman. Their PSCU1000 is a dual-channel unit with a 60-MHz bandwidth. Some of these units also contain a function generator.

National Instruments

National Instruments (NI) makes the myDAQ, a general purpose VI designed for college labs.

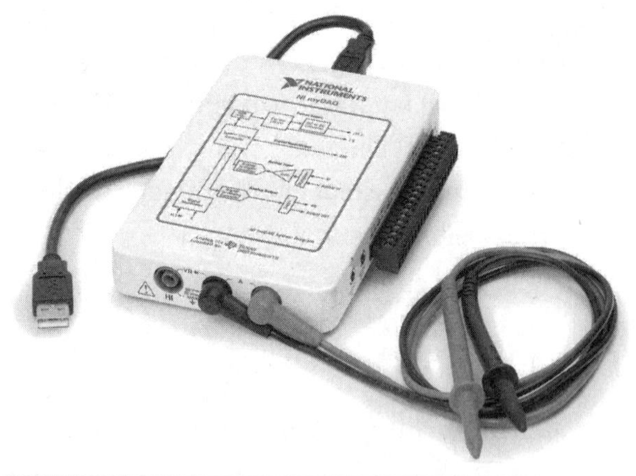

Figure 4.7 The National Instruments' myDAQ virtual instrument that includes power supplies, function generator, DMM, and an oscilloscope. Courtesy National Instruments.

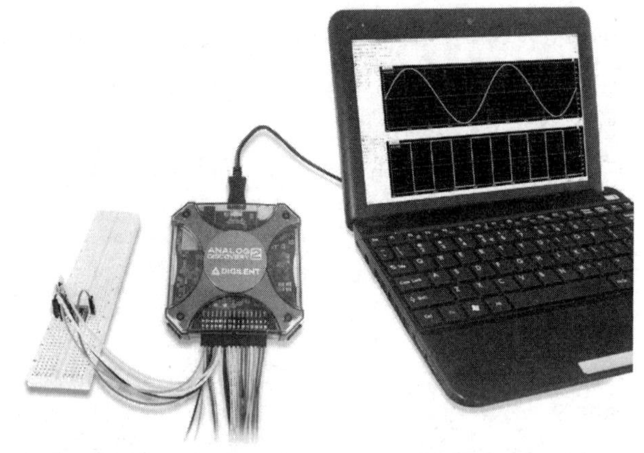

Figure 4.8 The Analog Discovery 2, a virtual instrument that includes power supplies, function generator, DMM, and an oscilloscope. Courtesy Digilent.

Figure 4.7 features a DMM with dc and ac voltage and current and resistance measurement capability. Built-in power supplies deliver +5 V and ±15 V dc. The oscilloscope has two channels with a bandwidth of dc to 400 kHz. The function generator delivers sine, square, and triangle waves from 0.2 Hz to 20 kHz. Other features are an arbitrary waveform generator (AWG or ARB), a Bode plotter for making frequency response graphs, and a dynamic signal analyzer that performs a fast Fourier transform (FFT) operation on an input and gives a frequency domain plot like a spectrum analyzer. The software is a subset of NI's popular LabVIEW. The myDAQ connects to a laptop or PC via a USB cable. The main limitation of this device is its limited scope bandwidth. Otherwise, it is a top-notch product.

Digilent

Digilent (a division of NI) makes the Analog Discovery 2 that incorporates a DMM, a scope, and a function generator. See Fig. 4.8. Its features are similar to the myDAQ but with some better specs. The DMM measures dc and ac voltages and currents and resistance. The built-in power supplies deliver ±0 to ±5 V dc. The oscilloscope has two channels with a bandwidth of dc to 30 MHz. The function generator delivers sine, square, and triangle waves from 0.2 Hz to 12 MHz plus the arbitrary waveform generator function. The network analyzer does Bode plots for making frequency response graphs and has a spectrum analyzer that performs a fast Fourier transform (FFT) operation on an input and gives a frequency domain plot like a spectrum analyzer. The software is called Waveforms, and the price is $279.

These VIs give you lots of instrumentation for the price, and if you are starting from scratch, it is the least expensive approach. I have used both myDAQ and the Analog Discovery 2, as well as the Velleman PC-based oscilloscope. All work great, and I recommend them highly.

Circuits for Testing

When you are building and testing circuits, you always need dc power as well as some test signals to apply or some outputs to monitor. If you get one of the VI units like the Digilent Discovery 2, you will have the power supplies and a function generator. For a low-cost start, I recommend that you build some test circuits you can use when you are breadboarding your design. Here are some simple circuits you can build to serve your input-output needs for some projects.

Power Supply

This topic was covered earlier. A good low-cost place to start is with batteries. Eventually you should consider a more substantial variable dual-voltage supply that can cover your needs for almost any circuit. Two 9-V batteries will take care of most projects until you can afford the power supplies. Buy the scope first!

Signal Sources

Both analog and digital sources are needed for testing the circuits and prototypes. Here are a few very useful sources.

Rectangular Wave Generator

Figure 4.9 shows a rectangular wave generator that you may use again and again. It uses the old faithful 555 timer IC. The frequency formula is

$$f = 1.443/(R_1 + 2R_2)C$$

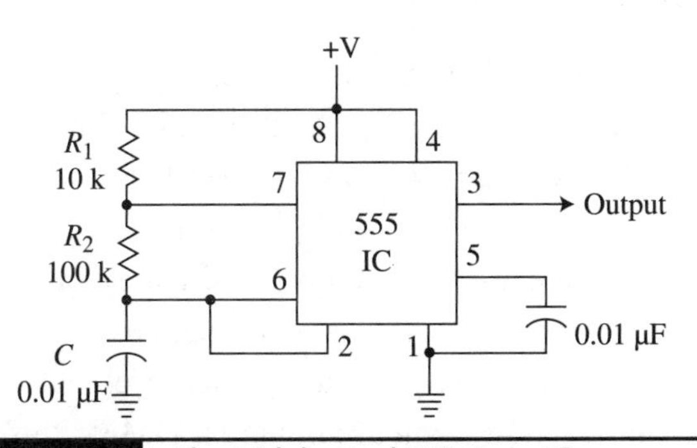

Figure 4.9 The widely used 555 timer IC connected as an astable multivibrator that generates rectangular waves.

The values shown produce an output frequency of about 600 Hz. Change C to change the frequency. In Chap. 8, some guidelines are provided to vary the frequency and duty cycle.

Select a supply voltage to match the circuits you are testing: 5 V for most digital logic and 12 V for analog circuits; 3.3 V is popular for some digital circuits; 6, 9, and 15 V are also common for analog circuits. All this explains why a good variable supply is very handy.

Wien Bridge Sine Generator

Some tests require a sine wave. Figure 4.10 shows a Wein bridge oscillator using an op amp. The output is a very clean sine wave. The frequency is set to approximately 1.6 kHz, but change it to something else by changing the R and C values. The frequency formula is

$$f = 1/2\pi RC$$

This circuit will be covered in more detail in Chap. 8.

Digital Circuits

Some tests will require logic signals. You can build an 8-bit binary word source with a DIP

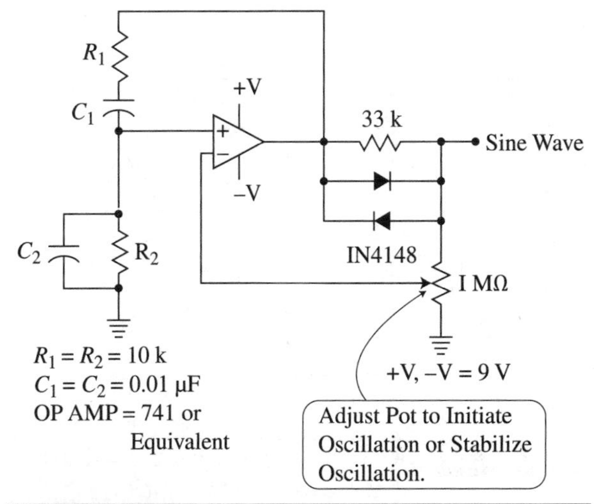

Figure 4.10 A Wein bridge oscillator that generates sine waves.

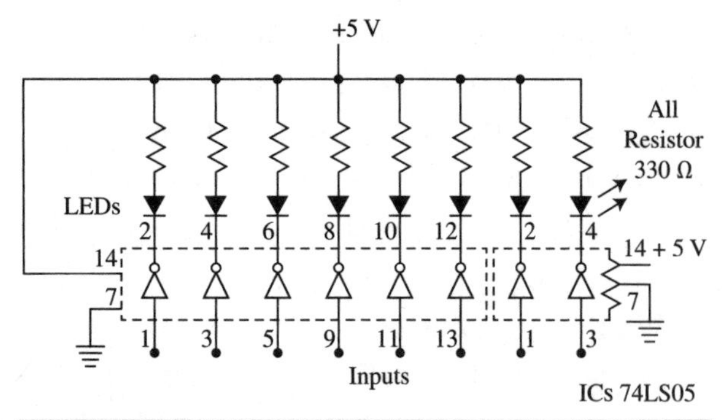

Figure 4.11 LED indicators for use in monitoring digital logic levels.

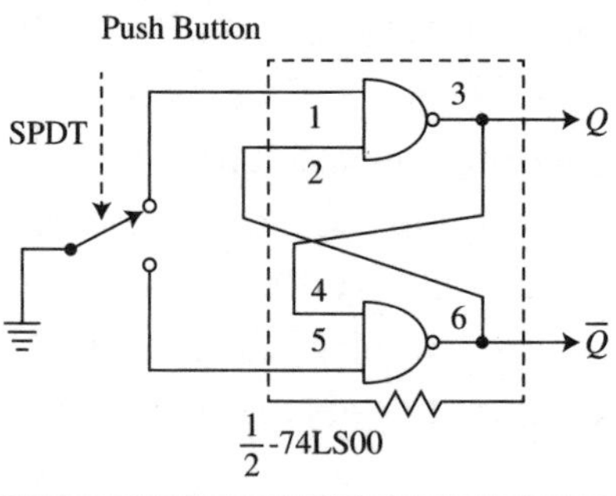

Figure 4.13 A 74LS00 NAND gate IC wired as a latch and used to debounce a push-button switch. If you are using CMOS circuits, you can use a 74HC00 here.

switch and some resistors as shown in Fig. 4.11. A DIP switch is a small package about the size of a common 14- or 16-pin (dual in-line package) DIP integrated circuit. By adding some resistors and a dc power source, you can set the switches to equivalent binary 0 or 1 levels for testing. A closed switch gives you a binary 0, and an open switch gives you +5 V or binary 1.

Another handy IO circuit is a group of LEDs for displaying logic levels. The circuit in Fig. 4.12 shows the circuit. It uses some TTL IC inverters as LED drivers. A binary 1 or +5 V input will turn the LED on.

Some momentary contact push-button switches are also handy in some projects. You can get some and make a test switch. The circuit is shown in Fig. 4.13. A logic circuit is used to "debounce" the switch. Most switch contacts bounce or chatter when the switch is pressed or released. The bounce produces multiple high-speed pulses that can falsely trigger a circuit. The debounce circuits produce a clean logic level transition when the button is pressed or released.

My recommendation is to get an extra breadboarding socket and build all of these input and output circuits on it. Keep the leads short and neat, and space the circuits out a bit so you will have easy access to the various inputs and outputs. All of your project design circuits should be breadboarded on a separate socket.

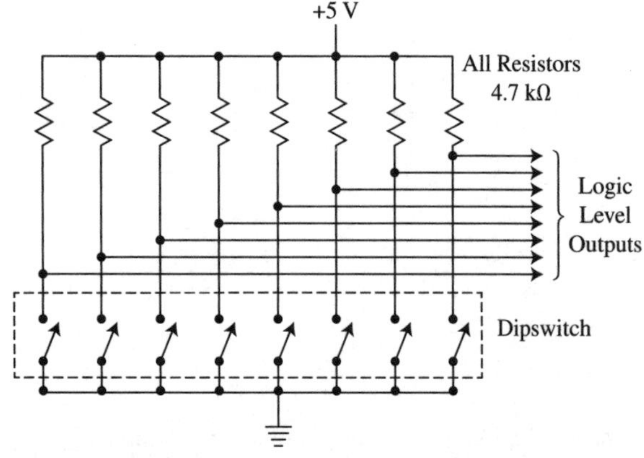

Figure 4.12 A switch register using a DIP switch that provides up to eight lines of binary outputs for testing and experimenting.

Making *L* and *C* Measurements

The ideal test instrument for measuring inductance and capacitance is called a Z meter or an LCR meter. It is way too expensive for the average experimenter, and you won't use it very often. There are other ways to measure *L* and *C*. First, recall that it was mentioned earlier that some DMMs have a capacitor measurement capability. When you are buying a DMM, this is a great feature to have, even if it costs a few dollars more.

Inductance is harder to measure, but here is a technique you can use to give you an approximate inductance measurement. It is not precise but will get you into the ballpark.

Figure 4.14*a* shows the basic measurement circuit. The unknown inductance (L) is connected in series with a known resistor value, and the circuit is driven by a sine wave signal source. The signal source should have a frequency close to what the inductor will encounter in its application. Use that frequency or close to it. Let's use 1 kHz.

For a resistor value, it should be approximately the same magnitude as the inductive reactance (X_L) at the chosen frequency (f). Remember that

$$X_L = 2\pi f L$$

You know the frequency but not the inductance. As a starting point, make a guess at the inductance. If you are making the inductor for a specific application and the value has been given, use that value. Then calculate X_L.

Suppose the desired value is 100 mH at 1 kHz. Then

$$X_L = 6.28(1000)(10 \times 10^{-3}) = 628 \text{ ohms}$$

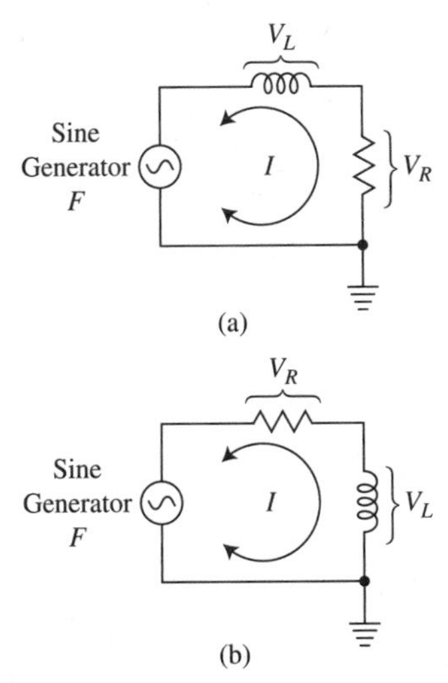

Figure 4.14 Inductance measurement circuits. (*a*) To measure I and (*b*) to measure X_L.

Then select an R value close to that. Anything in the 470 to 1500 range will work.

Once you have wired the circuit and connected it to the generator, follow the procedure below.

1. Set the generator to some modest voltage level range like 2- to 4-V peak-to-peak (pp). The frequency is 1 kHz.
2. Measure the voltage across the resistor (V_R) with your DMM set to ac or your oscilloscope. NOTE: Most DMMs have a very limited ac frequency range. Few will measure ac beyond about 5 kHz. Check your DMM manual for the specifications. If you are using a much higher frequency, use the scope to make the voltage measurements.
3. Calculate the circuit current (I). Knowing R and V_R, the current by Ohm's law is $I = V_R/R$.
4. Now rewire the circuit as shown in Fig. 14.14*b*. You are just swapping the positions of the L and the R. This is sometimes necessary if your generator, scope, or DMM are ac line powered. The internal grounds of these devices can create ground shorts when measuring the floating (nongrounded) component. Rearranging the circuit is simple and avoids the problem. The outcome will be the same.
5. Make sure the generator is still applying the same voltage as in the prior measurement.
6. Measure the voltage across the inductor (V_L).
7. Calculate $X_L = V_L/I$ using the previously determined current value.
8. Knowing X_L you can now find the inductance. $L = 2\pi(f)X_L$.

This method is not precise, but it will give you an approximate value to work with. This procedure also works with capacitors.

CHAPTER 5

Common Circuit Design Techniques

You will begin your actual design instruction in this chapter. Several basic devices and circuits occur again and again when you are designing something. This chapter is a collection of these fundamental devices and circuits. They include current limiting resistors, voltage dividers, pots, and transistor switches. At the same time, it is a good review of some basic electronic fundamentals, such as Ohm's law, Kirchhoff's law, and input and output impedances.

Drawing Circuits

There are lots of ways to draw the schematic diagrams for electronic circuits. However, there are some conventional ways that most engineers use. For example, Fig. 5.1*a* shows one way to show a battery powering an LED. The preferred alternative is given in Fig. 5.1*b*. The generally accepted rule is not to show the dc voltage source. Simply assume some source like a battery or power supply is connected and given the voltage and polarity as shown.

In more complex applications made up of multiple circuits, use a block diagram and arrange the circuits from left to right. Most engineers assume signal flow from input to output is from left to right. See Fig. 5.2. These conventions will be used in this book.

Series-Dropping Resistor

Also called a current-limiting resistor, this is a resistor that is connected in series with another component to achieve the desired current level and voltage drop across the component. This resistor is used in circuits where you want to operate a low-voltage device from a higher voltage source. A common example is powering a light bulb from a voltage source that is greater than the maximum bulb rating. Operating a light-emitting diode (LED) from some voltage source is another example.

Refer to Fig. 5.3. Note here that the LED is a semiconductor diode that must be forward biased to emit light. The anode connects to the positive side of the supply voltage. LEDs don't really have a voltage rating as such, but they do have a specific voltage drop across them when they are emitting light. This voltage drop varies with the color of the LED. The range is roughly 1.7 to 4 V. The following are some typical forward voltage drops for different color LEDs:

- Red, orange, yellow, and green: 1.9 to 2.1 V
- Pure green and blue: 3.0 to 3.4 V
- White and violet: 2.9 to 4.2 V

The LED also has a maximum current rating. The higher the current, the brighter the LED. A workable design rule of thumb is to assume a 2-V voltage drop for the average ubiquitous red or green LED. Then assume a current level in the 5 to 20 mA range. That will get you in the ballpark, and you can adjust later.

Figure 5.3 shows a white LED that normally has about 4 V across it in normal operation. It is to be powered by a 12-V battery voltage source (V_S). Connecting the LED directly to the

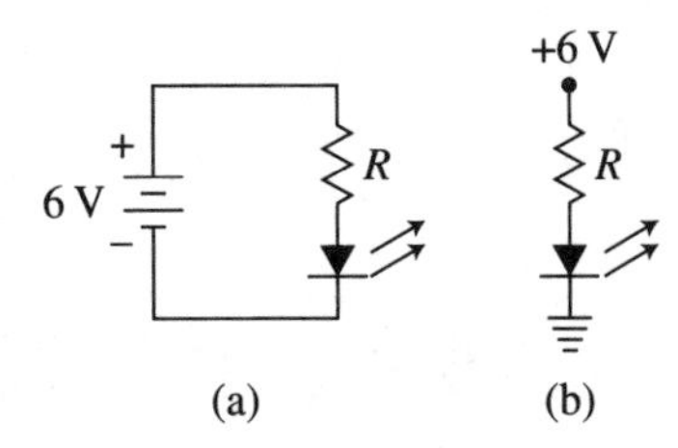

Figure 5.1 (*a*) One way to draw a schematic diagram and (*b*) the preferred way to draw a schematic diagram.

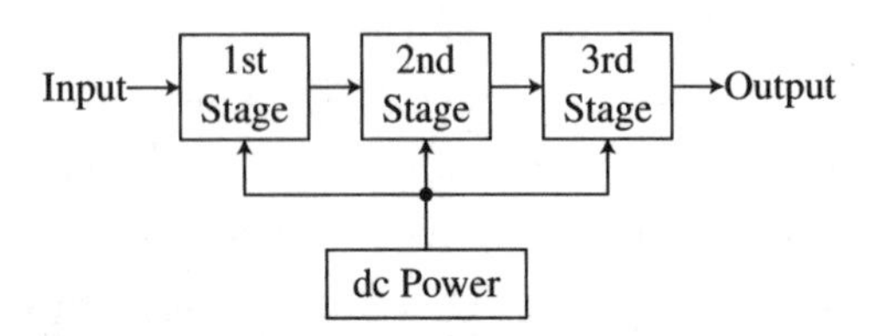

Figure 5.2 How to draw a block diagram showing signal flow from left to right.

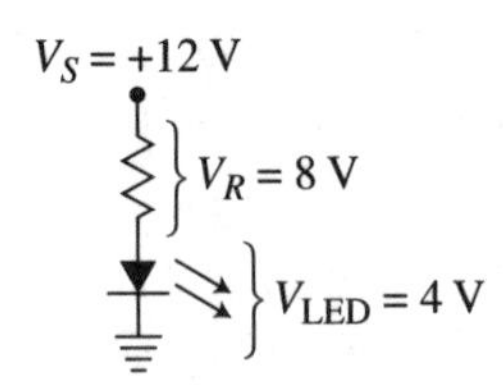

Figure 5.3 An example problem showing the circuit voltages. The sum of the series voltage drops equals the source voltage.

12-V source will exceed its rating and the LED will be destroyed. The resistor in series with the LED is used to absorb the other 8 V. What you need to know is the recommended current (I) in the LED. The current is usually in the 5 to 20 mA range, depending upon the brightness you want. Let's assume 10 mA (0.01 A). With this information, you can calculate the value of the resistor (R) value needed.

Remember that according to Kirchhoff's voltage law, the sum of the voltage drops around a series circuit is equal to the source voltage (V_S). In the circuit shown, the voltage of the LED (V_{LED}) and the voltage across the series resistor (V_R) adds up to the source voltage (V_S).

$$V_S = V_{LED} + V_R$$

When designing, you know the LED voltage and the source voltage. What you want to find is the resistor voltage. You rearrange the formula as follows:

$$V_R = V_S - V_{LED}$$

$$V_R = V_S - V_{LED} = 12 - 4 = 8 \text{ V}$$

What you want to find is the value of the resistor that will give you the desired current in the LED. For this calculation, you use Ohm's law. You now know the voltage across the series resistor and the desired LED current, so the resistor value is

$$R = V_R/I$$

$$R = 8/0.01 = 800 \ \Omega$$

Next, don't forget the power rating of the resistor. Remember that power can be calculated in several ways.

$$P = V_S(I)$$

$$P = I^2(R)$$

$$P = V_S^2/R$$

Since we know all the values, you can use any formula you want. Let's use the first one.

$$P = V_S(I) = 8(0.01) = 0.08 \text{ W}$$

Now you need to find a resistor with a value of 800 Ω and one that can handle 0.08 W of power.

As you may know, resistors come in standard sizes. You cannot usually get any specific desired value, but you can get close. The standard sizes are summarized in Chap. 14. Resistors with a tolerance of 5 percent are the most common. The closest values are 750 and 820 Ω. Choosing the 750-Ω resistor will allow a bit more current to flow so the LED will glow a little brighter. You probably should opt for the 820-Ω resistor that will absorb more voltage. This will decrease the current a bit, and the LED will be a little dimmer.

Resistors with a 1 percent tolerance offer many more resistance choices if you want or need to get a value closer to your design.

As for power rating, resistors come in several standard wattage ratings or ⅛, ¼, ½, 1, and 2 W. A common ¼ -W rating resistor is a good choice.

The whole design procedure is summarized below.

1. Identify the voltage and current rating of the component or device to be operated.
2. Identify the source voltage.
3. Calculate the amount of voltage to be dropped across the resistor. $V_R = V_S - V_{LED}$.
4. Calculate the resistor value. $R = V_R/I$.
5. Select the closest standard resistor value. See Chap. 14.
6. Calculate the power dissipation of the resistor. $P = V_R(I)$.
7. Select the closest higher power rating resistor.

Remembering Ohm's Law

Ohm's law says that the current (I) in a circuit is proportional to the applied voltage (V) and inversely proportional to the resistance (R). Mathematically, this is the voltage divided by the circuit resistance (R).

$$I = V/R$$

This basic formula can be rearranged to solve for voltage or resistance.

$$V = IR$$

$$R = V/I$$

Voltage Dividers

Another basic circuit that shows up repeatedly is the voltage divider. It has an input or source voltage (V_S) and output voltage (V_O) that is lower. Figure 5.4*a* shows the basic circuit. The total divider resistance (R_D) is the sum of R_1 and R_2.

$$R_D = R_1 + R_2$$

The output is the voltage drop across the lower resistor (R_2), so it is always lower than the input. A voltage divider attenuates an input signal or makes it smaller.

There are two types of dividers: unloaded and loaded. Unloaded means that there is no load connected to the output that draws current. Loaded means that there is a load on the output that will draw current from the source. A voltage divider with no load is a hypothetical case. Most dividers will have a load of some kind. However, if that load is negligible, then you can basically ignore it. Negligible here means that the load resistance should be at least 10 or more times higher than the total divider resistance. A good design goal is to make the divider resistance less by a factor of 10 to a 100 of the load resistance. Then you can call that an unloaded divider.

Another way to look at this is by considering current. The load is considered negligible if the load current is less by a factor of 10 or more than the current in the divider. A factor of 100 is better.

Unloaded Voltage Divider

Using the definition described earlier, let's summarize the procedure for designing an unloaded or lightly loaded voltage divider. Refer to Fig. 5.4*b*. The basic formula for calculating the output voltage is

$$V_O = V_S R_2/(R_1 + R_2)$$

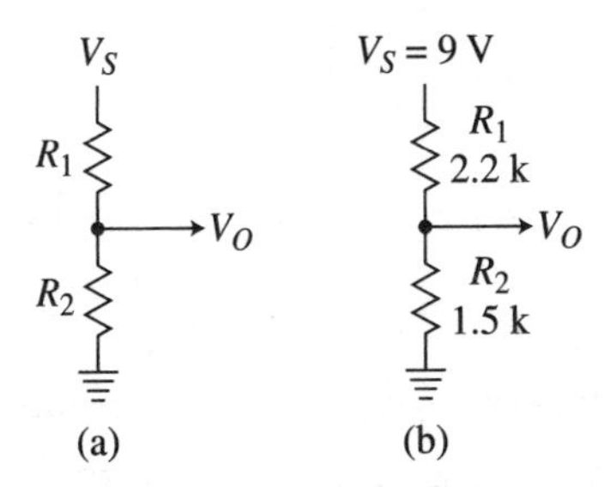

Figure 5.4 (*a*) An unloaded voltage divider and (*b*) an unloaded voltage divider with specific resistor values.

The output voltage is then:

$$V_O = V_S(R_2/R_D)$$

With the values shown in Fig. 5.4b, the output voltage is

$$V_O = 9(1500/2200 + 1500) = 9(0.405) = 3.65 \text{ V}$$

To design a voltage divider, you need to know the input or source voltage (V_S), the desired output voltage (V_O), and the total series resistance of the divider (R_D) = $R_1 + R_2$. You will need to provide a value for the total. That can vary from a few ohms to many megohms. The actual value will depend somewhat on where and how it will be used. Remember, we are discussing an unloaded voltage divider here but if you know the load resistance (R_L), then make the divider resistance (R_D) lower by a factor of 100.

$$R_D = R_L/100$$

If you know the load current (I_L), you can find R_D with this:

$$R_D = V_S/100I_L$$

Once you know R_D, you can rearrange the formula to solve for R_2. Then you can calculate R_1 from that.

$$V_O = V_S(R_2/R_D)$$

$$R_2 = V_O R_D/V_S$$

$$R_1 = R_D - R_2$$

Assume a load of 100 kΩ. The supply voltage is 12 V and the desired output is 3 V. The total divider resistance is

$$R_D = R_L/100 = 100\text{k}/100 = 1 \text{ k}\Omega$$

Assume a total resistance of 1 kΩ. The input or source voltage is 12 V and the output is 3 V. R_2 then is

$$R_2 = V_O R_D/V_S = 3(1\text{k})/12 = 250 \ \Omega$$

$$R_1 = 1\text{k} - 250 = 750 \ \Omega$$

The standard resistor values are 240 and 750 Ω. Using these values, the output voltage is

$$V_O = V_S R_2/(R_1 + R_2)$$
$$= 12 \times 240/(750 + 240) = 2.9 \text{ V}$$

Not exact but close.

The percentage error is calculated with the following expression:

$$\%\text{Error} = 100(\text{Actual voltage} - \text{Specified voltage})/\text{Specified voltage}$$
$$= (2.9 - 3)/3 = -0.1/3$$
$$= -0.033333 \times 100 = 3.33\%$$

Ignore the negative sign as this indicates that the actual voltage is less than the specified voltage. To eliminate the negative value, just reverse the actual and specified values. Multiply by 100 to get the percentage or 3.333 percent.

That may be close enough for many applications. A good objective is ±5 percent. You can use 1 percent resistors if you need greater precision.

Loaded Voltage Dividers

A loaded voltage divider is shown in Fig. 5.5. A resistive load is connected to the output. The load appears in parallel with R_2, so this modifies the relationship between the input and output voltages. To compute the output, you need to know the resistance value of the load R_L or how much current is drawn by the load I_L at the desired output voltage.

One general rule of thumb in design is that if the load is light, meaning a resistance value that is much larger than R_2, then you just use the unloaded design process. For example, if the load is at least 100 times R_2, then you can essentially ignore the effect on the output. The output will be a little less than desired but close. For a load 10 times (×10) R_2, the error will be less than 10 percent.

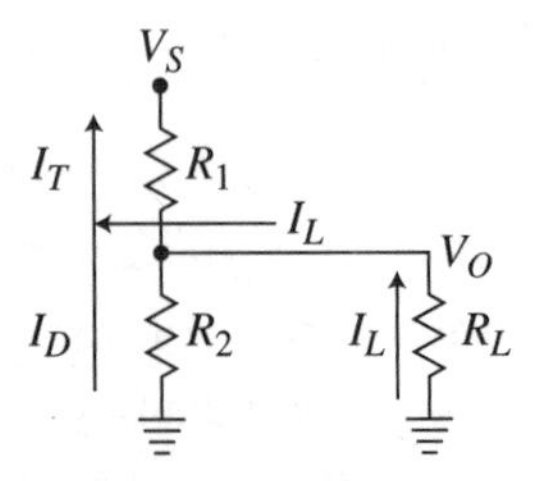

Figure 5.5 A loaded voltage divider showing current flow.

The first step to designing a loaded divider is to specify a value of divider current (I_D). This is an arbitrary selection since the value is not usually critical. Check to ensure that the source voltage supply can furnish the load current as well as a divider current. A commonly used approach is to make the divider current I_D about 10 times the load current.

$$I_D = 10I_L$$

This makes the divider a low-impedance source compared to the load. Such a divider is called a "stiff" divider, since its output voltage will have minimum variations with load changes. Refer to Fig. 5.5. The divider current I_D flows through R_2. The total circuit current (I_T), consisting of the divider current (I_D) and the load current (I_L) flows in R_1. The total circuit current is given as

$$I_T = I_D + I_L$$

A simpler approximation is to make R_D a factor of 10 lower than R_L.

$$R_D = R_L/10$$

Even better, use the previously mentioned guideline of RD = RL/100. Then just ignore the load current.

Next, calculate the value of R_2. Use the following expression:

$$R_2 = V_O/I_D = V_O/10I_L$$

Now calculate the value of R_1. This is given as

$$R_1 = (V_S - V_O)/I_D + I_L = (V_S - V_O)/I_T$$

If you used the $R_D = R_L/10$ approach, then find I_D.

$$I_D = V_S/R_D$$

Knowing the values of R_1 and R_2, you can specify the closest standard 5 percent, or 1 percent resistor values, depending on the desired accuracy.

Now calculate the true output voltage with the specified resistor values. Use the actual values specified for R_1 and R_2 in the equation.

$$V_O = V_S R_2/(R_1 + R_2)$$

If you want, calculate the percentage of error.

%Error = 100(Actual voltage − Specified voltage)/Specified voltage

Here is an example. Assume a source voltage of 6 V and a desired output voltage of 2 V. The load current is 10 mA.

$$I_D = 10I_L = 10(10) = 100 \text{ mA}$$

$$I_T = I_D + I_L = 100 + 10 = 110 \text{ mA} = 0.11 \text{ A}$$

$$R_2 = V_O/I_D = V_O/10I_L = 2/0.11 = 18\ \Omega$$

$$R_1 = (V_S - V_O)/I_D + I_L = (V_S - V_O)/I_T$$
$$= (6 - 20)/0.11 = 4/0.1 = 40\ \Omega$$

Standard ±5 percent resistor values are 18 and 39 Ω. Calculate the actual output voltage.

$$V_O = V_S R_2/(R_1 + R_2) = 6(18/39 + 18)$$
$$= 6(18/57) = 1.89 \text{ V}$$

The error is

%Error = 100 (Actual voltage − specified voltage)/Specified voltage

Error = 100(1.89 − 2)/2 = −0.11/2 = 5.5%

Special Sensor Resistors

Most of the voltage dividers you will design will use standard resistors. But some applications may call for a sensor to sense temperature or light intensity. Special resistors are available for such applications. Two of the most popular resistive sensors are thermistors and photocells.

Thermistors

A thermistor is a resistor that changes value over a wide range depending upon the temperature to which it is exposed. Most thermistors are of the negative temperature coefficient (NTC) type whose resistance varies inversely with the temperature. As heat increases, the resistance decreases. The variation is nonlinear. Such resistors can be used to measure temperature or to detect a specific temperature as a warning or set point. Positive temperature coefficient (PTC) thermistors are also available.

Photocells

A photocell is also known as a photosensitive resistor, photoconductive resistor, or a light-dependent resistor (LDR). Most are made of cadmium sulfide (CdS). As the intensity of the light falling on the cell varies, its resistance changes over a wide range. In the dark, the resistance can be as high as a few megaohms. With maximum brightness, the resistance can drop to 1 k or so. The resistance goes down as the light intensity increases.

Both of these special resistive sensors are often used as one element in a voltage divider to produce an output voltage that varies with the temperature or light intensity. Figure 5.6 shows four possible configurations. In Figure 5.6*a* and 5.6*b*, notice the special symbol for a thermistor. The photocell (LDR) symbol is shown in Fig. 5.6*c* and 5.6*d*.

Consider the circuit in Fig. 5.6*a*. As the heat on the thermistor is increased, the resistance goes down. More current will flow, and the output voltage across standard resistor will increase. In Fig. 5.6*b*, the opposite operation will occur. Increasing the temperature decreases the current in and the output voltage across the thermistor.

Now, figure out how the output voltage varies with light intensity from the photocell dividers in Fig. 5.6*c* and 5.6*d*. See App. B for the solution.

Potentiometers

A potentiometer or pot is basically just a variable voltage divider. If you absolutely must have a precise voltage setting from a voltage divider, use a pot. The pot has a fixed resistance element along with a slider arm that taps into the element. It continuously varies the R_1 and R_2 values while maintaining a constant total resistance. As R_1 goes down, R_2 goes up; or as R_1 goes up, R_2 goes down. See Fig. 5.7*a*. It is still a voltage divider so is subject to the design rules covered earlier. This circuit will deliver an output voltage anywhere between 0 and +5 V.

Most pots have a rotary slider. A knob is usually attached to a shaft that moves the slider. Some smaller pots have a screwdriver adjustment slot to rotate the slider. There are also linear pots that implement the slider in a straight line along the resistive element.

A pot can also be used as a variable resistor. Refer to Fig 5.7*b*. You use one end of the resistive element and the slider to produce a variable resistor with a range of 0 to the maximum resistance value of pot. For example, using a 100 k pot, you can set the resistance to any value between zero and 100 kΩ. Variable resistors are sometimes referred to as rheostats.

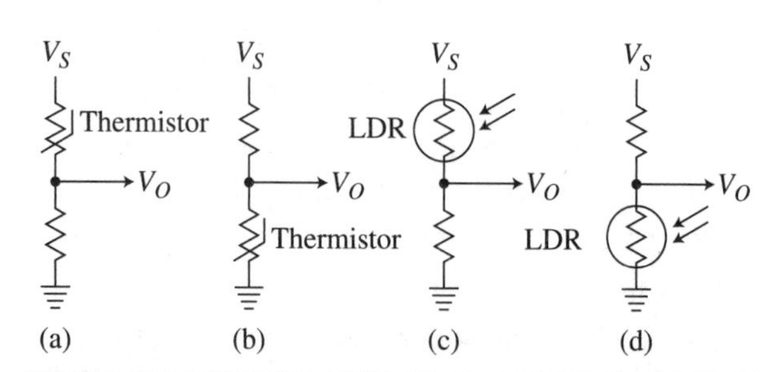

Figure 5.6 (*a* and *b*) Voltage dividers made with thermistors and (*c* and *d*) photocells.

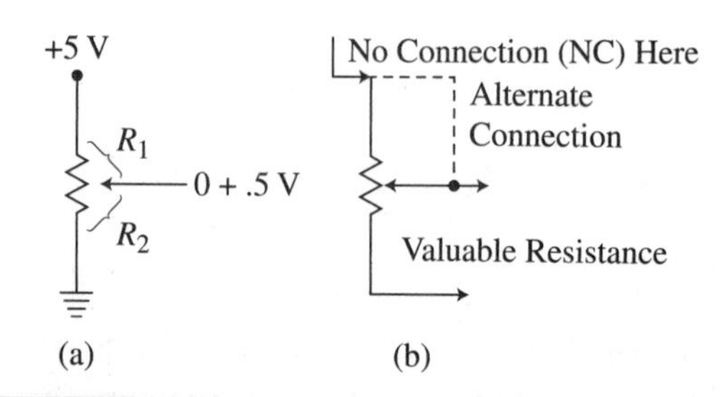

Figure 5.7 (*a*) A pot is an adjustable voltage divider and (*b*) using a pot as a variable resistor.

Just So You Know

Voltage source tolerances plus resistor values slightly different from the calculated values give you output voltages that are not dead-on your design target. For many designs, precision is not needed. If you are close (less than 10 percent variation), things will probably work. Don't worry about the actual value.

Not all designs will tolerate such leeway. In that case, you can specify 1 percent tolerance resistors. These are a little more expensive, but that may be necessary if you must achieve exact values.

Error and Accuracy

Here is something every designer should know and understand. Keep this in mind as you are dealing with voltage and current measurements as well as component values.

Error is the amount by which a calculated or measured value differs from the true, desired, or correct value. It is usually expressed as a percentage.

%Error = 100(True value − Measured or calculated value)/True value

If the measured or calculated value is greater than the true value, simply reverse their position in the formula to eliminate the negative value that results. Or just ignore the negative value.

Accuracy is a value that indicates how close a measured or calculated value is to the true or desired value. It, too, is expressed as a percentage.

%Accuracy = 100(Measured or calculated value)/True value

If the measured or calculated value is more than the true value, simply reverse their positions in the formula.

Error and accuracy are related by these expressions.

%Error = 100 − %Accuracy

%Accuracy = 100 − %Error

Variable Voltage Dividers

Variable voltage dividers are made with a potentiometer. Using a pot alone, you can vary the output voltage from 0 to the maximum input voltage and anything in between. However, you may encounter a need to vary the voltage over a narrower range. This can be done with a voltage divider like that shown in Fig.5.8*a*. When the pot wiper is at its lower limit, the voltage will be V_{min}, and with the pot set to its maximum upper limit, the output voltage will be V_{max}. The goal is to find the values of R_1, R_2 (the pot), and R_3. Two other variations are illustrated in Fig. 5.8*b* and 5.8*c*. The version in Fig. 5.8*b* is very useful when you need to develop a very small output voltage from a much larger input or source voltage. The design rules given earlier apply here.

The only tricky part is selecting a pot value. Obviously, it needs to be less than the total divider resistance (R_D). A good rule of thumb is make the pot value about one-third to one-half of R_D. Select the closest standard value available.

Transistor Switches

When you are working with digital circuits, you will almost always need to connect a signal to another device to turn it off or on. You may need to operate an LED, a light bulb, a relay,

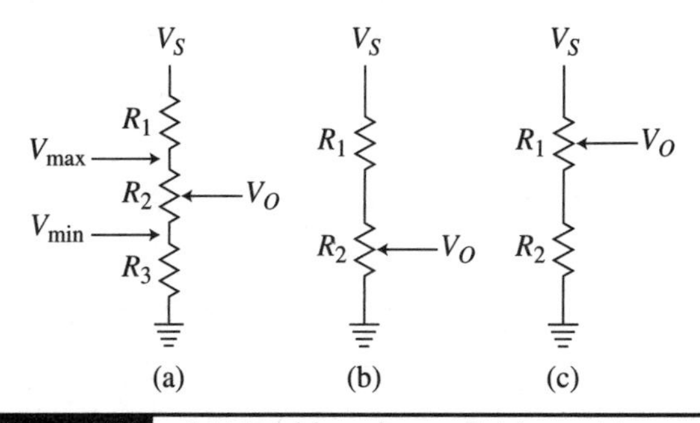

Figure 5.8 (*a*) Variable voltage dividers with upper and lower voltage limits, (*b*) an upper limit, and (*c*) a lower limit.

a solenoid, or a small motor. The circuit or IC output will usually not be capable of driving the component of choice. The way to do this is to use an external transistor switch. The digital signal source, whether it is simple logic circuits or the output from a microcontroller, supplies a signal to the transistor. These switches are sometimes referred to as device drivers.

Bipolar Junction Transistor

Figure 5.9 shows the basic configuration. When the digital/logic input is near zero volts, the emitter-base junction is not forward biased, so the transistor does not conduct. It acts as an open switch so that the lower end of the load is floating. No collector current flows. When the input goes high to its standard voltage level, the emitter-base junction is forward biased. Base current I_B flows, and the transistor conducts.

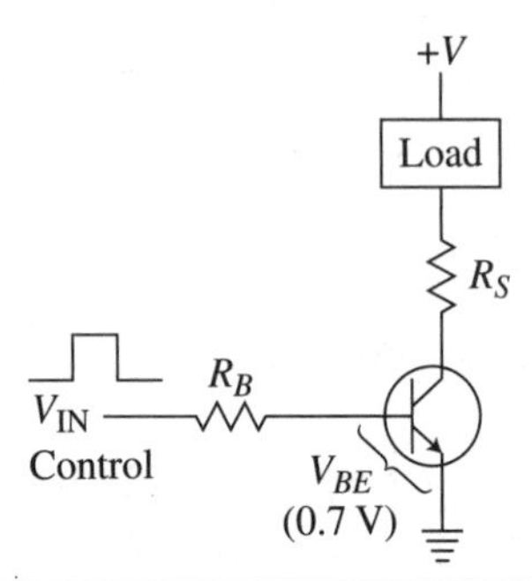

Figure 5.9 A bipolar transistor used as a switch to turn a load off and on.

Ideally, the base current is high enough to saturate the transistor. Collector current flows. At saturation, the emitter-collector voltage drop is typically less than 0.2 V. You can almost ignore this voltage and assume it is zero. Therefore, the transistor acts as a closed switch. The lower end of the load is effectively connected to ground. The full supply voltage, less a few tenths of a volt, appears across the load and R_S, causing it to activate.

Figure 5.10 shows a basic bipolar junction transistor (BJT) switch with several different loads connected in series with the collector. Note that when a load is an inductance like a relay, solenoid, or motor, you must connect a diode across it. These three loads are inductive. When the transistor switch turns off, the current in the inductance collapses and induces a short-duration, very high reverse voltage across the load that will destroy the transistor. The addition of a diode prevents the high voltage from damaging the transistor. When the transistor switches off, the induced voltage still occurs, but the diode conducts and clamps the voltage at the collector to a very low value. A common rectifier diode like a 1N4001 is a good choice.

The main design criterion for this circuit is to ensure that the transistor saturates when it receives a turn-on input signal at the base. You determine this by first knowing the collector current (I_C)—the amount of current needed to

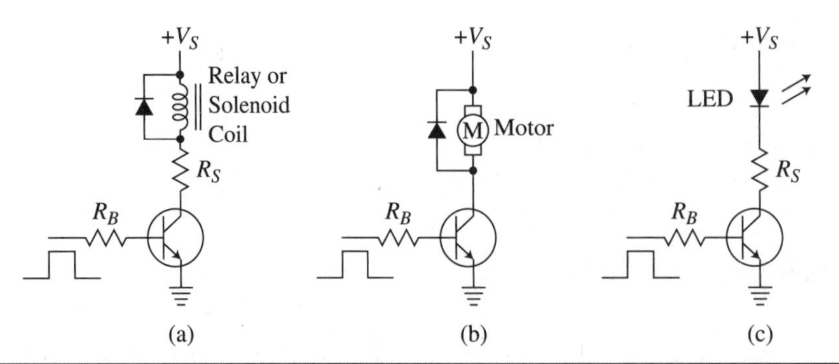

Figure 5.10 Bipolar transistors (*a*) driving a relay or solenoid coil, (*b*) a motor, and (*c*) an LED. Note protective diodes on the circuits driving inductive loads.

operate the device. It may only be a few mA for an LED, but it could be many hundreds of mA for a relay or solenoid or up to an ampere or more for a small motor.

Once you have that value, you can calculate how much base current (I_B) you need to turn the transistor on and cause it to saturate.

The main characteristic of the BJT of interest is the dc current gain, referred to as a beta.

$$(\beta) \text{ or } h_{FE}$$

$$\beta = h_{FE} = I_C/I_B$$

You can find the value of β or h_{FE} in the transistor data sheet. It varies with collector current and temperature. You want to know the minimum value that is usually specified. Minimum values may be as low as 10 for a power transistor or in the 50 to 100 range for an ordinary NPN or PNP transistor. With this value, you can find the needed base current and the value of the base resistor.

As you can see in Fig. 5.10, some of the collector loads are connected directly between the supply voltage (V_S) and the collector of the transistor. When the transistor turns on, the full supply voltage is connected to the load. In other cases, like the LED, a series resistor is used to drop the excess voltage and set the current level. When a series resistor is used, the basic procedure used earlier for finding the series-dropping resistor is used.

Here is the basic design procedure.

1. Define the load characteristics.
   a. LED. Assume a 2-V drop and determine the amount of current to flow.
   b. Relay or solenoid. These devices are designed to operate at a specific voltage level, such as 5 or 12 V. Using the data sheet, determine the amount of current that flows at the rated voltage.
   c. For a motor, identify the stated operating voltage and current.
2. Select or identify the supply voltage. Typical levels are 3.3, 5, or 12 V. This voltage must be the same or higher than the relay, solenoid, or motor voltage.
3. If the supply voltage is higher than the load voltage, you will need a series-dropping resistor to limit the voltage. Use the procedure given earlier to determine that value.
4. Identify the collector current (I_C). This is the current that will flow in the load.
5. Identify the input signal specifications. This will usually be a logic signal that switches between some low value near ground, usually a few tenths of a volt. For most designs, we just assume this level is zero. The upper logic level will usually be the supply voltage level. If the supply voltage is 5 V, the input signal may be 5 V or at least in the 3- to 5-V range.
6. Select a transistor. Any common NPN transistor designed for switching service will work. Be sure the transistor can handle the collector current required by the load. For LEDs and small relays, almost any device will work. For heavy-duty solenoids or a motor, high currents may be present. The transistor maximum collector rating from the data sheet will define the limitations. There are probably dozens of transistors that can be used in this application. It is recommended that you use common popular devices for low cost and availability. The 2N3904 is a good one. So is the 2N2222A if higher current is needed. Other popular choices are 2N4124, 2N4401, and 2N5134. Most of these come in a TO-92 plastic package. Be sure to acquire the data sheet.
7. Calculate the base current needed to saturate the transistor. Base current is found with the following expression:

$$I_B = I_C/h_{FE}$$

Divide the needed collector current by the h_{FE}. Use the minimum value of h_{FE} from the data sheet. Then divide it by two for a safety factor. If the minimum value is 80, use 40 for the value of h_{FE}.

8. Calculate the base resistor value. The voltage across the base resistor R_B is the input logic level V_{IN} less the emitter-base voltage drop (V_{BE}) at the base that is usually 0.7 V for a common silicon transistor. Then divide that by the base current.

$$R_B = (V_{IN} - V_{BE})/I_B$$

Design Example 5.1

1. The load is a 5-V relay that is activated by a current of 40 mA.
2. The supply voltage is 12 V.
3. You will need to get rid of 12 − 5 = 7 V across the series-dropping resistor. With a current of 40 mA or 0.04 A, the resistor value is 7/0.04 = 175 Ω. Standard sizes are 160 and 180 Ω. Selecting the 160-Ω value will ensure enough current to operate the relay.
4. The collector current is 40 mA.
5. The input signal has logic levels of 0 and +5 V.
6. The 2N3904 is selected. The minimum h_{FE} is 70.
7. Using half the h_{FE} of 35, the base current is

$$I_B = I_C/h_{FE} = 40 \text{ mA}/35$$
$$= 1.14 \text{ mA} = 0.00114 \text{ A}$$

8. Assuming that the V_{BE} is 0.7 V, the base resistor then is

$$R_B = (5 - 0.7)/.00114 = 3772\ \Omega$$

The closest standard values are 3300, 3600, or 3900. Any of these will work OK.

MOSFET

More and more MOSFETs are replacing BJTs in basic device drivers. These enhancement mode devices turn on and act as a closed switch when the gate threshold voltage (V_{TH}) is exceeded. With low or zero gate voltage, the MOSFET is off and acts like an open switch. To turn the device on, the gate to source voltage must be greater than the gate threshold value for that transistor. This is a voltage in the 1.5- to 8-V range. When that gate voltage is applied, the device conducts heavily with its current being determined by the applied drain voltage and the load resistance. The on-resistance of the MOSFET is super low, from a few milliohms to several ohms, depending upon the transistor.

A couple of examples are shown in Fig. 5.11. You still need to use a series-dropping resistor if the load voltage is less than the supply voltage. You also should still use the suppression diode across any inductive load.

Here is a summary of the design procedure.

1. Define the load characteristics.
   a. LED. Assume a 2-V drop and determine the amount of current to flow.
   b. Relay or solenoid. These devices are designed to operate at a specific voltage level, such as 5 or 12 V. Using the data sheet, determine the amount of current that flows at the rated voltage.
   c. For a motor, identify the stated operating voltage and current.

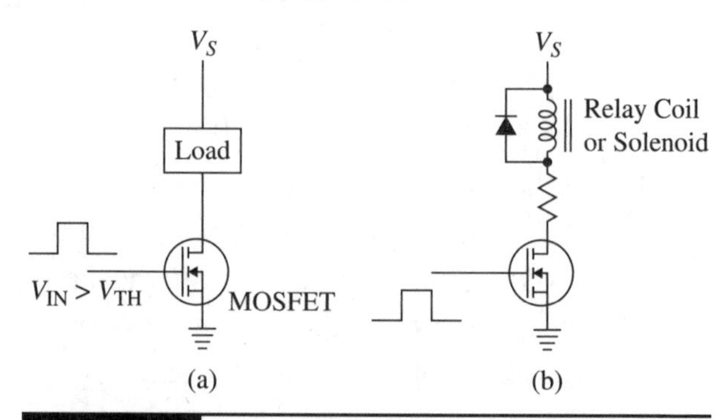

Figure 5.11 (*a*) A MOSFET driving a load and (*b*) a relay coil.

2. Select or identify the supply voltage. Typical levels are 5 or 12 V. Higher levels are sometimes needed. Common values are 15, 24, or 48 V. This voltage must be the same or higher than the relay, solenoid, or motor voltage.
3. If the supply voltage is higher than the load voltage, you will need a series-dropping resistor to limit the voltage. Use the procedure given earlier to determine that value.
4. Identify the drain current (I_D). This is the current that will flow in the load.
5. Identify the input signal specifications. This will usually be a logic signal that switches between some low value near ground, usually a few tenths of a volt. For most designs, we just assume this level is zero. The upper logic level will often be the supply voltage level. If the supply voltage is 5 V, the input signal may be 5 V or at least in the 3- to 5-V range.
6. Select a transistor. Any common N-channel enhancement mode MOSFET transistor designed for switching service will work. Be sure the transistor can handle the drain current required by the load. For LEDs and small relays, almost any device will work. For heavy-duty solenoids or a motor, high currents may be present. The transistor maximum drain rating from the data sheet will define the limitations. There are probably dozens of transistors that can be used in this application. It is recommended that you use common popular devices for low cost and availability. The 2N7000 is a good one and so is the BS170. If higher current is needed, try any of those devices with an IRFxxx number. Most of these come in a TO-92 plastic package, but others may be in a TO-220 package. Be sure to acquire the data sheet.
7. Determine the gate threshold voltage (V_{TH}). It could be from 2 to 7 V or so. In any case, the input logic level must be at least that high to turn the transistor on.

Design Example 5.2

1. Assume that the load is a solenoid. It operates from 12 V with a current of 0.6 A.
2. Use a supply voltage of 12 V.
3. No series-dropping resistor is needed.
4. The transistor must handle a drain current of 0.6 A.
5. Input signal will be zero for off and 12 V for on.
6. The selected transistor is the IRF510. It can handle up to 100 V and 5.6 A. It is housed in a TO-220 package.
7. The gate threshold voltage is in the 2- to 4-V range. That value varies with the device. As a result, the input level should be greater than 4 V to be sure that the device turns on.
8. A suppression diode is needed. A common 1N4001 silicon rectifier diode will work.

Design Project 5.1

Now you try it. Assume a source voltage of 9 V and a standard green LED. Design for a current of 15 mA. What is the series-dropping resistor value? After you make the calculations, you may want to simulate the circuit using one of the simulators given in Chap. 3. This basic circuit will help simplify your introduction to the software. Finally, you should breadboard the circuit and test it for real to be sure it works. See App. B for the solution.

Design Project 5.2

Use the design procedures described earlier in this chapter.

1. Design a voltage divider to produce 1.2 V output with a source voltage of 5 V. The load is a 10 kΩ resistor. Use standard resistor values.
2. Simulate the circuit to check your work.
3. Build the circuit. Adjust resistor values as needed to achieve an output within ±5 percent of the desired output voltage.

See App. B for the answer.

Design Project 5.3

Design a voltage divider using a CdS light sensor and a standard resistor. Use Fig. 5.6*d*. With normal ambient lighting, the divider output should be about half the supply voltage. Use a 12-V supply. When shining a bright light from a flashlight, how does the output voltage vary?

1. Simulate the circuit to test its operation. Most simulation software will have a light-sensitive resistor sensor.
2. Build the circuit and test it.
3. Repeat steps 1 through 3 and use a 10 kΩ thermistor. Heat the thermistor with a soldering iron or hair dryer and describe how the output varies.

Design Project 5.4

1. Design a variable divider to produce an output voltage that covers the range from 2 to 7 V. The supply voltage is 9 V, and the load is 100 kΩ.
2. This is a good one to simulate first until you get it right.
3. Build the circuit and test it.

Design Project 5.5

1. Design a BJT driver for a red LED with a supply voltage of 6 V and a LED current of 20 mA. The logic drive signal on the base is 6 V from the power supply. Use a 2N3904.
2. Design a MOSFET driver for the same LED and same circuit conditions. Use a 2N7000 MOSFET.
3. Simulate both circuits to check your work.
4. Breadboard and test each circuit.

Refer to App. B for the solutions.

CHAPTER 6

Power Supply Design

Every electronic product needs a power supply. All electronic circuits rely on one or more dc voltage sources for power. As a result, any design effort should start by acknowledging the need for a power supply. You should design or specify the end product first to determine the voltage and current requirements of the supply.

Power Supply Choices and Specifications

Your options are batteries, ac power, or solar. If the product is to be portable, mobile, or handheld, a battery supply is mandatory. Figure 6.1 shows the basic configuration. In most designs, a regulator is needed to maintain a steady output to the load. Another addition may be a dc-dc converter. This is a separate IC or module that translates the battery dc into another needed dc voltage, either higher or lower. In some designs, a built-in battery charger circuit is needed.

If the user has a place to plug the product into the ac mains, an ac to dc supply is the best choice. Figure 6.2 shows the main components of such a supply. AC line voltage is applied to a low-pass line filter that passes the 50/60 Hz sine wave line voltage but keeps out any high-frequency transients, harmonics, or other interfering signals. The clean line ac is then stepped down to a lower voltage by the transformer. Diodes forming a rectifier convert the ac sine wave input into a pulsating dc voltage. A capacitor filter smooths the dc pulses into a constant dc. This is usually followed by one or more regulators and/or dc-dc converters.

A special case is the solar alternative. More and more solar supplies are being incorporated into some designs. One common configuration is shown in Fig. 6.3. The solar panel supplies dc to a charge controller that manages the charging of a battery. The battery can then be used as needed. However, in most cases the battery dc operates an inverter. An inverter is a power supply that converts dc into line voltage ac, such as 120 V 60 Hz. The inverter output may be a sine wave or a special more efficient stepped version of a sine wave.

Your first design decision then is to select the power source, battery, ac, or solar.

As a next step, you need to specify the power supply features and characteristics. These include the following:

1. Input voltage, ac or dc. You may need both in some designs.
2. DC output voltages needed. Voltage tolerance: Most ICs require a specific voltage with a tight tolerance of less than about ±5 percent or ±10 percent.
3. DC current requirements. Current needs of the devices being used.
4. Need for regulation. Regulation percentage required.
5. Need for dc-dc converters.

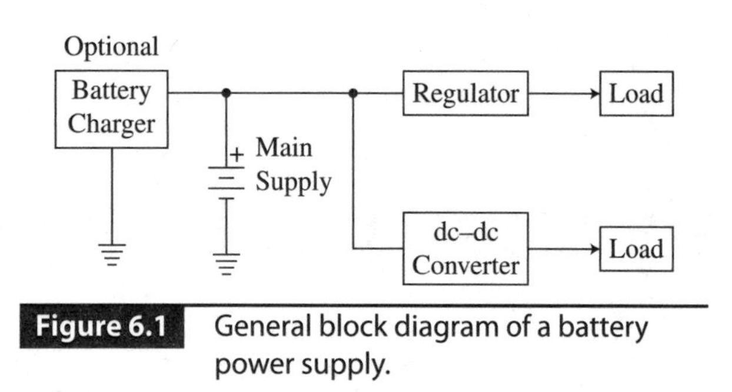

Figure 6.1 General block diagram of a battery power supply.

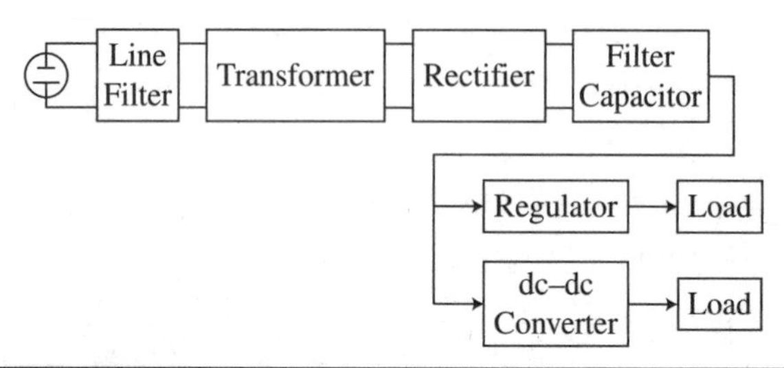

Figure 6.2 Block diagram of an ac power supply.

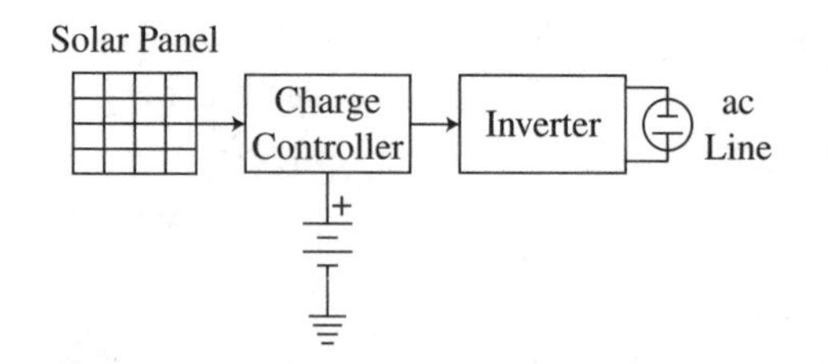

Figure 6.3 The main components of a solar power supply.

6. Efficiency concerns. Efficiency is power out divided by power in. Power supply efficiency ranges from 0 to 40 percent to over 90 percent for some types.
7. Packaging details. Highly variable.

The Make vs. Buy Decision

Because power supplies are so common, there are hundreds of commercial power supply products to satisfy your need. The buy option should be considered because it saves time and effort. However, it may be a more expensive option to designing your own. The trade-off is your design time and the attendant cost versus the extra cost of an immediately available commercial supply.

If you are making a one-off, one-of-a-kind product, a commercial supply makes sense. But if you are designing for a high-volume product, you will typically save more if you design a supply specifically for that product.

Common Voltages

Most circuits are designed to operate from several standard supply voltages. You must stay with these standard values. The most widely used voltages are: 3.3, 5, and 12 V. However, 3, 6, 9, 15, and 24 V are also used. Some special devices use 0.9-, 1.8-, and 2.5-V supplies. The 3.3- and 5-V supplies are used for digital logic and microcontrollers. The 12- and 15-V supplies are common for analog or linear circuits like op amps. You can deviate if you need a special value, but it will complicate the design and may require a special separate supply. It is a costly choice. You can identify your voltage needs from the requirements of the ICs and modules you are using.

Designing a Battery Supply

It is best to use the most common and widely available batteries to ensure a good supply, low cost, and reliability. The recommended types follow:

- Alkaline manganese dioxide cells. These are the popular AA, AAA, C, and D primary cells. Their nominal output is 1.5 V. You connect them in series to get closer to a desired voltage like 3, 6, 9, or 12 V or in-between voltages like 4.5 or 7.5 V. An interesting and still available choice is the 9-V battery, also an alkaline. These were originally designed for small transistor radios. They have been around for decades and are still widely used where low-current devices are to be powered.

All of these cells have a long life and can withstand heavy current draws. Ampere hour (Ah) ratings are in the 90 mAh to 20 Ah range. It should be obvious that the larger C and D cells have a greater current capability than the AA and AAA cells because they contain more chemicals to make current flow. Note that these are primary cells that are not rechargeable—a great choice for many designs. See the section on the Ah rating.

- Lead-acid. These come in a variety of sizes, and they are made up of six 2.1-V cells in series to supply 12.6 V. Small sizes are implemented with a special sealed enclosure and are referred to as gel cell batteries. Ampere-hour (Ah) ratings run from about 3 to 200. Higher Ah ratings can be obtained from what we all know as car batteries. These are available with an Ah rating of 100 A or more. Most electronic products do not require larger batteries like this. Remember, too, this is a secondary type of battery, so you will need a charger. Many models are available for purchase.
- Lithium. These batteries are the current choice for most portable and mobile devices today. Lithium batteries come in multiple variations and may be rechargeable in secondary or primary form. They are typically smaller, lighter, and more expensive than other types, but their energy storage capacity for a given size is better than almost any other type. That is why you find them in smartphones, laptops, electric vehicles, and hybrid vehicles. There is an amazing variety of forms and types. The basic cell voltages run in the −3 to 3.7-V range. A popular example is the coin-sized lithium manganese dioxide ($LiMnO_2$) CR2032 that is not rechargeable. Lithium ion (Li-ion) batteries are rechargeable.
- Nickel cadmium (NiCd) and nickel metal hydride (NiMH) rechargeables are still available but not recommended here at least for common designs. They are more expensive and have a lower cell output voltage of about 1.2 V, so you often must use more of them in series to get to the desired voltage, and they require a charger.

Just So You Know—Cells and Batteries

What most of us call batteries are really just cells. Batteries are multiple cells connected in series to achieve a higher voltage. A 12-V car battery is made up of six 2.1-V lead-acid cells. A common 9-V battery is six 1.5-V alkaline cells connected in series. A battery we call AA is really an AA cell. Anyway, it is good to know the appropriate terminology even though we do not use it.

Ampere-Hour Ratings

The energy capacity of a cell or battery is expressed in ampere-hours, the current in amperes (A) time in hours (h) or Ah. It is often shortened to amp-hours. Some ratings are given in milliampere-hours or mAh. What it means is that you can use a given cell for a specific amount of time at some current level before it is exhausted. A rating of 50 Ah tells you that you can use a cell for one hour at 50 amperes or 50 hours at one ampere, 2 amperes for 25 hours, and so on. With this rating, you can get a feel for what size cell or battery will best match your application. Table 6.1 shows Ah ratings for common cells.

Table 6.1 Ah Ratings

Type of Cell/ Battery	Voltage	Ampere-Hour Rating	Rechargeable?
Alkaline (AA)	1.5	2.2 Ah	No
Alkaline	9	565 mAh	No
Lead-acid	2.1	3–300 Ah	Yes
Lithium	3.6–3.7	90 mAh–6 Ah	Yes and some no

Battery Supply Design Procedure

Batteries are a huge subject. There is plenty of information available to answer any questions you may have. Do an internet search and see for yourself. Go to the battery manufacturers for in-depth material and data sheets on the battery you choose.

Here is a basic step-by-step design procedure for a battery-powered device:

1. Identify your voltage needs. What voltages do the ICs and other circuits operate from? What are the voltage requirements of any modules or subassemblies you may be using? Ideally, if you know you are going to use battery power, it is best to design everything to operate from one battery source. Try to make that voltage one that is readily available from a battery. Common battery voltages are 1.5 and its multiples of 3, 4.5, 6, and 9 V. 12 V is popular. Higher voltages can be obtained by putting batteries in series. If you need multiple voltages, it is usually possible to add regulators or dc-dc converters to develop the extra voltages. For example, you may design most circuits to work from 12 V. You may also need a 5-V supply for some digital logic. You can add a 5-V regulator to the battery output to obtain the 5 V. See Fig 6.4a.

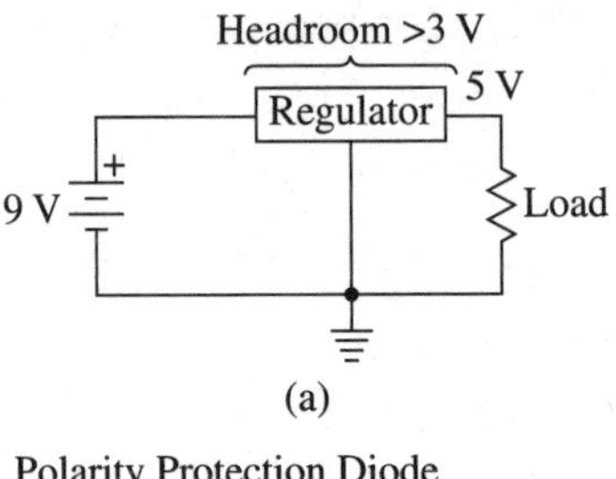

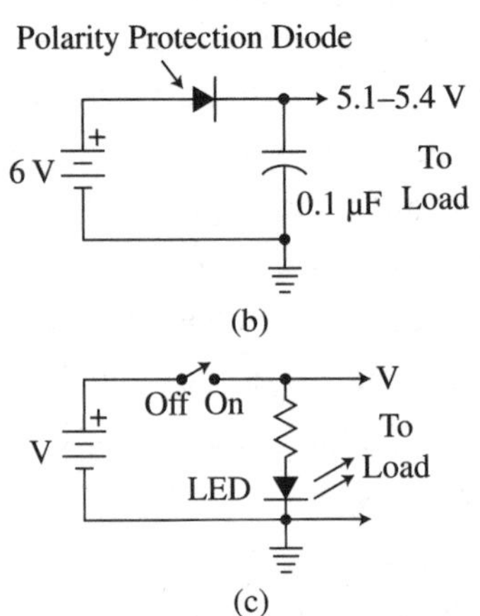

Figure 6.4 Two ways to obtain the desired dc output from a battery supply *(a)* IC regulator, *(b)* series diode, *(c)* an off-on switch, and an LED power on indicator enhancements to a battery supply.

2. Also estimate the current needs of each voltage source. As a safety factor, increase your current drain estimates by 10 to 20 percent.

3. Select either primary or secondary batteries. Primary batteries are not rechargeable. When they die you have to replace them. That is the best choice for some low-priced products. Secondary batteries are rechargeable. That means you will need a charger. Will the charger be an external unit or built into the product? Chargers add considerable extra cost but ultimately save on battery replacement costs over time.

4. Select the battery type. For primary batteries, the widely available AA, AAA, C, or D alkalines are popular. Lithium primary batteries of various types are also available. For rechargeable types, there are multiple choices. The lead-acid is the obvious choice for a 12-V supply. Car batteries are usually too large to be practical, but smaller sealed gel cell lead-acid batteries are also available. Lithium ion rechargeables are also available. Select only popular sizes and types for lowest cost and highest availability from multiple sources.

5. Estimate battery life. Batteries are rated in ampere-hours (Ah). That rating tells how long you can draw current from them before they are exhausted. See related sidebar.

Battery Supply Design Example

1. Initial analysis of the desired product indicates that most circuits will operate from 5 V that must be regulated. There are no 5-V batteries. You may also reconsider the 5-V circuits, and use alternative circuits that operate from 3.3 V.
2. Estimated current maximum is 80 mA. Increase that to 100 mA for safety factor.
3. Primary batteries are acceptable. Secondary batteries require a charger that adds considerable extra cost.
4. AA alkalines are a good choice. No charger is needed.
5. Desired operational time before replacement is 10 h. Ten hours at 100 mA gives a rating of $10 \times 0.1 = 1$ Ah. This is a good match to the AA type whose Ah is 2.2.

One solution is a standard 9-V battery. This can be regulated down with an IC to 5 V. See Fig. 6.4*a*. It may seem reasonable to use four AA cells to get 6 V. However, most 5-V linear regulators do not have enough headroom to work at 6 V; therefore, a higher voltage input is needed for the regulator to function. One solution is to use a low-dropout (LDO) IC regulator with a 6-V battery. See LDO sidebar. Another solution is to use 5 AA cells to get 7.5 V, giving the 5-V regulator more headroom. The 9-V battery is probably the best choice.

What Is Headroom?

Headroom is a term that refers to the voltage across a series regulator measured from input to output. See Fig. 6.4*a*. Most regulators must have enough headroom voltage to function properly. For linear regulators, that is a minimum of about 2 to 3 V. Any lower than that and the IC regulator will not work. The amount varies with the specific regulator IC. Generally allotting 3 V or more is sufficient.

Refer to Fig 6.4*b*. Assume that a 6-V battery supply is available. One simple way to get 5 V is to put a silicon rectifier diode in series with the output. When current is flowing, the voltage drop across the diode is in the 0.6 to 0.9 V. The output will then be in the range from $(6 - 0.6) = 5.4$ V to $(6 - 0.9) = 5.1$ V. This will usually work because it is within a 2 percent tolerance. A bypass capacitor is a good idea for this arrangement.

You will need a battery holder to contain the batteries. You can select from a wide range of choices from manufacturers. Some holders require the individual cell holders to be wired together in series. Some holders are already prewired and just have the + and − output wires or terminals. Holders are usually available for 2, 4, 6, or 8 AA cells. A popular model is that shown back in Chap. 4, Fig. 4.2*a*. Check your distributor Web site.

Battery Supply Enhancements

One desirable addition to a battery supply is a diode to prevent damage to the powered circuits in the event that the batteries are installed with the wrong polarity. Reversing the polarity can cause total failure of many ICs. A series diode like that in Fig. 6.4*b* conducts when the polarity is correct. If the batteries are reversed, the diode is reverse biased so no current flows, protecting the load. One thing to keep in mind is that when the circuit is drawing current, there will be a 0.6- to 0.9-V drop across the diode that takes away some of the available output voltage.

An on-off switch is a common need. It lets you stop the battery drain when you are not using the device. Longer battery life is the result.

A handy addition is a pilot light showing that the circuits are on. An inexpensive LED with a series-dropping resistor can be used as shown in Fig. 6.4*c*. Keep in mind that the LED will take extra current from the battery. A typical LED will draw about 5 to 20 mA. Use the design procedure from Chap. 5.

Specifying a Linear Supply

A linear supply as defined here means one that is operated from the ac mains. Linear also refers to the type of regulator used. Regulators may be linear or switching. The primary supply for many devices or products is a linear supply. It is usually built into the product but could be external if required. Some part of the supply could also be external.

The specification process follows:

1. Identify your voltage needs. State a desired range with maximum and minimum values. If multiple voltages are needed, state each and the desired range.
2. Identify the current requirements for each supply. Increase the maximum value required by 10 to 20 percent for a safety factor.
3. Determine the need for regulation.
4. What are the space requirements?
5. What is the input ac voltage range?
6. Is there a ripple specification?
7. What is the most useful form factor?

Available Form Factors

Form factor refers to the way the power supply is mechanically packaged. Here are the most common arrangements.

- Integral to the product. Assembled on the same printed circuit board of the devices to be powered.
- Open frame. Power supply is built as a stand-alone unit on its own PC board that will be built into the finished product.
- Tabletop. This is a supply that is typically housed in an enclosed package external to the product. A standard power cord connects it to the main ac socket. A separate cable connects the supply to the product itself. Many laptop computers use this form. A standard 2.1, 2.5, or 3.5 mm circular connector is used to connect the supply to the product.
- Wall adapter. Wall adapters are complete or partial power supplies built into a housing that contains the ac power line plug. A cable connects it to the product. Wall adapters come in three basic forms:
    - ac to ac transformer only. The remainder of the supply is inside the product to be powered.
    - ac to dc unregulated. The output is a filtered dc. Linear or switching types are available. Its output connects to the product where additional filtering may be added along with regulators. There is no regulation, so the output voltage will be over a given range. Some applications do not need a specific precise voltage.
    - ac to dc regulated. Most of these use switch-mode regulators for good efficiency and small size.

The first decision is make or buy. It is usually easy to find a commercial supply that does the job. Price and availability are the main considerations. Go to the catalogs or do an internet search to find something that will work for you. In the meantime, while you are deciding upon a power supply, use some batteries to get you going faster.

Power Supply Regulation

Most ICs and other electronic circuits require a fixed power supply voltage for proper operation. That voltage must remain constant. This is the purpose of a regulator circuit. The regulator will maintain a constant output voltage despite changes in the ac line voltage or the load on the regulator. The ac power line varies depending upon the time of day and the load on the line. These line voltage variations cause the rectified voltages to vary.

Other output variations are caused by changes in the load. The greater the load, the larger amount of current drawn from the supply. This causes the output voltage to drop because of the increased current and internal voltage drops in the transformer windings, rectifier diodes, etc. A regulator senses these output changes and corrects them so that the output remains constant. A bonus feature is that the regulator also corrects for ripple variations resulting from insufficient filtering of the rectifier output. Most power supplies use an IC regulator to provide a clean fixed voltage to the product.

The amount of regulation can be determined with the following formulas:

ac line regulation = (change in dc output voltage/change in ac input voltage) × 100%

Load regulation = (No-load dc output voltage − Full-load dc output voltage)/(Full-load dc output) × 100%

Suppose the ac line voltage varies from 124 to 117 V, causing an output voltage variation from 12.8 to 11.6 V. The line regulation then is

$$\text{Line regulation} = (12.8 - 11.60)/(124 - 117) \times 100 = 1.2/7 = 0.17 \times 100 = 17\%$$

Assume the output voltage changes from a no-load voltage of 5 to full-load voltage of 4.75 V. The load regulation is

$$\text{Load regulation} = (5 - 4.75)/4.75 = 0.25/4.75 \times 100 = 5.26\%$$

In both cases, a good regulator at the output of the filter capacitor will fix both problems. Most regulators are of the feedback control type where the output voltage is monitored and compared to a fixed standard. Any variations are automatically corrected for.

Types of Regulators

There are three basic types of regulators commonly used:

- Zener stabilizer—Uses a single zener diode to limit the output voltage range over a very narrow range. It does not use feedback correction. It is only used in cases where some minimum output voltage variation can be tolerated.
- Linear IC—Uses a series power transistor whose conduction is varied to provide correction for any variations. Many IC regulator chips are available to meet fixed or variable voltage needs. Typical efficiency is in the 30 to 50 percent range.
- Switch-mode regulator—Uses pulse-width modulation (PWM) where the duty cycle is varied to maintain a constant output voltage average. More complex than any other regulator. Uses a main switching IC plus external capacitors, inductors, diodes, and MOSFETs to maintain a fixed output voltage. These supplies are very efficient, in the 80 to 95 percent range. Furthermore, they operate at a higher frequency, so they are usually much smaller and lighter. The downside is higher cost than a comparable linear supply.

Zener Stabilizer Design Procedure

The basic zener stabilizer circuit is shown in Fig. 6.5. This circuit is only used where some minor voltage variations can be tolerated and the load current is low, typically, less than 100 mA. It uses a single resistor and a reverse-biased zener diode. Zener diodes come in fixed voltage values. The most common values are 3.0, 3.3, 4.7, 5.1, 5.6, 6.2, 9.1, 12, and values up to

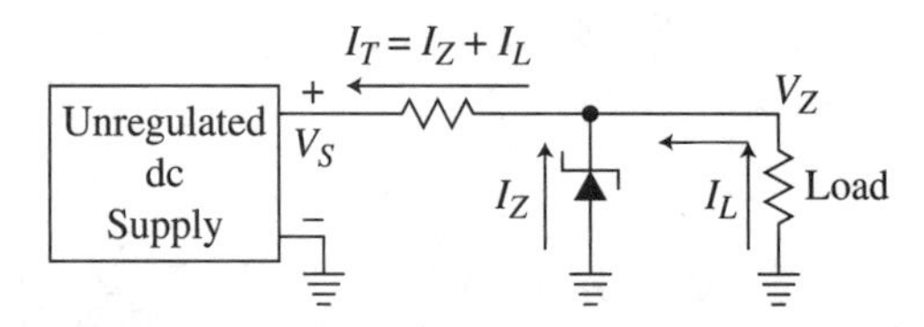

Figure 6.5 Deriving the desired dc output with a zener diode.

about 200 V. Wattage ratings of 0.5 and 1 W are the most common. Some 2- and 5-W units are also available.

Design a zener stabilized power supply to furnish 9 V at 20 mA. Refer to Fig. 6.5. Note that the zener is reversed biased with cathode to +.

1. Make sure that the output from the unregulated power supply (V_S) is at least 1.5 to 2 times the zener voltage (V_Z). More is OK. Let V_S be 12 V.
2. Now, choose the zener with the closest value to the voltage you need. In this case, 9.1 V is available.
3. Confirm the load current I_L. Assume 20 mA.
4. Set the zener current (I_Z) to at least 5 to 10 mA. The goal is to keep the zener into breakdown but to ensure that if the load is removed, the zener can withstand the total current. Let $I_Z = 10$ mA.
5. Calculate the total current through the resistor: $I_T = I_Z + I_L = 10 + 20 = 30$ mA (0.03 A)
6. Now calculate the value of the series-dropping resistor R_S.

$$R_S = (V_S - V_Z)/(I_Z + I_L)$$
$$= (12 - 9.1)/(0.01 + 0.02)$$
$$= 2.9/0.03 = 96.7\ \Omega$$

   Standard values are 91 or 100 Ω. Either of these will work.

7. Calculate the power dissipation of the resistor. The voltage across the resistor is

$$12 - 9.1 = 2.9 \text{ V}$$
$$P = VI = 2.9(0.03) = 0.087 \text{ W}$$

   Choose a ¼ W resistor.

8. Calculate the zener power rating. Use the total current of 0.03 A.

$$P = VI = 9.1(0.03) = 0.273 \text{ W}$$

   A ½ -W zener would work.

Linear Supply Design Procedure

The design process starts with the selection of the rectifier. There are several widely used and proven circuits to choose from. These are illustrated in Fig. 6.6. In all cases, the transformer is used to step down the 120/240-V ac from the power line to a lower voltage closer to what the dc output will be. The half-wave design in Fig. 6.6*a* is the simplest but requires a larger filter capacitor to achieve a desired low-ripple voltage. It is still widely used because of its simplicity and low cost.

The full-wave design in Fig 6.6*b* doubles the ripple frequency, thereby decreasing the filter capacitor value needed for a given ripple. The configuration requires a center-tapped (CT)

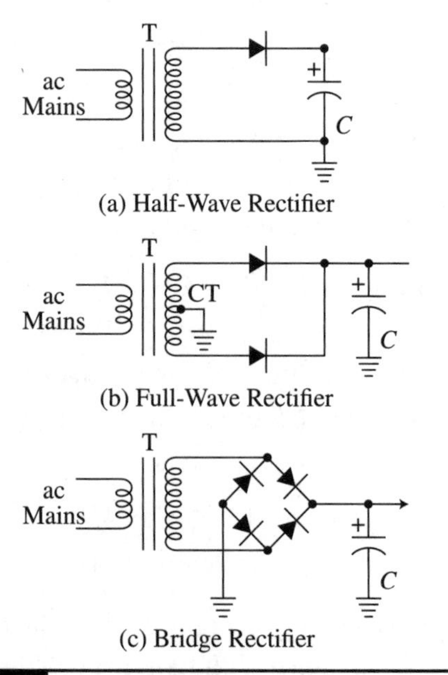

Figure 6.6 Summarizing the main rectifier configurations (*a*) half-wave, (*b*) full-wave, and (*c*) bridge.

secondary winding on the transformer that may be harder to find and a bit more expensive. The full-wave alternative in Fig 6.6*c* is the bridge rectifier. That is probably the optimum choice for most designs.

One good design approach is to work backward from load to power source. Refer to Fig 6.7. Start with the desired output voltage (V_O) and current requirements. Then identify the need for regulation. Most modern circuits require some regulation, so assume you will need it, too. Then choose an IC linear regulator for that voltage and current capability. There are dozens to choose from, but try to stay with popular available ICs.

Linear regulators are available for standard voltage values, including 3.3, 5, 6, 8, 9, 12, 15, and 24 V. These produce a positive output voltage. Negative output regulators are also available for −5, −6, −12, −15, and −24 V. You can also select a regulator whose output voltage is adjustable from 1.2 to 37 V. You set the output voltage with external resistors.

Next, assume that the regulator needs at least 3 or more volts of maneuvering room, called headroom, to regulate properly. Add that voltage to the output voltage to determine the minimum voltage input to the regulator. This is also the voltage across the filter capacitor that will need to be at least this much.

At this point, you can select the filter capacitor. This is usually a polarized electrolytic. Its value is determined by the desired ripple voltage upper limit, the ac line frequency (50 or 60 Hz), and the current being drawn by the load. The formula is

$$C = I/fV_r$$

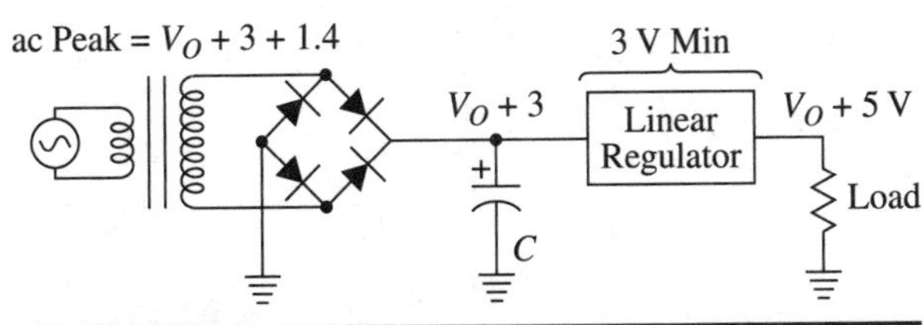

Figure 6.7 A complete typical ac-dc power supply showing the main components.

C is the capacitance in farads (F), I is the load current in amperes, f is double the ac power line frequency in Hz (usually 120 Hz for a full-wave rectifier), and the V_r is the peak-to-peak ripple voltage. The formula gives an approximation, but it will produce a value of C that will work. The actual value will be some larger standard value.

Unless you have been given a specification for a desired ripple level, a good estimate is less than 5 to 10 percent of the output voltage. You can also specify the ripple across the filter capacitor. It can be higher because the regulator will compensate for it and produce a lower output from the regulator. You can start with anything 1 V or less ripple across the capacitor. You can always increase the capacitor value to reduce it.

Select the standard value of capacitance higher than the computed value. Be sure that the selected capacitor will withstand the voltage level.

Assume you require no more than 2 V of ripple across the capacitor. Load current is 200 mA.

$$C = I/fV_r = 0.2/120(2)$$
$$= 0.0008333 \text{ F or } 8333\ \mu\text{F}$$

The closest higher standard value is 10,000 μF. That would work and would give even less ripple. If a regulator is to be used, you can probably withstand a bit more capacitor ripple, so the capacitor can be smaller. Be sure the capacitor voltage rating is higher by at least 20 percent more than you need.

Continuing the design of the supply in Fig. 6.7, the secondary voltage of the transformer can now be determined. Add to the capacitor voltage the drops across the rectifier diodes: one for the half-wave and full-wave rectifiers and two for the bridge rectifier. That voltage drop is in the 0.6- to 0.9-V range for common silicon rectifier diodes. Use 1.4 V for two series rectifiers in the bridge rectifier.

There will also be a voltage loss or drop across the internal resistance of the transformer. It varies widely, so for an initial design just ignore it.

The resulting voltage is what the minimum peak ac voltage should be. The peak voltage is 1.414 times the rms voltage. Transformer voltage ratings are usually given in rms values. Calculate the desired rms secondary voltage.

$$V_{rms} = V_p/1.414 = 0.707V_p$$

Then choose a transformer that will give a minimum of the rms value. This is the value that is produced with an ac input voltage of usually 120 or 240 V from the mains. The transformer may be an open-frame type that you will mount inside the product or a wall-adapter type.

It is not likely that you will find a transformer with your computed voltage. Use standard available voltages when possible. Simply find the transformer secondary of the nearest voltage greater than your computed value. This will give you more peak voltage across the filter capacitor and more headroom for the regulator. However, a disadvantage is that the regulator will be subjected to a higher headroom voltage across the internal series-pass transistor. The result is a higher power dissipation in the regulator and a higher operating temperature. A heat sink may be needed to dissipate the excess heat. One possible solution is to use a low-dropout regulator. See sidebar.

One final word about the transformer. The secondary voltage specified occurs at the maximum current rating of the winding. For example, a transformer with a 24-V secondary winding has a current rating of 2 A. If your design does not draw that much current, the secondary voltage will be higher. It could be over 30 V. The reason for this difference is the resistance of the transformer winding where some voltage is lost when current flows. This higher voltage will also cause the voltage across the filter capacitor and the input to the regulator to be greater. This difference must be factored into the design.

Low-Dropout Regulator

A low-dropout regulator is a special linear design that requires only a small headroom voltage, usually in the 0.1 to 1 V range for proper operation. That means that the source of dc need only be about 0.1 to 1 V more than the output to provide regulation. LDO ICs dissipate less power for a given load current than standard linear regulators. They are available for 3.3, 5, and 12 V outputs.

Linear Supply Design Example

Assume you need a supply that will deliver 5 V at 125 mA. Ripple must not exceed 400 mV peak-to-peak across filter capacitor at the regulator input. You select the bridge rectifier circuit of Fig. 6.6*c*. Now refer to Fig. 6.7 for the following steps.

Select Regulator

First, select an IC regulator. A good choice is the popular 7805. It can handle current up to 1 A. An alternative is the LM340. Both come in the popular TO-220 package. See Fig. 6.8. A smaller version of the 7805 housed in a TO-92 transistor package is also available but with an upper

Figure 6.8 The popular TO-220 package used for power transistors and IC regulators.

current limit of 100 mA. That would not work for this example. You can also use a regulator whose output is set by a pair of resistors. The LM317 is a popular choice. Be sure to go online and get the data sheet for the regulator IC you choose.

Add 3 V headroom to the 5-V output to get the minimum regulator input voltage of 8 V. The voltage across the filter capacitor must be 8 V minimum.

Add to that the rectifier diode drops. An average may be 0.7 V per diode that indicates a total of 1.4 V for a bridge rectifier. The peak voltage at the input to the rectifier should be a minimum of 8 + 1.4 = 9.4 V.

Select Rectifier Diodes

You also need to specify a diode rectifier. It must have a current rating suitable for the desired load current. It also must have a high enough peak inverse voltage (PIV) rating. PIV is the peak ac voltage that the diode sees when it is reverse biased and off. It is usually the peak ac secondary winding voltage. Silicon diodes are preferred. A popular choice is the 1N4000–1N4007 series. All of them can handle current up to 1 A. PIV ratings range from 50 to 1000 V. The 50-V version should work here, so choose the 1N4001. Don't forget the diode data sheet.

An alternative to four individual diodes is to choose a prewired bridge rectifier device. The four diodes are already interconnected inside the package. Again, be sure it can handle the load current and has a PIV to survive the transformer peak voltage.

Select Transformer

Next, define the transformer. The transformer secondary must put out at least 8 + 1.4 = 9.4 V peak. The transformer secondary voltage in rms is

$$V_{rms} = V_p/1.414 = 9.4/1.414 = 6.6 \text{ V}$$

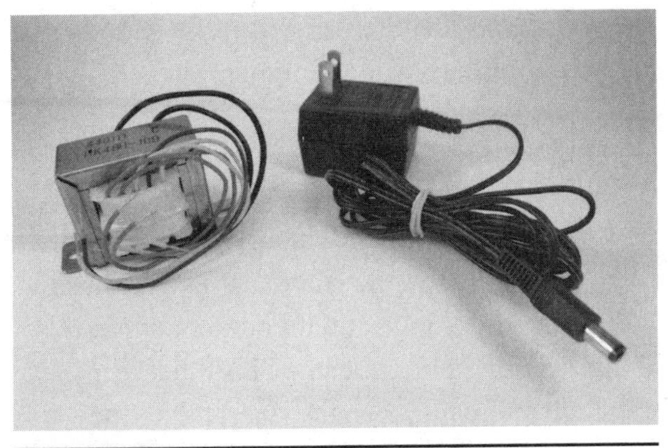

Figure 6.9 Types of commonly used power transformers.

Select a transformer that gives at least this much secondary voltage. Figure 6.9 shows both open-frame and wall-adapter transformers. If you have chosen an external wall-adapter transformer for 120 V 60 Hz, you typically have common secondary voltage options of 5, 9, and 12 V. Choose the 9-V version. Be sure the transformer is rated to supply the desired current.

Open-frame transformers are also available. Typical secondary voltages are 6.3 and 12.6 V. Eighteen- and 24-V windings are also available. The 6.3-V secondary may provide enough voltage. Its peak value would be 6.3(1.414) = 8.9 V. Your calculations said 9.4, but 8.9 may work.

Now work in the other direction to see what the real input to the regulator and capacitor voltage will be. The capacitor will charge up to the peak of the transformer secondary voltage stated in rms voltage. Recall that a 9-V secondary was chosen. That is an rms value, so you need to convert to a peak value.

$$V_p = 1.414 V_{rms} = 1.414(9) = 12.7 \text{ V}$$

Subtract the 1.4 V diode drops.

That gives 12.7 − 1.4 = 11.3 V

Remember that if the full current rating of the transformer secondary is not used, this voltage will be higher. Plan for a higher value. This is the voltage on the filter capacitor. Be sure the capacitor voltage rating is higher.

AC Power Line Filter

Most new power supplies incorporate an ac line filter. This is a low-pass filter that passes the 50/60 Hz line voltage but attenuates any signals that are higher in frequency. This filter keeps any high-frequency signals generated in the equipment from getting on the power line and interfering with some other device. The filter also keeps out high-frequency signals that may be lurking on the power line from getting into the power supply. Select the filter by its mechanical mounting method and the current it will handle.

Calculate the Filter Capacitor

The capacitor voltage rating should be higher than the peak voltage you calculated. Common capacitor voltage ratings are 16, 25, 50, or 100 V. A 16-V unit should work, but to be on the conservative side, select the 25- or 50-V versions to accommodate the higher secondary voltage.

As for the capacitor value, it is calculated from the formula given earlier. It is based upon the current draw and the desired ripple level: I is 125 mA or 0.125 A, f is 120 Hz with 60 Hz input, and maximum V_r is 400 mV or 0.400 V across the filter capacitor.

$$C = I/fV_r = 0.125/120(0.4)$$
$$= 0.002604 \text{ F or } 2604\ \mu\text{F}$$

A 2200 μF capacitor is available and would work but produce a bit more than 400 mV or ripple. A better choice is a 3300 μF aluminum electrolytic capacitor that costs about the same but would give you a ripple less than 400 mV. To calculate it, you can rearrange the basic formula to compute ripple for a given capacitor size. Using a value of $C = 3300\ \mu\text{F}$ or $C = 0.0033\text{F}$,

$$V_r = I/fC = 0.125/60(0.0033)$$
$$= 0.3156 \text{ V} = 316 \text{ mV}$$

Remember, electrolytic capacitors are polarized, so be careful in connecting it.

Regulator Ripple Suppression

Since the regulator automatically compensates for voltage variations due to ac line or load changes, it is also fast enough to correct for ripple variations. In other words, the regulator greatly reduces the ripple. This means that the ripple across the filter capacitor can be higher. This ripple reduction factor is commonly given in the regulator data sheet. That specification is given in dB. The minimum value for the 7805 regulator is 62 dB, and a typical value is 73 dB. Use the 62 dB version for a conservative design.

Remember that voltage decibel calculations are of the form

$$\text{dB} = 20 \log(V_O/V_{\text{in}})$$

V_O is the ripple out of the regulator, and V_{in} is the ripple input to the regulator. We use the previously stated input ripple of 400 mV or 0.4 V. With that we can calculate the maximum ripple at the output of the regulator V_O. Rearranging the dB formula gives

$$\text{dB} = 20 \log(V_O/V_{\text{in}})$$
$$\text{dB}/20 = \log(V_O/V_{\text{in}})$$
$$\text{antilog (dB/20)} = \text{antilog} \log(V_O/V_{\text{in}})$$
$$V_O/V_{\text{in}} = \text{antilog (dB/20)}$$
$$V_O = V_{\text{in}}(\text{antilog})(\text{dB}/20)$$
$$= (0.4)/\text{antilog } (-62/20)$$
$$= (0.4)\text{antilog } (-3.1)$$

Since the ripple reduction factor is an attenuation rather than a gain, we use a −62 dB.

$$V_O = V_{\text{in}}(\text{antilog (dB/20)}) = 0.40/\text{antilog } (-62/20)$$
$$= (0.4)\text{antilog } (-3.1) = 0.000317 \text{ V}$$

or about 317 μV peak-to-peak. A significant reduction.

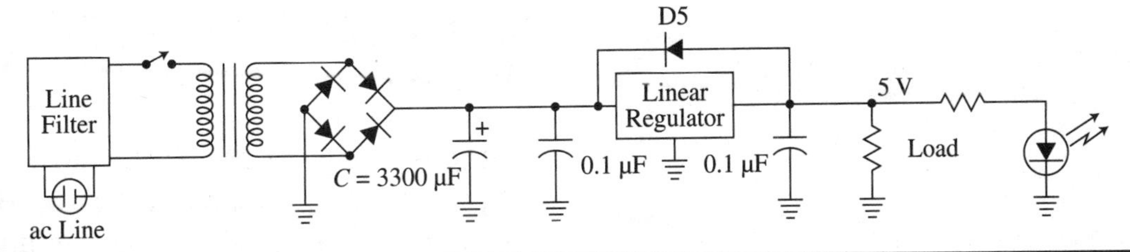

Figure 6.10 A popular ac-dc power supply configuration.

Design Details

There are a few other items to include. These are some additional filter capacitors and protection diodes as recommended in the 7805 data sheet. See Fig. 6.10. If you followed the data sheet, you should be aware to these features. An additional capacitor is added in parallel with the electrolytic filter capacitor. The filter capacitor has a relatively high internal resistance (equivalent series resistance—ESR) that prevents it from filtering out some high-frequency noise or transients. It is traditional to use a 0.1 to 0.33 μF ceramic capacitor across the filter capacitor to correct this problem. In addition, a 0.1 μF capacitor across the output to eliminate noise and transients. Diode D5 protects the regulator in case its input gets shorted to ground for some fault or if the filter capacitor becomes shorted. The popular 1N4001 diode will work here. That concludes the electrical design.

A useful next step would be to simulate this supply on your PC. If it seems to be working correctly, go ahead and build your prototype.

Testing the Design

After you have built a prototype of the supply, now is the time to test it. First, apply ac to the circuit. If no smoke arises from the circuit, measure the output voltage. It should be 5 V ± the percentage stated in the data sheet. The typical range should be 4.75 to 5.25 V. The supply is working.

Finally, measure the ripple, first across the filter capacitor and then at the regulator output. Use an oscilloscope and determine the peak-to-peak value. The ripple across the capacitor should be roughly what you calculated earlier. As for the ripple at the regulator output, it should be significantly less. The regulator action of maintaining a constant output voltage also compensates for any changes in the input and that includes the ripple variations across the filter capacitor. The output ripple will probably be less than that calculated earlier.

Here is something to watch out for. Since the input to the regulator is likely higher than planned, you need to be sure that the regulator can handle the power and heat. Measure the voltage between the output and input of the regulator and multiply it by the load current. That gives you the power dissipation. The maximum power allowed is specified by the manufacturer. Look it up in the data sheet. It depends upon the ambient temperature. With the filter capacitor voltage of 11.3 V, the power dissipation is

$$P = (V_{in} - V_O)/I = (11.3 - 5)0.125$$
$$= 6.3(0.125) = 0.7875 \text{ W}$$

The data sheet says that the maximum power dissipation is about 2 W without a heat sink for the TO-220 package. Use a heat sink (see sidebar) if necessary to dissipate any excess power.

Heat Sinks

A heat sink is a metal surface to which the regulator IC or a power transistor is attached in an effort to help divert heat generated by the series transistor in the IC. If the voltage across the regulator or transistor current is too high and the load current is excessive, the power dissipated as heat may burn up the IC. The heat sink absorbs most of that heat and radiates it into the surrounding air. You can also attach the IC to any nearby metal surface to use as a heat sink. More common is a commercial heat sink made of black anodized aluminum. Get one designed for the TO-220 package of the IC regulator. Figure 6.11 shows what they look like. Most heat sinks are for the TO-220, but others are available for the TO-3 power transistors. Check with your distributor for the many variations that are available.

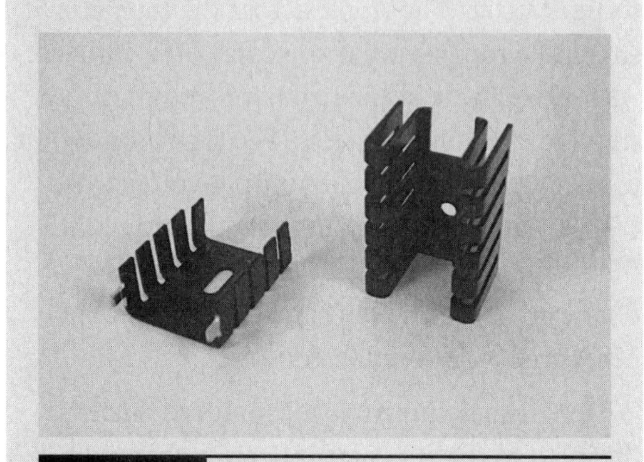

Figure 6.11 Heat sinks are metal devices with expanded surface area that help dissipate heat from an IC regulator or power transistor.

If you are attaching the IC to a metal surface that also serves as ground to your circuit, it may be necessary to insulate the IC from the heat sink. The metal mounting tab of the TO-220 package (Fig. 6.8) is connected to one of the three leads. In the case of the 7805, the ground lead connects to the mounting tab. In this case, you can attach the transistor directly to the chassis that serves as the housing for the product. That chassis is also usually the common ground for most of the circuitry. It is also a pretty good heat sink.

However, some TO-220 packages have either the input or output lead connected to the mounting tab. Connecting the TO-220 directly to the chassis will short the circuit and burn it up. The solution is to get an insulator that isolates the IC output from ground but effectively transfers the heat to the metal chassis. Heat sink mounting kits are available. These include the insulator, screws, and bolts and other mounting hardware as needed. Thermal grease may be needed to ensure good transfer of heat from IC to heat sink.

Switch-Mode Power Supplies (SMPS)

The main reason for choosing a switching supply is greater efficiency. Remember that efficiency is the ratio of the power output (P_O) to the power input (P_I) expressed as a percentage.

$$\%\text{Efficiency} = P_O/P_I \times 100$$

Not all the input power gets to the output. A great deal of it gets converted to heat, usually in the series-pass transistor of the regulator. Diodes and transformer windings also consume power. Less power dissipation means less heat.

The efficiency of a linear supply is no more than 30 to 50 percent, but a switching supply can be as efficient as 90 percent or more. Switching supplies are also usually smaller because they work at a higher frequency (50 kHz to 4 MHz or higher), so transformers and capacitors can also be smaller. Because the transistors switch rather than conduct all the time, efficiency is higher. An on transistor has a very low voltage across it, so power dissipation is minimum. An off transistor has no current flowing in it, so it dissipates no power.

The downside of a switching supply is that it generates noise that can interfere with not only the circuits it is powering but also nearby equipment. High-speed switching circuits produce transients and harmonics that are difficult to eliminate. However, with good design practices, this problem can be mitigated.

If you absolutely must have high efficiency, go with a switching design. They are more complex, but for a volume product, it is probably the best choice. Otherwise, linear supplies are simpler, cheaper, and easier to design.

Types of SMPS

There are two major categories of SMPS: dc-dc converters and regulators. Both circuit types use the same fundamentals. As it turns out, both do the dc-to-dc conversion and both regulate. The circuitry is more complex, but many good ICs are available to simplify that to some extent. Some of the basic and widely used SMPS circuits are covered in Fig. 6.12.

Figure 6.12*a* shows a buck converter. Its output is always less than the input. The input dc voltage from a rectifier and filter capacitor or another regulated dc is chopped into pulses by the MOSFET. When the MOSFET is on, the dc input charges the output capacitor and supplies current to the load. Current in the inductor causes energy to be stored as a magnetic field. When the MOSFET switches off, the current is interrupted and the magnetic field in the inductor collapses, inducing a voltage across the inductor. This voltage polarity is such that the Schottky diode becomes forward biased. Therefore, the capacitor is charged and current flows in the load.

A boost converter is shown in Fig. 6.12*b*. Its output is higher than the input. When the MOSFET is on, the input voltage causes current to flow in the inductor. With the MOSFET off, the magnetic field around the inductor collapses and induces a voltage into it. That voltage adds to the input voltage. The diode conducts and charges the capacitor to the higher voltage and produces current flow in the load.

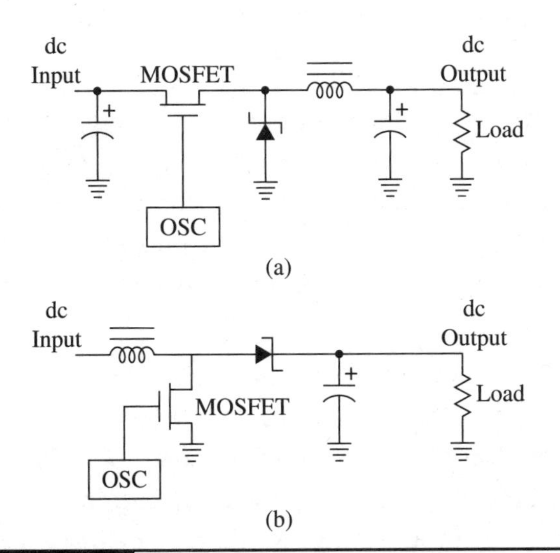

Figure 6.12 Two popular SMPS regulators (*a*) buck, and (*b*) boost.

You may notice that the diodes are Schottky types. Why? Mainly because the circuit is switching at a much higher frequency than the 60 Hz of a linear supply. The range is about 50 kHz to 4 MHz. Standard diodes like the 1N4001 just don't switch fast enough. Schottky diodes do.

There are a half dozen or more variations of these circuits. While the MOSFET is widely used, the transistor switch may also be a bipolar. This switch is driven by an oscillator or clock at a frequency in the 50-kHz to 4-MHz range. Pulse-width modulation (PWM) is used to maintain a constant output voltage. See PWM sidebar. Most of these configurations are implemented in IC form. There are literally dozens of choices from many semiconductor manufacturers. These devices may be used in regulators or dc-dc converters.

Another Form of SMPS

Figure 6.13 shows a complete SMPS. This configuration is called a flyback converter. Note that no input transformer is used. Instead, the bridge rectifier converts the ac line of 120 V into a +170 V dc that is filtered by the capacitor. A MOSFET switch driven by a high-frequency clock in the 50-kHz to 4-MHz range chops the dc into pulses and sends them to the transformer. This transformer steps the voltage down to a high-frequency stream of ac pulses. These are rectified and filtered into dc and applied to the load. The duty cycle of

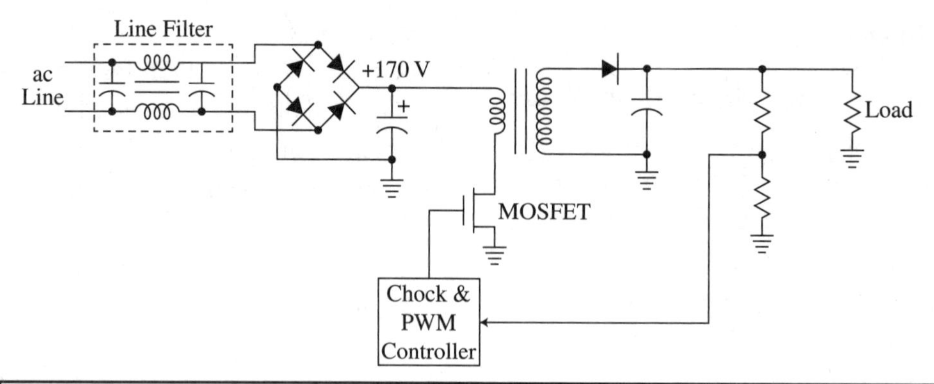

Figure 6.13 A flyback-type of SMPS.

the pulses varies the average on and off time to maintain a constant output voltage. PWM is summarized next.

Pulse-Width Modulation

Pulse-width modulation (PWM) is a method of using variable duty cycle pulses to vary the average dc voltage across a load. Duty cycle is the ratio of the pulse on time to the period of the pulses. Refer to Fig. 6.14.

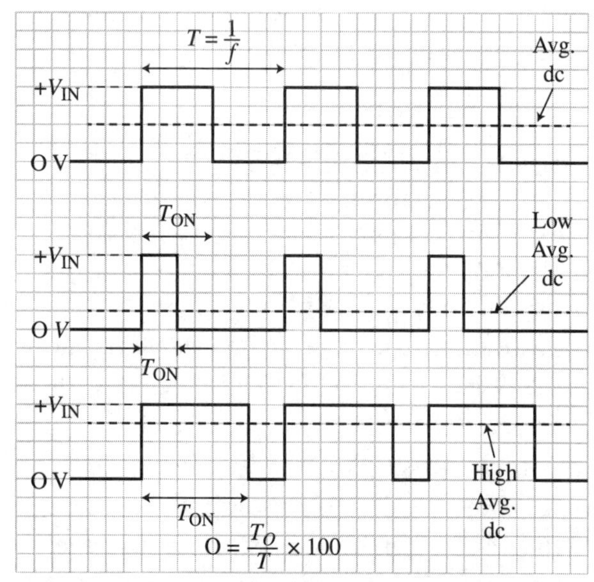

Figure 6.14 Illustration showing the concept of pulse-width modulation (PWM) that varies the duty cycle to correct for output changes.

A clock generates a fixed frequency (f) output with a constant period (T) where $T = 1/f$. The pulse on time (T_{on}) can be varied. The duty cycle (DC) is

$$= T_{on}/T$$

This is a fractional value, but many times you will see it multiplied by 100 and expressed as a percentage.

For example, if the clock frequency is 100 kHz, the period is

$$T = 1/f = 1/100{,}000 = 0.000010 = 10\ \mu\text{s}.$$

If the on time is 5 μs, the duty cycle is

$$\text{DC} = 5/10 = 5 \text{ or } 50\%$$

Now, if the PWM wave is applied to a load, the average voltage across the load will be

$$V_L\ (\text{avg.}) = \text{DC}\ (V_{IN})$$

Here V_{IN} is the peak pulse voltage and DC is the fractional duty cycle. The average voltage across the load is what a voltmeter would read if connected across the load. The pulses are often filtered in a low-pass RC filter to get the average value.

Now, if you vary the duty cycle, you vary the average DC across the load. Figure 6.14 illustrates this. If the input voltage is 9 V peak and the duty cycle (DC) is 22 percent, the average output voltage would be

$$V_L\ (\text{avg.}) = \text{DC}\ (V_{IN}) = 9(0.22)$$
$$= 1.98 \text{ V or about 2 V.}$$

SMPS Components

Most regulator/dc-dc converter ICs contain mostly the circuitry including the MOSFET switch, clock, and the PWM control circuitry. The only external components are the inductor, diode, and capacitors—or the transformer, if used. The inductor and transformer are critical components in a SMPS. They store energy in a magnetic field and then release it to provide current to a load or boost the voltage level. These "magnetic" components add a complex dimension to SMPS design. You may be able to buy some of them, but in some instances, you may have to design the magnetic components yourself.

Another SMPS Example

A popular buck mode (step-down) switch-mode IC is the LM2576HV. See Fig. 6.15. This IC has been around for years, but it is relatively easy to design with. It is available with fixed voltage outputs of 3.3, 5, 12, and 15 V. An adjustable output voltage version is also available. It can supply up to 3 A to a load. The design uses a minimum of external components and is simple to use. It is also one of the few switch-mode controllers that is available in a TO-220 package. Most other switch-mode ICs are housed in smaller surface mount packages. The main switch is bipolar and contained within the IC (LM2575). Also inside is the PWM circuitry for maintaining a constant output voltage. The oscillator runs at 52 kHz. This circuit could easily be substituted for the 7805 linear supply shown in Fig. 6.10, producing a much more efficient supply that runs cooler.

Design Assistance

It is generally recommended that you buy a switching supply. Their design is highly complex, and many options, configurations, and design issues are involved. It is a major engineering specialty. However, you can make your own switching supply by simply following the linear design procedure described earlier for linear supplies then using an available switch-mode controller IC.

Figure 6.15 shows the fixed-voltage circuit using the LM2575. The design is pretty much given to you in the data sheet along with all component values. Download the data sheet for this IC and see for yourself. A similar IC is the Texas Instruments TPS62150.

Another method is to use a switch-mode reference design available from such vendors as Analog Devices, Maxim Integrated, and Texas Instruments. Magnetic components, such as transformers and inductors, are usually needed, and those designs are beyond the scope of this book. Contact vendors like Coilcraft and Bournes because both make standard values of inductors that work in many designs.

To get a feel for this, test out one popular software design tool from Texas Instruments called WEBENCH. Go to www.ti.com/power-designer/switching-regulator. This tool will give

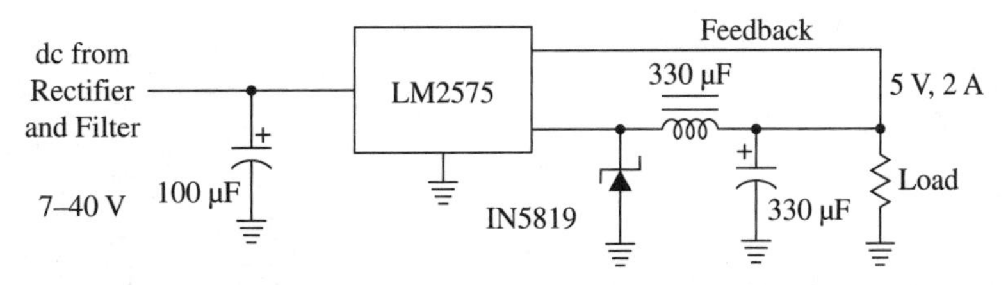

Figure 6.15 A typical SMPS with a commercial regulator IC (LM2575) and necessary external components.

you multiple designs using TI regulators and dc-dc converters. This will give you a feel for the complexity of the switch-mode designs.

Design Project 6.1

If you have become familiar with some circuit simulation software, you should try simulating the power supply in Fig. 6.10. It will reveal any errors in the design. Then you can build a working prototype.

Design Project 6.2

Design a dual-voltage supply with both positive and a negative dc output voltage that can be varied over the 2- to 16-V range. Current capability should be up to 1 A. Use the popular LM317 positive regulator and the LM337 negative regulator ICs in this supply. Use an available transformer and build the circuit on perf board. Be sure to get the data sheets for both ICs.

If this design is too challenging for you, one good alternative is to build this supply from a kit. Jameco sells the JE215 kit that provides the desired output voltages. (See Fig. 4.5.) Elenco Electronics sells the XP-620 power supply that effectively implements this design. It also has a separate 5-V supply. Whatever options you choose, you will end up with a good bench power supply that will power almost any of your projects.

Design Project 6.3

Design a solar supply that will provide at least 100 W of 120 V 60 Hz ac. Use a 12 V gel cell battery. Build and test. Use existing commercial products. This is a good example of a systems design with no circuit design needed. To find all these pieces, use your catalogs, go to local electronics stores, or search the internet. Then assemble and test your emergency supply.

Design Project 6.4

Design an LED night-light using two or three white LEDs. Power it directly from the 120 V 60 Hz line, meaning no step-down transformer. **BE CAREFUL! If you build a prototype, be sure to keep any 120 V wires isolated to avoid shock. Also, do not work on the circuit while it is plugged in.**

See App. B for solutions.

CHAPTER 7

Amplifier Design

Amplifiers are the second most common type of electronic circuit. There are far more digital logic circuits in the world than any other kind thanks to large-scale semiconductor integration, microcontrollers, and computers. There are hundreds of different types of amplifiers that are categorized according to their application. It is just not possible to cover them all in one chapter. Therefore, this chapter discusses only the most common and popular types that you may end up using in a design project.

It is generally unnecessary to fuss with individual transistor amplifier circuit design since it is possible to implement most new designs with available ICs. However, some basic discrete component amplifier design procedures are covered in App. C. If you are interested in more advanced transistor amplifier design, I refer you to the many texts available with that nitty gritty detail (see App. A). That said, the main emphasis here is on operational amplifiers (op amps). Because of their versatility, they can cover probably 90 percent of all your amplifier needs. A few special IC amplifiers will also be covered.

Amplifier Types

The following is a list of basic amplifier types:

- Voltage amplifiers. Boost low-level signals to larger values as needed.
- Power amplifiers. Drive heavy loads like speakers, motors, long cables, antennas, etc.
- Low-noise amplifiers (LNA). Use for very small signals that may be masked by noise.
- Audio amplifiers. Amplify sound signals in the 10-Hz to 30-kHz range.
- Video amplifiers. Amplify video signals up 10 MHz or more.
- Radio frequency (RF) amplifiers. Amplify radio signals in wireless applications over the 1-MHz to 30-GHz range.
- Instrument amplifiers. High-gain differential amplifiers for sensor signal conditioning.
- Op amps. A versatile all-purpose IC amplifier that can be configured or specified for many of the previously mentioned categories.

In summary, most amplifiers are designed based upon the frequency range they cover and whether or not they amplify voltage or power.

Specifying Amplifiers

The main specifications are

- Frequency range or frequency response. Highest and lowest frequencies to be amplified.
- Voltage or power amplification.
- Gain. How much must the input signal be amplified to fulfill the application? The factor by which an input signal is amplified. The ratio of the output voltage or power to input voltage or power. May be expressed in decibels (dB).

- Input impedance. The load presented to the input signal source. A resistive value but may include a shunt capacitive value as well.
- Output impedance. The internal impedance of the amplifier as a driving source.
- Output voltage swing limits or output power to the load.
- Input signal amplitude range.
- dc power supply needed, voltage and current requirements.
- Single-ended or differential input signals.
- For audio amplifier, total harmonic distortion (THD). Waveform distortion indicating heavy harmonic content. For RF power amplifiers, the specification is called linearity, the lack of harmonics and distortion.

Understanding the Specifications

Here is a closer look at several key specifications. The universal symbol for an amplifier is the triangle. See Fig. 7.1.

Gain

The gain of the amplifier is the ratio of the output to the input. For a voltage amplifier, gain is

$$A = V_O/V_I$$

For a power amplifier, gain is

$$A = P_O/P_I$$

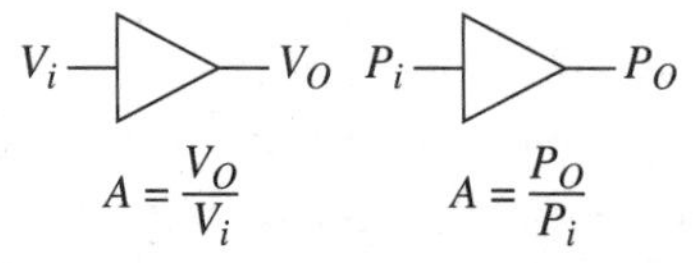

Figure 7.1 The universally recognized symbol for voltage and power amplifiers is the triangle.

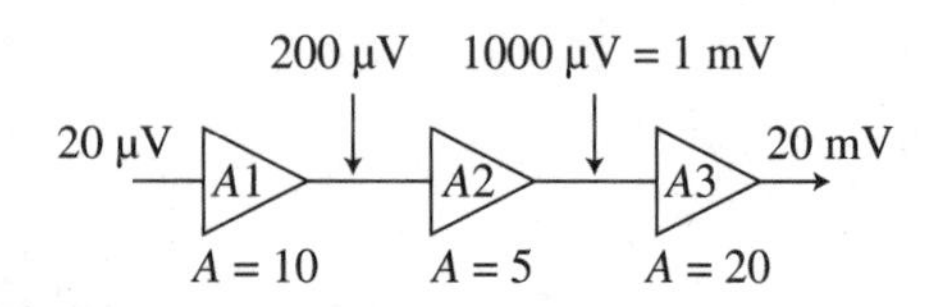

Figure 7.2 The total gain of cascaded amplifiers is the product of the individual amplifier gains.

Sometimes a single amplifier (amp) does not have enough gain to meet the specifications. The solution is to cascade two or more amplifiers. That means stringing the amplifiers together, where the output of one is amplified by the next. Refer to Fig. 7.2. Note that each amp has a gain factor $A1$, $A2$, and $A3$. A signal of 20 μV is first amplified by 10 to get 200 μV, then by 5 to get 1000 μV or 1 mV, and then by 20 to get 20 mV. The total gain is the product of the amplifier gains or

$$A_T = A1 \times A2 \times A3 = 10 \times 5 \times 20 = 1000$$

Rearranging the gain formula, you can see that the output voltage is just the input voltage multiplied by the gain.

$$V_O = AV_I = 1000(20) = 20{,}000\ \mu V = 20\ mV$$

In the real world, the actual gain of each amplifier is not realized because the input to one amplifier loads the previous stage resulting in a lower overall gain. This is caused by the amplifier input and output impedances forming voltage dividers that decrease the gain.

Decibel Measure for Gain and Attenuation

You may see gain expressed as the ratio of the output to the input of an amplifier.

$$A = V_O/V_i$$

$$A = P_O/P_i$$

There is another more common method. It is referred to as decibel (dB) expression of gain or attenuation. It takes the voltage or power ratio and translates it into a dB expression using logarithms. The following expressions show the dB calculations:

$$dB = 20 \log(V_O/V_i)$$
$$dB = 10 \log(P_O/P_i)$$

You can find the logarithm of a number with your scientific calculator.

Some examples follow:

An amplifier with a voltage gain of 500 has a gain of

$$dB = 20 \log(500) = 20(2.7) = 54 \text{ dB}$$

A power amplifier with a gain of 70 has a gain of

$$dB = 10 \log(70) = 10(1.845) = 18.45 \text{ dB}$$

Decibel measure is also used for attenuation or loss calculations.

If the input is 8 V and the output 6 V, the attenuation is

$$dB = 20 \log(6/8) = 20 \log(0.75) = 20(-0.125) = -2.5 \text{ dB}$$

The logarithm of a number less than 1 is negative.

Note that the dB value is negative, signifying loss or attenuation.

If the power input to a filter is 5 W and the output is only 3.5 W, the attenuation is

$$dB = 10 \log(3.5/5) = 10 \log(0.7) = 10(-0.155) = -1.55 \text{ dB}$$

The negative dB value signals a loss.

You will see gain and attenuation more commonly expressed in dB than just a ratio.

Input and Output Impedances

Every amplifier has an input and an output impedance. The input to the amplifier represents a specific amount of resistance. This is a resistance that is usually some complex composite of the amplifier input circuits. It can range from a few hundred ohms to many megaohms—usually the higher the better.

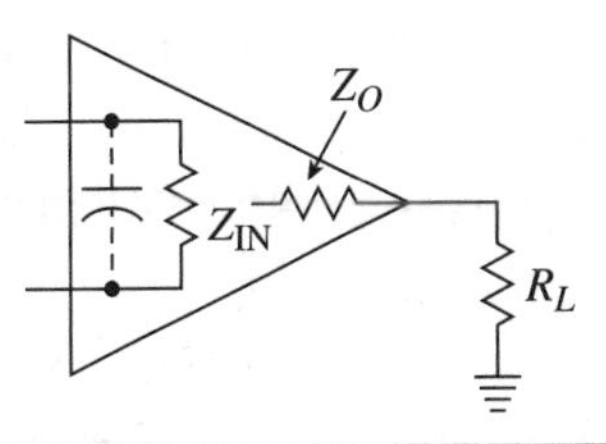

Figure 7.3 A simplified view of the input and output impedances of an amplifier.

The output impedance is also a complex amalgam of circuit values that essentially appear in series with the amplifier output. Figure 7.3 shows a simplified representation. The input impedance is represented by the resistor R_{IN} or Z_{IN}. Most amplifiers also have a parallel capacitance associated with the input resistance. It is usually small (picofarad range). At low frequencies of operation, the capacitance is usually ignored. However, at higher frequencies (greater than several MHz), it begins to act as a low-pass filter, thereby attenuating the higher frequencies.

The amplifier output voltage acts as a kind of signal generator in series with the output impedance R_o or Z_O. The most desirable output impedance, of course, is 0 Ω or as low as possible.

The importance of this is that the input and output impedances appear as part of a voltage divider when the amplifier is connected to an input voltage source and a load. See Fig. 7.4.

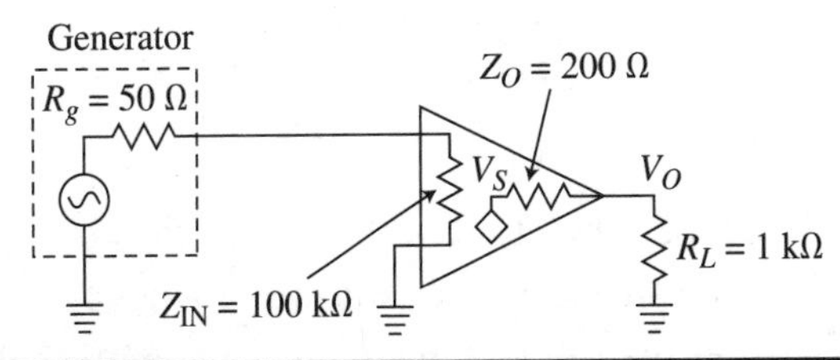

Figure 7.4 How the input and output impedances make up voltage dividers that attenuate the signals being amplified.

The generator has an output impedance R_g that forms a divider with Z_{IN}. The amplifier output impedance Z_O forms a voltage divider with the load R_L. This also occurs when amplifiers are cascaded to get higher gain. The voltage dividers formed actually attenuate the signal, the result being an overall gain factor less than expected. Figure 7.4 illustrates this.

Figure 7.4 shows a signal generator with an output impedance R_g of 50 Ω sending a signal to an amplifier with an input impedance of 100 kΩ. With no load, the generator output is 1.5 V. The generator output impedance is so much lower than the input impedance of the amplifier, the voltage divider effect is minimal. Therefore, we would just assume that most of the 1.5 V from the generator gets to the amplifier input.

To calculate the actual values, use the familiar voltage divider formula.

$$V_{IN} = V_g Z_{IN}/(R_g + Z_{IN})$$
$$= 1.5(100{,}000)/(50 + 100{,}000)$$
$$= 150{,}000/100{,}050 = 1.499 \text{ V}$$

The input voltage then is essentially 1.5 V.

The amplifier has a gain of 4, so it generates a no-load output voltage of

$$V_O = AV_I = 4(1.5) = 6 \text{ V}$$

Theoretically, the output should be 6 V if there is no load or a light load. The output impedance in this case is 200 Ω and the load is 1 kΩ. So we do get some attenuation from the voltage divider made up of the 200-Ω output impedance of the amplifier and the 1 k load.

The actual output across the load then is

$$V_O = V_S R_L/(R_L + Z_O)$$
$$= 6(1000)/(1000 + 200)$$
$$= 6(0.8333) = 5 \text{ V}$$

The amplifier output voltage V_S is the voltage output from the amplifier without a load. It appears across the voltage divider made up of the output impedance of the amplifier Z_O (200 Ω) and the load R_L (1 K). The input is 6 V, and the output is 5 V. The amplifier gain of 4 is reduced by the voltage divider effect. The total net gain then

$$V_O = AV_I$$
$$A = V_O/V_I = 5/1.5 = 3.33$$

If you need more output, you need to adjust the amplifier gain upward to offset the attenuation effect here.

Just So You Know—Rule of Thumb

A design rule of thumb for amplifiers is to keep the input impedance as high as possible and the output impedance as low as possible. In this way, the attenuation is minimum and the gain is maximum. Most circuits are designed this way, but there are still instances where real values are involved, and the voltage divider effect comes into play.

Maximum Power Transfer

In some circuits, especially power amplifiers for audio or radio frequencies (RF), the goal is maximum power transfer. In the previous examples of loading, the goal is to get maximum voltage transfer. You do that with low output impedance and high input impedance.

With power it is different. The condition for maximum power transfer is to have the load impedance match the output impedance. See Fig. 7.5*a*. The output impedance of the

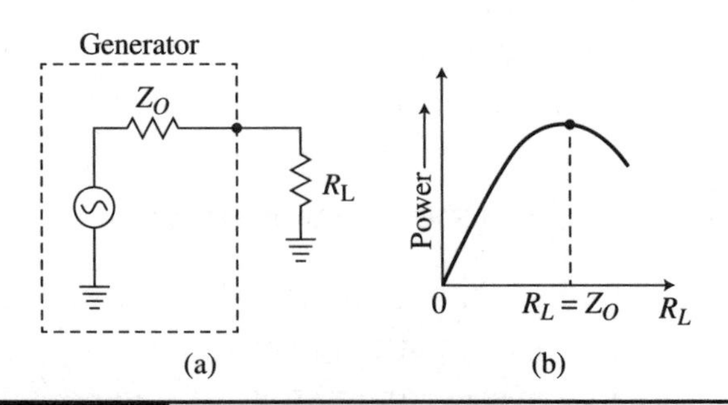

Figure 7.5 (*a*) The circuit configuration that produces maximum power transfer from source to load and (*b*) a plot of power output for different load values.

generator is Z_O. The load is R_L. To transfer maximum power, $R_L = Z_O$. If you plot power delivered to the load versus the value of the load, the curve looks like Fig. 7.5*b*. If load resistance is equal to the output impedance of the driving circuit, you get maximum power. It does not seem logical, but it is true. Under those conditions, the same amount of power is dissipated in the driving circuit (the amplifier) as is dissipated in the load, but that is when maximum power is delivered to the load.

In RF circuits, special impedance-matching circuits made up of inductors and capacitors are widely used to make the load impedance "look like" the output impedance so that maximum power is transferred.

A Microphone Amplifier

Some audio projects require a microphone. A good choice is the low-cost electret microphone. It is made in a small (0.5 in diameter) circular package. The main element of this microphone is a capacitor with one moveable plate. The capacitor is precharged. When you speak into it, the capacitor plate moves in response to the sound waves. This changes the capacitance, and a voltage is developed across the capacitor that is the electrical version of the voice. This signal is very small, so an amplifier is needed to boost the small voice signal up to a useable level.

The electret microphone module contains a built-in JFET amplifier stage. It serves as the interface between the electret element and an external amplifier. This internal amplifier needs a bias voltage to operate. It is usually some dc voltage in the 3- to 12-V range. Refer to Fig. 7.6. The resistor value is in the 1- to 10-k range. Follow the manufacturer's recommendations as presented in the data sheet. The microphone signal is very small, so it is usually sent to another amplifier for a further boost. An op amp is a good choice.

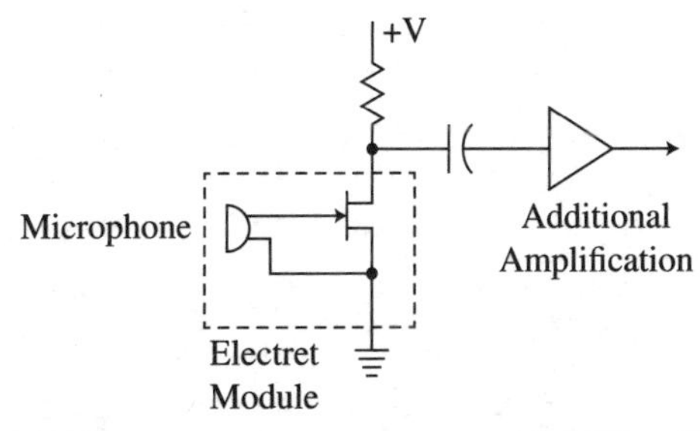

Figure 7.6 A popular microphone for many projects and products is the electret. It features an internal FET that interfaces the high impedance of the microphone element to a following amplifier.

Designing with Op Amps

Perhaps the most useful analog or linear integrated circuit is the operational amplifier (op amp). An op amp is a very high-gain differential amplifier that amplifies dc or ac signals. The versatile IC op amp practically eliminates your need to design discrete component amplifier circuits. High-quality IC op amps are available for every application. For most designs, all you need to do is connect a few external passive components to the op amp. Most linear amplification applications can be implemented with an op amp. IC op amps can be used in any application requiring the amplification or processing of dc and ac signals up to about 1 GHz. Most uses are at audio frequencies and the low MHz range. Output voltage limits are approximately 30-V peak-to-peak maximum, depending upon the power supplies, load value, and frequency of operation. Lower voltages are more common. In this chapter, you will learn to design op amp circuits. Most popular circuit configurations are covered with design examples.

Designing with IC op amps is similar to the generic procedures outlined earlier.

1. Define the need by writing out the functions to be implemented. What will the circuit do?
2. Write down a set of specifications as best you can define them. Define gain, frequency

range, input and output impedances, output voltage levels, and load.

3. Choose an op amp. There are literally hundreds of choices, but it is recommended that you stick to popular proven devices. The old bipolar 741 is probably the most used and still a viable choice. Some widely used bipolar devices have JFET inputs. These are the LF411, TL071/2, and TL081/2. There are also fully CMOS op amps. One good choice is the Texas Instruments TLC2272, a CMOS op amp. The major op amp suppliers are Analog Devices, Maximum Integrated, NXP, and Texas Instruments, so check their product lines for other choices. Don't forget to get the data sheet once you have made your choice. Check for any related documentation or applications notes. Make sure the specifications are compatible with the requirements you defined earlier in step 2.
4. Designate the dc power sources and connections: bipolar or single supply.
5. Select a circuit configuration to fit the application.
6. Calculate the external component values.
7. Determine the need for offset compensation if required to minimize errors.
8. Adjust gain, frequency, and phase compensation as needed.
9. Simulate the circuit. (Optional)
10. Build a working prototype and test it.

DC Power Connections

Most op amps operate from dual positive and negative dc supplies in the 3- to 15-V range. ±12- and ±15-V supplies used to be the most common, but today ±5 is more popular. The dual supply configuration is illustrated in Fig. 7.7. The two supplies are referenced

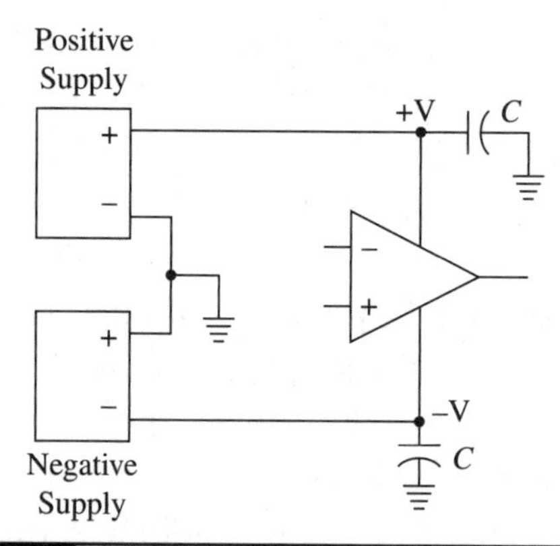

Figure 7.7 The widely used two power supply connections to an op amp. There is no ground on the IC.

to ground, and each is usually decoupled with a 0.001 to 0.1 μF ceramic capacitor that helps reduce the possibility of oscillation and minimizes noise from the power supplies.

Key Point

Note that the op amp itself does not have a ground pin. Ground is established at the junction of the positive and negative power supplies.

You can also use a single supply in some applications. The single supply connection is shown in Fig. 7.8. A dc voltage equal to half the supply voltage is applied to one of the inputs, in this case, by way of R_1 and R_2. R_2 is usually bypassed to ground with a large capacitor. This sets the no-signal output level to a value of half the supply voltage. The output then is centered around this value. If ac operation is the target, you can always use input and output coupling capacitors that block the dc but let the ac through.

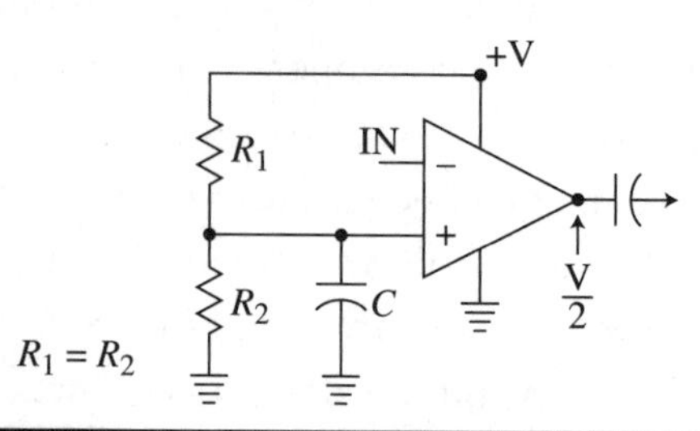

Figure 7.8 A voltage amplifier with a single supply uses a voltage divider to bias the op amp.

The power supply voltages set the maximum output voltage variation of the amplifier. With both positive and negative supplies, the output can vary from negative to positive. The output is zero with no input signal. The peak-to-peak swing for bipolar junction transistors (BJT) op amps is usually 5 to 10 percent less than the sum of the two supply voltages. With dual 12-V supplies, the maximum potential swing will be 24 V. In reality, the maximum output is about 11.4 V. Total output swing is about 5 percent less than 24 or 22.8 V peak-to-peak. This output swing is also determined by the load resistance and operating frequency. The power supply voltages, or power rails as they are called, set the upper and lower limits. Some available all-CMOS op amps can have nearly a full rail-to-rail swing.

Open Loop Operation

An IC op amp is rarely used open loop, that is without feedback. A look at its open loop behavior will help you better understand the operation of this versatile circuit. In most designs, external components are used to provide a substantial amount of negative feedback that sets the circuit characteristics. There are a number of standard circuit configurations that fit most applications, and we will cover them here. The characteristics of each circuit will be given along with recommended applications.

IC op amps have two inputs: an inverting input (−) V_1 and a noninverting input (+) V_2 making this a differential amplifier. The open loop gain, A_{OL}, is very high, typically between 10^4 and 10^9. The amplifier amplifies the difference between the two inputs V_1 and V_2 to get the output V_O. This relationship is

$$V_O = A_{OL}(V_2 - V_1)$$

The open loop gain (A_{OL}) is so high that even with tiny input voltages, the output goes to saturation at one of the output limits. Consider an op amp with ±12-V supplies and a full swing output between the two limits. A_{OL} is 100,000. If $V_1 = 3$ V and $V_2 = 2.5$ V, the input difference is 0.5 V. The output voltage will be

$$V_O = A_{OL}(V_2 - V_1) = 100{,}000(2.5 - 3)$$
$$= 100{,}000(-0.5) = -50{,}000 \text{ V}$$

Obviously, this amplifier cannot deliver that kind of output, so it goes as far as it can and produces an output of almost −12 V at the negative limit.

Other open loop characteristics feature an input impedance in the 100 k to 3 MΩ with bipolar op amps. Using JFETs or MOSFETs to implement the input stage, input impedances of 10 MΩ or more can be achieved. The typical open loop output impedance is usually less than 200 Ω. This is reduced to a few ohms or less when negative feedback is applied.

Frequency Response

See Fig. 7.9. This plot shows the amplifier gain over a range of 0 Hz (dc) to 10 MHz. The open loop gain A_{OL} of op amps is usually over 100 dB or 100,000 and as high as 1,000,000. That gain is good from dc to 10 MHz. The 3 dB down cutoff frequency is 100 Hz, and the gain response rolls off at 6 dB/octave or 20 dB/decade. Depending upon the IC used and its upper frequency, as well as the amount of feedback used, the frequency response can be as high as 10 MHz.

Op amps have a fixed gain-bandwidth (GBW) product, meaning that as gain goes down, the upper cutoff frequency becomes higher. Maximum bandwidth is at unity gain.

$$\text{GBW} = A(BW)$$

BW is the frequency range or bandwidth that is between 0 Hz (dc) and upper frequency limit.

With a GBW of 10,000,000 and a gain of 100, the bandwidth is

$$\text{GBW} = A(BW)$$
$$BW = \text{GBW}/A = 10{,}000{,}000/100$$
$$= 1000 = 100 \text{ kHz.}$$

Verify this yourself in Fig. 7.9.

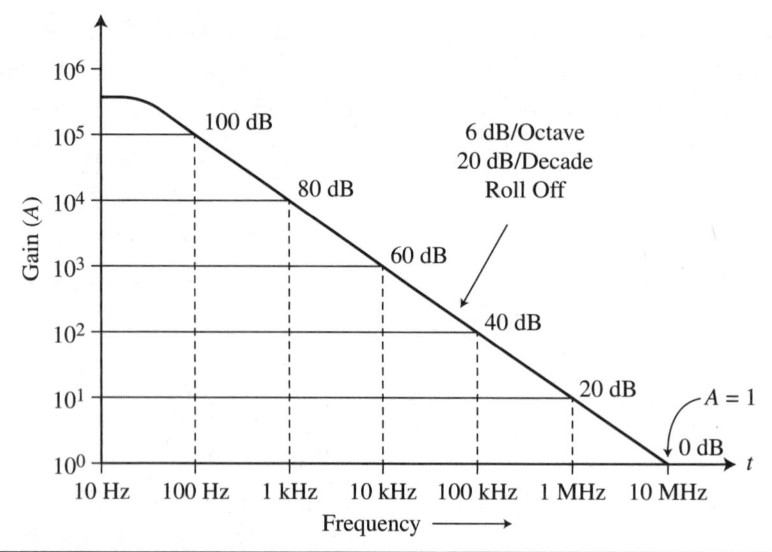

Figure 7.9 An open loop gain plot of an op amp vs. frequency.

Slew Rate

Slew rate is the maximum rate of change of the output voltage over time that an op amp can achieve. It is usually expressed as volts per microsecond (V/µs) or volts per nanoseconds (V/ns). It is one of the basic specifications of an op amp. It's importance is related to the maximum signal frequency possible for a desired output voltage level. The slew rate determines the upper frequency limit of amplification without signal distortion. The slew rate (SR) is a function of the frequency of operation (f) and the maximum peak output voltage (V_O) possible.

$$\mathrm{SR} = 2\pi f(V_O)$$

Usually, you want to know how much output voltage you can get without distortion with a given op amp.

$$V_O = \mathrm{SR}/2\pi f$$

Suppose that the op amp slew rate is 3 V/µs and you want to operate at 250 kHz. What is the peak voltage possible without distortion?

NOTE: To compute with 3 V/µs, use $3/10^{-6} = 3{,}000{,}000$

$$V_O = \mathrm{SR}/2\pi f = 3\ \mathrm{V/\mu s}/6.28\ (250{,}000)$$
$$= 3{,}000{,}000/6.28(250{,}000) = 1.91\ \mathrm{V}$$

Another possibility is, what is the maximum sine wave frequency you can amplify without distortion to achieve a 5-V peak output? SR = 3 V/µs.

$$f = \mathrm{SR}/2\pi V_O = 3\ \mathrm{V/\mu s}/6.28(5)$$
$$= 3{,}000{,}000/6.28(5) = 95.5\ \mathrm{kHz}$$

Primary Op Amp Application Circuits

Here are the most widely used op amp circuits.

Inverting Amplifier

The most popular op amp circuit is probably the inverter shown in Fig. 7.10*a*.

Resistor R_f applies negative feedback to input inverting (−) input. R_i is the input resistor. This arrangement produces a virtual ground at the junction of R_i and R_f at the inverting input. It is not an actual physical ground, but it acts like one. The basic circuit characteristics are

- This circuit amplifies both dc and ac signals.
- This amplifier inverts the input signal by 180°.
- Gain: $A = R_f/R_i$
- Input impedance: $Z_i = R_i$

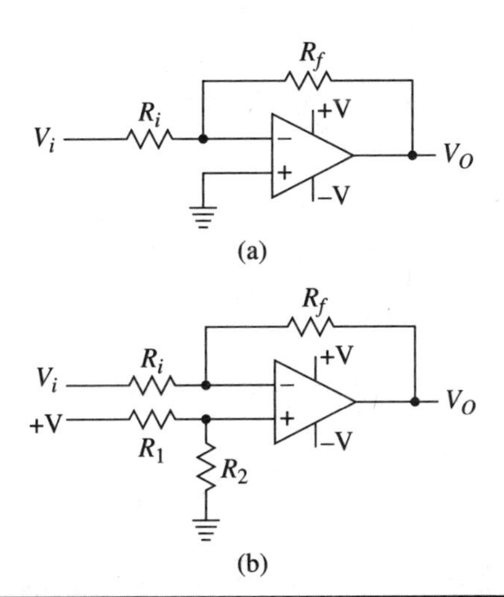

Figure 7.10 (*a*) An op amp inverter, probably the most widely used configuration and (*b*) voltage divider resistors on the + input set the output voltage to about one-half the supply voltage.

- The output impedance, if you really need to know it is to divide the open loop output impedance value in the data sheet by the open loop gain. It is usually a fraction of an ohm. Of course, the ideal is zero, but a fraction is very good.
- The input (V_i) and output (V_O) are related by the following:

$$V_O = -V_i(R_f/R_i)$$

  The negative sign indicates phase inversion.
- Bandwidth range is from dc to the upper 3 dB down cutoff frequency f_{co} that depends upon the op amp and the closed loop gain. Design tip: Virtually any gain up to the open loop gain value is possible by selecting the proper resistor values. Generally, the gain is less than 100, but you can get higher values if you need it. The best design method is to state your desired input impedance then make R_i equal to that value. Then calculate the feedback resistor value knowing the desired gain. $R_f = A(R_i)$.
- Select the power supply voltages based upon the desired output voltage swing. Select the dc supply values 5 to 10 percent higher than the desired output swing for bipolar op amps. For CMOS op amps, you can almost get an output from rail-to-rail.
- For single dc supply operation, connect the circuit as Fig. 7.10*b* shows. The positive supply voltage +V is applied to a voltage divider $R_1 = R_2$ so that the voltage at the noninverting (+) input is one-half the positive supply voltage +V. The output signal will be referenced to that voltage (+V/2).

Just So You Know—Single-Supply Op Amp Circuits

Having to supply two voltages to an op amp is a pain. So if offered, a single-supply circuit will be most likely to be selected over the two-supply configuration. Just keep in mind that the single-supply circuits have the following disadvantages:

- Limited output voltage swing. One-half the maximum of the two supply circuits.
- More likely to clip and introduce distortion in the output.
- A dc output that may complicate the connection to other circuits.
- A more complicated circuit with extra resistors.

Some problems can be eliminated by using capacitor coupling on the input and output that will block the dc but readily pass the ac signals you want to amplify.

Noninverting Amplifier

The circuit is used when you do not want signal inversion; otherwise, its performance is similar to the inverter.

See Fig.7.11*a*. The circuit characteristics are

- The circuit amplifies both dc and ac signals.
- The output is in phase with the input.
- Gain: $A = (R_f/R_i) + 1$
- The input impedance is very high. It is typically higher than the input impedance given in the data sheet. The value is usually many megaohms.

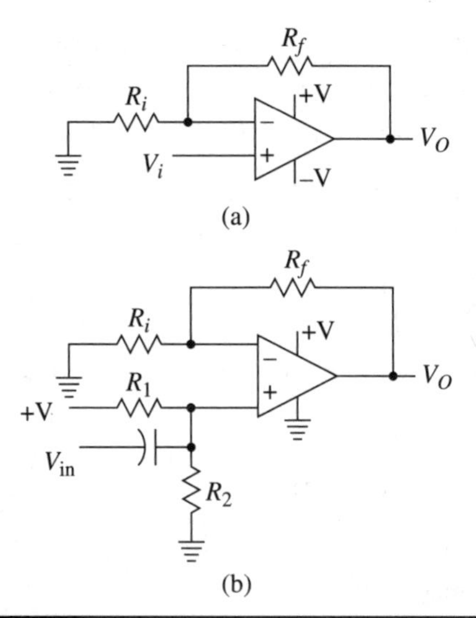

Figure 7.11 (*a*) The often used noninverting amplifier (*b*) can also be set up for single power supply operation by using the bias voltage divider on the input.

- The frequency of operation depends upon the op amp and its GBW.
- The single-supply circuit is in Fig. 7.11*b*. $+V$ is the single supply voltage. R_1 and R_2 are equal and bias the output to $+V/2$. Capacitive coupling gets rid of the dc.

Follower

The op amp follower is used to provide isolation between circuits, buffering, impedance translation, and power amplification. The circuit is shown in Fig. 7.12*a* and is a simple modification of the noninverting amplifier. The single-supply version is given in Fig. 7.12*b*.

The following are the circuit characteristics:

- Unity gain.
- Output in phase with the input.
- Very high input impedance of many megaohms and very low output impedance of a fraction of an ohm.
- Wideband frequency response determined by the op amp.

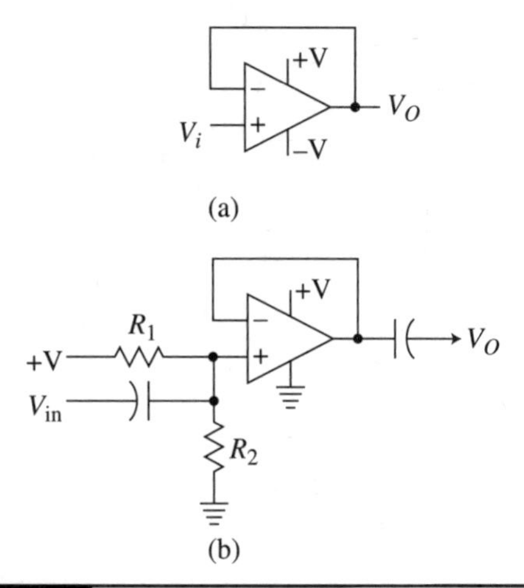

Figure 7.12 (*a*) An op amp follower (*b*) and an ac coupled single-supply version (*b*).

Summing Amplifier

The summing amplifier is an inverting amplifier but with multiple input resistors. See Fig. 7.13*a*. This circuit algebraically adds the input signals. The output is the inverted sum of the input voltages. The inputs may be ac signals or dc. Each input is multiplied by a gain factor that is set by the feedback and input resistors. You can use as many inputs as needed for your design goal.

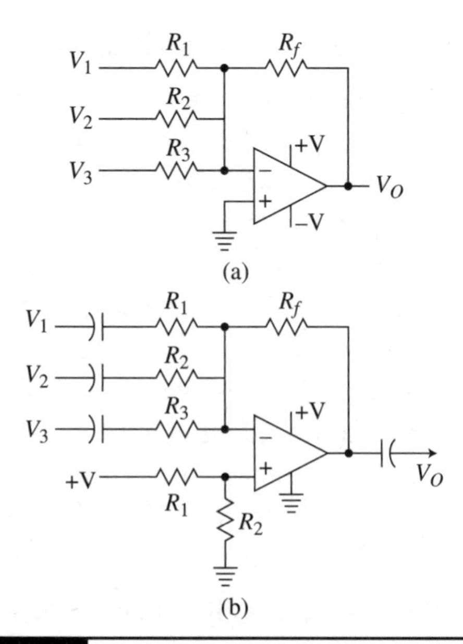

Figure 7.13 (*a*) A summing amplifier (*b*) and its single-supply version.

Typical applications include audio mixers, level shifting, and analog computing.

1. Both dc and ac inputs may be accommodated.
2. Gain = R_f/R_i where R_i is the individual input resistor.
3. $Z_i = R_i$
4. The output is calculated with this expression.

$$V_O = -[V_1(R_f/R_1) + V_2(R_f/R_2) + V_3(R_f/R_3)]$$

The single-supply version with capacitive coupling is shown in Fig. 7.13*b*.

Integrator

An integrator is a circuit that produces an output that is the mathematical integral of the input. The circuit is given in Fig. 7.14. Notice that the feedback component is a capacitor instead of a resistor. Specifically, the output is

$$V_O = -(1/RC)\int V_i\,dt$$

The integrator gain is $1/RC$.

Integrators are used for a wide range of functions. Besides actual mathematical integration, the integrator is used as a low-pass filter, a ramp generator, a triangle generator, and a phase shifter.

With sine wave input signals, the integrator produces phase shift and low-pass filtering. An accurate 90° phase shift occurs because of the mathematical relationship between cos(ωt) and sin(ωt). Note: ($\omega t = 2\pi f$.) See Fig 7.14*b*.

Sine and cosine signals are inherently 90° out of phase with one another. An additional phase shift of 180° occurs since the amplifier is an inverter.

While the integrator is occasionally used with sine/cosine waves, it is more common to process dc voltages. When you apply a fixed dc input voltage (V) to the integrator, its output will be a voltage ramp that will increase or decrease linearly (straight line) with respect to time. This is expressed by the following mathematical formula, according to the expression:

$$V_O = -V_i(t)/RC$$

The dc input voltage is V_i, and t is the time in seconds that the circuit integrates. While the input is applied, the feedback capacitor charges or discharges in a straight line. It is assumed that the initial charge on the capacitor is zero. The capacitor will continue to charge until it is limited by the power supply voltages. Here is an example:

Assume supply voltages of ± 12 V. The outputs saturate at 11.5 V. Let the input resistor be $R = 100$ k, $C = 0.1$ µF, $V = -5$ V, and $t = 8$ ms. The output voltage will rise to:

$$V = -V_i(t)/RC = -(-5 \times 0.008)/(100{,}000 \times 0.1 \times 10^{-6})$$

$$V = 0.04/.01 = 4\text{ V}$$

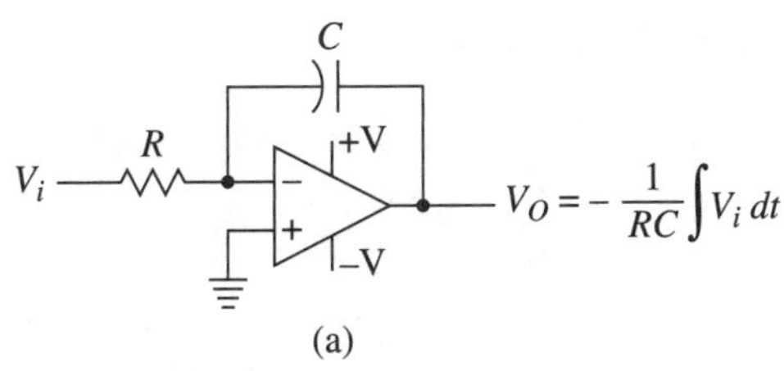

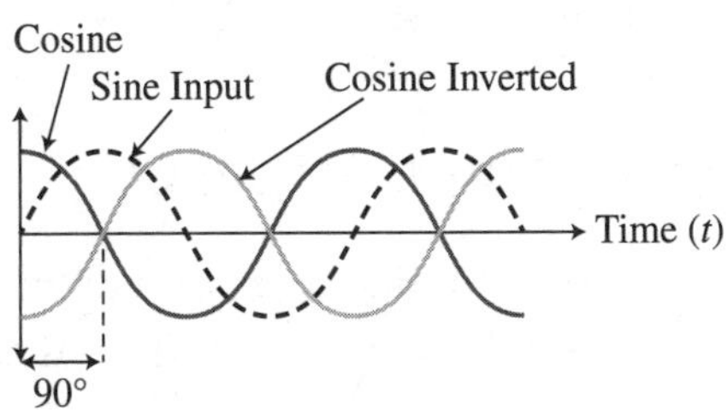

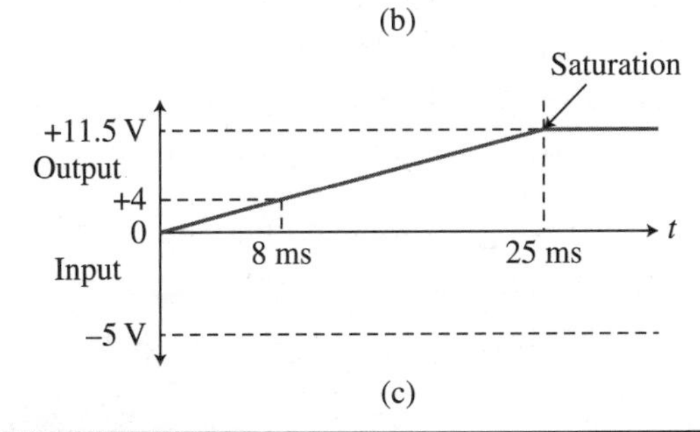

Figure 7.14 (*a*) An op amp inverter with capacitor feedback, (*b*) the sine and cosine inputs and outputs, (*c*) and the + dc output for a negative input.

Refer to Fig. 7.14*c*. Note that the output is going positive because the integrator inverts the polarity of the input.

If you let the integrator continue to run, say for a total of 25 min, the output voltage will hit its +11.5 V limit.

The integrator gain ($1/RC$) and the input voltage value determine the rate of rise of the output voltage. The higher this gain and input voltage, the faster the output will rise or fall.

One useful modification to the integrator is a resistor in parallel with the feedback capacitor. This resistor gives some dc feedback that assists in reducing errors caused by input offset current and voltage. It may not be needed if the op amp specifications for offsets are minimal. If you use the resistor, a typical value for R_f is greater than $10R_i$.

Differential Amplifier

Op amps are all differential amplifiers, but they are rarely used as such. However, there are applications where the differential mode is desirable. A differential connection can reject common-mode signals and noise. The basic differential configuration is shown in Fig. 7.15.

The output is the difference between the two input signals, V_1 and V_2, that is multiplied by the gain that is the ratio of the feedback to input resistors. Generally, the two feedback resistors (R_f) are made equal, and the two input resistors (R_i)

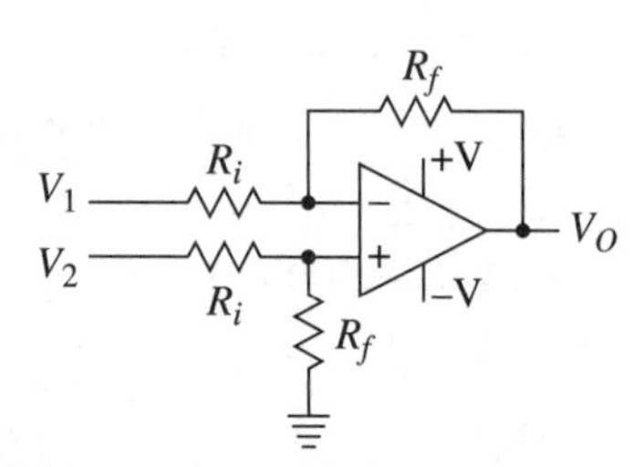

Figure 7.15 A differential op amp circuit.

are made equal. The inputs are usually from single-ended sources. The gain is

$$V_O = (V_2 - V_1)(R_f/R_i)$$

The input can also be a true differential signal that is connected to the two inputs with no ground reference. This causes the output to be two times the normal gain calculation.

$$V_O = 2(R_f/R_i)V_i$$

A common application of a differential amplifier is to help cancel common-mode noise. For example, many circuits are exposed to high levels of 60 Hz power line signals. This signal is induced into both op amp inputs. Since the interfering signals are equal, the amplifier will just subtract one from the other and cancel them out. To ensure the effectiveness of this action, the resistor values must be carefully matched to get optimum cancellation. Typically, 1 percent tolerance of better resistors are needed. The following is a summary of key circuit characteristics:

- Gain = R_f/R_i (single-ended)
- Gain = $2R_f/R_i$ (differential)
- $Z_{in} = R_i$ (for V_1)
- $Z_{in} = R_f$ in parallel with R_i (for V_2)

Instrumentation Amplifier

An instrument amplifier is a special form of differential amplifier widely used to signal condition various sensors in industrial applications. It features very high input impedances and gain plus a single resistor to set the gain. The basic circuit is shown in Fig. 7.16.

The noninverting input amplifiers have an input impedance of 10^9 or more. The gain A is set by R_g. If all other resistors (R) are equal, the gain is

$$A = 1 + 2R/R_g$$

It is possible to achieve gains of up to 1000 or more with this type of amplifier. All of

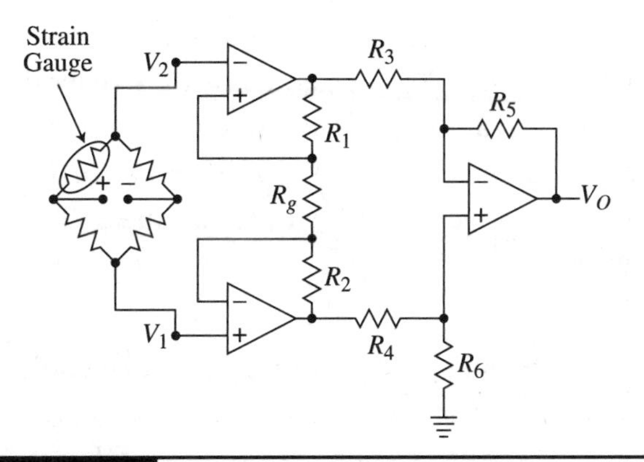

Figure 7.16 An instrumentation amplifier.

the circuitry is made on a single chip, so it is possible to obtain good matching of resistor values and other components to achieve a very high common-mode rejection ratio. The circuit in Fig. 7.16 is used to amplify the very small voltage generated by the bridge circuit containing a strain gauge. The strain gauge's resistance changes with any stress or strain put upon it. That unbalances the bridge, producing a voltage across V_1 and V_2. The voltage variation is sometimes only microvolts, so high gain is needed to boost the signal to a useable level. Good noise rejection is also needed so that this small signal will not be masked by any surrounding noise. Many good commercial IC instrumentation amplifiers are available, so you do not have to build this yourself. An example is Analog Devices AD620 with a −3 dB bandwidth of 800 kHz and a CMRR of 95 dB.

Error Source Compensation

Op amps are not perfect. All of the configurations and their specifications are not ideal, but almost, making the previous descriptions 99.99 percent correct. Op amps suffer from some problems caused mainly by unequal characteristics in the differential amplifier inputs. Differences in the transistor characteristics and resistor values introduce errors into the output. Some of the errors are also affected by temperature. The most detrimental errors are input offset current and input offset voltage. Their presence causes unwanted dc outputs. These unwanted variations hurt you most when you are amplifying dc signals. As a result, you may need offset correction circuits.

These error-causing problems were a major problem in early bipolar op amps. However, with the development of FET input amplifiers and CMOS op amps, these offsets are minor and may not need correction. If you are using an older op amp, you need to determine your need for correction.

For example, if you are amplifying low-level dc signals, minor amplifier offsets will interfere with the small input signal and will be amplified along with the input. The result may be a large error in the output. These offsets that also drift with temperature may actually mask the original input signal to be processed. These offsets must be canceled. The problem is especially acute in an integrator. These minor offsets will get integrated. A huge dc offset will result in an output that has no relationship to the input. Offset compensation is required. The proper adjustment of the offset correction pot is to ground the inputs then adjust for zero or minimum dc output.

Please note that such input current and voltage offsets are extremely small in the newer model op amps. Most do not generally require compensation. But if you are using an older op amp, you can adopt offset compensation methods. Add a few external components to the op amp circuit and solve the problem. Your job will be to determine whether your application requires it. Then implement the circuit in Fig. 7.17.

Older op amps like popular 741 have extra input pins for adding offset correction. The figure shows a 10 k pot with its arm connected

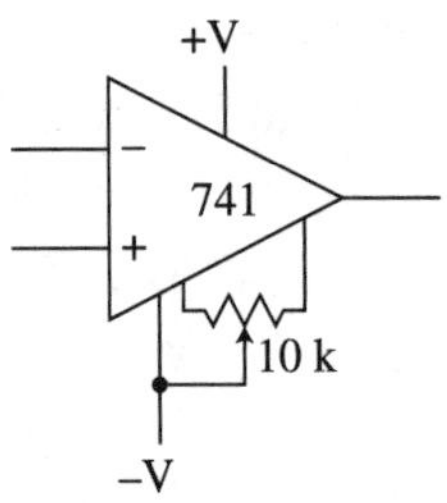

Figure 7.17 A pot used to correct input offset voltage.

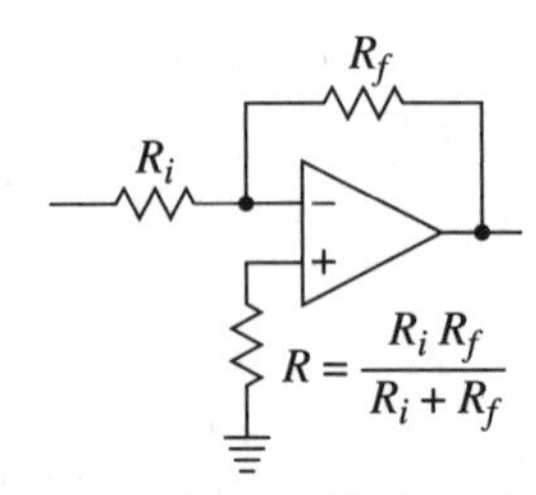

Figure 7.18 A resistor to correct for input offset current.

to the negative power supply. This arrangement usually corrects all offset current and voltage problems. Check the data sheet of the op amp you are using to see the recommended arrangements for offset correction.

One special case involves input offset caused by different input bias currents. One simple arrangement usually minimizes the problem. Figure. 7.18 shows that a resistor of the proper value connected in series with the noninverting input lead can help equalize and balance out the input bias currents. The value of this resistor should be the parallel combination of the input and feedback resistors ($R_f \parallel R_i$). This resistor can eliminate or greatly minimize the offset current problem.

Comparators

A comparator is an op amp-like circuit that compares an input signal to a reference voltage and produces an output voltage that indicates whether the input is less than, more than, or equal to the reference. It is a great decision-making circuit. You can actually use a general-purpose op amp as a comparator. Because of the very high open loop gain, the difference between the inputs is very small but still produces saturation of the output to one power rail or the other.

As a general guideline, op amps make really bad comparators because their output changes too slowly. Furthermore, noise causes false switching and inaccuracies. In addition, the most desirable output is a digital logic level like 0 and +5 V that switches rapidly when the two inputs are equal. Luckily, this problem resulted in a unique comparator circuit that is based upon op amp theory but is optimized for fast response, high speed, and a digital output. Special IC comparators are available when a comparator is needed.

Examples are the popular LM311, LM393, and a four-circuit version called the LM339. The power supply is a single supply like 5 V referenced to ground. These do not use any feedback. And the output is set with an external pull-up resistor in the 1- to 10-k range. See Fig. 7.19. A reference voltage is applied to the inverting input. The voltage divider makes a good reference, but it can be some other source, such as a pot or other op amp circuit. The noninverting input is then compared to the reference. If the input is greater than the reference, the output switches low or near zero. If the input is less than the reference, the output

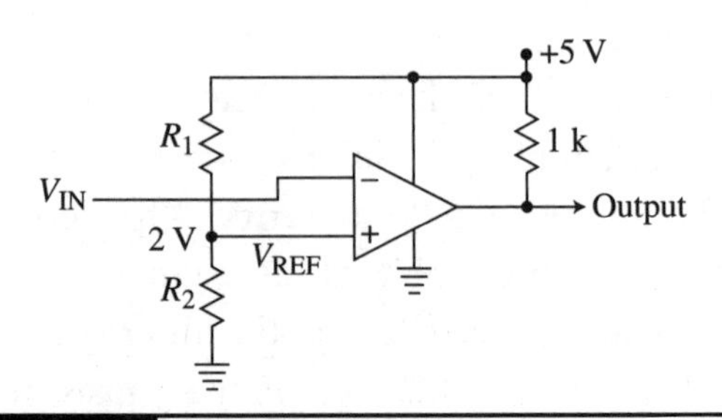

Figure 7.19 A comparator.

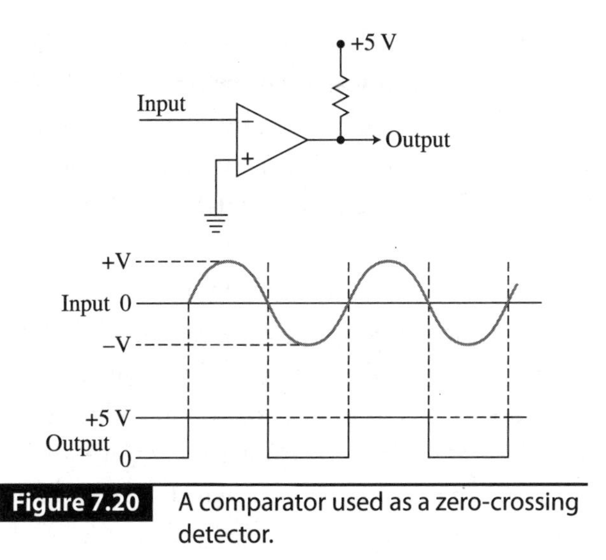

Figure 7.20 A comparator used as a zero-crossing detector.

goes high, in this case set by the pull-up resistor connected to +5 V.

One popular use of the comparator is converting an analog input signal like a sine wave into a compatible digital signal. See Fig. 7.20. Note that the reference voltage is ground or 0 volts. The comparator switches when the sine wave crosses zero. This is called a zero-crossing detector.

Power Amplifiers

You should be able to cover most of your design needs with op amps, but you may encounter some special needs like power amplifiers. A common example is an amplifier to drive a speaker. Again, there is no need to design one yourself since audio power amplifiers are available in several forms. There are power amp ICs, power modules, and available kits and assembled amplifiers. Here are some popular examples.

Op Amp Follower

A follower is a power amp. The voltage gain is only one, but with its high input impedance and low output impedance, it can handle some power amplification needs.

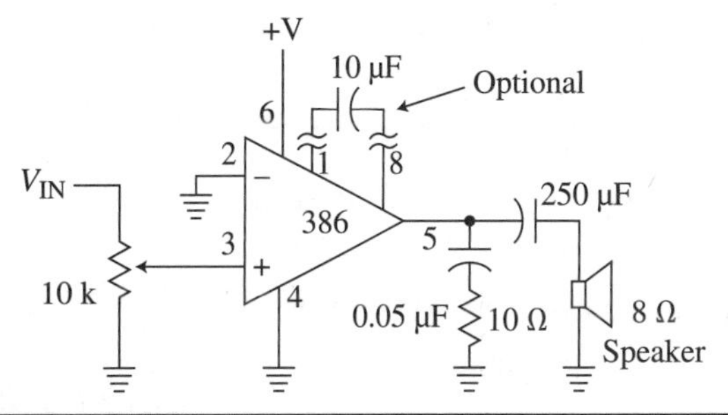

Figure 7.21 The popular 386 power op amp used for audio applications.

IC Power Amp

One of the most widely used audio IC power amp is the LM386. This IC has been around for years, but you will still find it in new equipment. The LM386 is essentially an audio power op amp. Its primary application is to drive a speaker. Figure 7.21 shows the basic arrangement. The load is a common 8 Ω speaker. Note that the speaker is ac coupled with a 250 μF capacitor. The input is applied to a pot that is used as a volume control. The pot value is not critical and should be selected so as not to be a heavy load on the signal source.

The gain of the amplifier is set at 20 by an internal feedback resistor. You can boost the gain to 200 by putting a 10 μF capacitor across pins 1 and 8. Putting a resistor in series with this capacitor will let you set the gain to any value between 20 and 200.

As for dc power, any supply voltage in the 5- to 18-V range will work. The higher the supply voltage, the greater the potential output power. This IC can deliver an output power from about 250 mW to 1 W depending upon the dc supply.

You can also tailor the audio frequency response with an external series RC network at the output as Fig. 7.21 shows. Before you do your design, don't forget to get the data sheet with more details and application info.

If you need more power to drive a larger speaker, you can find ICs that can deliver power levels up to 20 W or so. These usually require heat sinks to get rid of the heat they produce. Some choices include the LM380 that will get you up to about 2 W and the TDA2040V that will deliver up to 20 W. Beyond that, you typically have to use discrete power transistors to deliver 50+ W or more.

Class D Amplifiers

Standard linear (class A, AB, or B) power amplifiers are very inefficient. Much of the power they produce is wasted as heat, yet their benefit is very low distortion. However, there is a way to get low distortion, power amplification, and good efficiency at the same time. Use class D amplifiers. These are switching amplifiers. Like switching regulators, class D amplifiers chop the input signal and develop a switching signal that uses PWM. Figure 7.22 shows the basic process.

The output transistors are MOSFETs, Q_1 a p-type and Q_2 an n-type. The gate driver alternatively turns Q_1 and Q_2 off and on. The output switches from V_{DD} to ground. When a transistor is off, it does not generate power or heat. When a transistor is switched on, its source to drain resistance is only a few ohms or less,

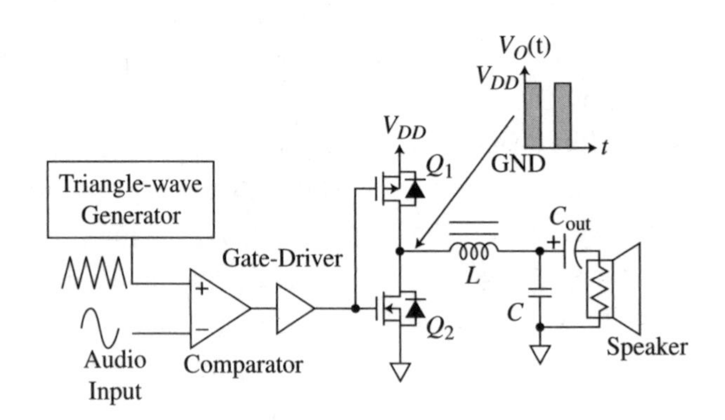

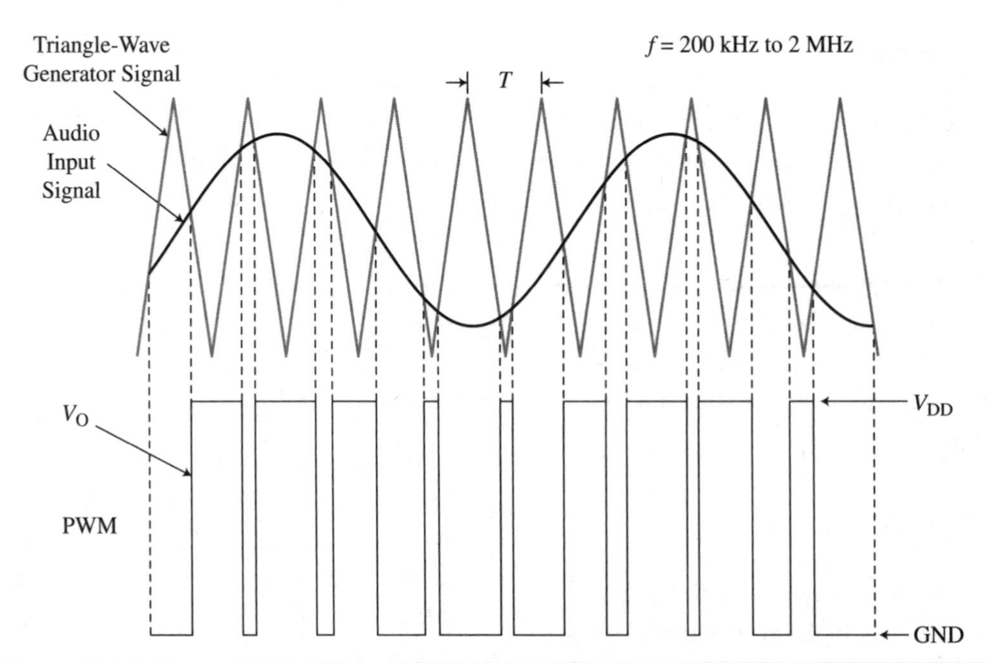

Figure 7.22 General schematic of a class D switching amplifier and how it works using pulse width modulation (PWM).

so little power is consumed. A small amout of power is lost during the transistor switch on and switch off transitions. Overall the efficiency of a class D amplifier is better than 80 percent.

The PWM gate drive signal is developed by comparing a triangle wave to the audio input to be amplified. The comparator output is a PWM signal with a switching speed in the 200-kHz to 2-MHz range. The switching frequency should be higher than the highest frequency signal to be amplified. A factor of five times the highest input frequency of, say, 15 kHz would give a switching frequency of 75 kHz. This signal is sent to the MOSFET switches. An LC low-pass filter between the tranistors and speaker smoothes the PWM pulses back into a high-power version of the input. The pulses are averaged in the filter, and the original audio signal is reproduced in the speaker. Capacitor C_{OUT} removes the dc component.

To obtain a true ac signal at the output and to avoid the need for both positive and negative power supplies, a circuit arrangement called a full H-bridge is used to produce the output. Figure 7.23 shows four MOSFETs connected in an H-bridge configuration.

The speaker is connected between the half-bridge outputs. This arrangemenet is also called a bridge-tied load. There is no ground connection for the speaker as the H-bridge

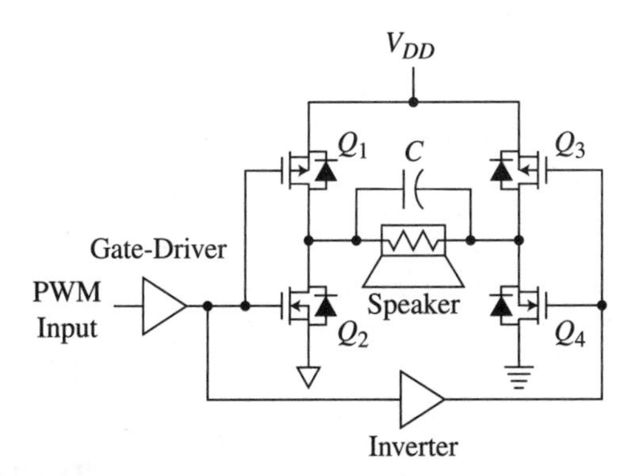

Figure 7.23 One form of bridge-connected speaker in a class D amplifier.

allows current to flow in both directions in the speaker. When Q_1 and Q_4 conduct, electrons flow from right to left. Then Q_2 and Q_3 conduct, current flows from left to right through the speaker. The capacitor across the speaker is the low-pass filter that averages the pulses in the speaker. This connection also eliminates the need for a large coupling capacitor, C_{OUT}, to pass ac but block dc.

Most audio power amplifiers in the 10- to 400-W range use class D. While their harmonic distortion is greater than that of a linear class AB, it is low enough to be acceptable. If you are needing to add an audio amp and speaker to a project, class D is the way to go. Some of the companies making class D ICs are International Rectifier, Maxim Integrated, and Texas Instruments.

Design Projects

Solutions to these design projects are in App. B.

Design Project 7.1

Design and build or acquire an audio amp with speaker to use in projects and other experiments. A recommendation is to build one with the popular LM386. Alternately, design and build the amplifier from a class D chip. Get the data sheet and any app notes for the device selected. Build the amplifier so that you can easily attach it to other audio projects you may be working on. You can also build a kit. A number of kits are available. Several vendors make audio amp kits or prewired modules with power outputs of 5 to 20 W: All Electronics (Velleman K4001) and Jameco (EBST-11). Prices are usually in the $10 to $30 range. These ICs or kits can save you time in developing your end product because you will not be reinventing older circuits. Get one (and a speaker) for your workbench so you will have one to test audio circuits and other projects.

Design Project 7.2

Design an audio op amp with a gain of 15, an input impedance of 10 k, and an upper cutoff frequency of at least 20 kHz. Phase inversion is not necessary or detrimental. The output swing should be at least ±6 V.

Design Project 7.3

Design a two-channel audio mixer for a microphone and a guitar pick-up. Buffer each input with a 1 MΩ input impedance buffer and provide a gain of 10 for each input with individual volume controls. Test the circuit by combining two signals to the inputs, one of them 1 kHz and the other lower, about 100 Hz. Set both inputs to 1 V peak-to-peak. Draw and/or describe what you see.

Design Project 7.4

Using one of the IC comparators described earlier (LM393), design and build a night-light that turns on when the light level drops to some low level. Use a CdS photocell (Chap. 5) as the sensor that activates a white LED for illumination. It might be useful if this one was battery operated so it could be used anywhere needed.

See App. B for typical solutions.

CHAPTER 8

Signal Source Design

Signal sources generate some kind of signal like a sine wave or square wave. Such signal sources are called oscillators, multivibrators, clocks, timers, or frequency synthesizers. Some of these circuits are available as ICs you can use as needed. Others use an IC that requires external components. Some calculations are usually involved. Then there are the circuits you need to design from scratch.

Oscillator design is particularly challenging, as one example in this chapter will illustrate. In many design projects, you can solve the need of a signal source by simply adopting one of the popular circuits summarized in this chapter. No design is needed. Just acquire the ICs and add the external components that set the frequency and other characteristics. Then modify as needed.

Signal Source Specifications

Here are the features and specifications you need to know about signal sources:

- Type of waveform output, sine, rectangular pulses, sawtooth, etc.
- Frequency of operation or range of frequencies.
- Frequency variation if needed. Single frequency only. Continuously variable, frequency steps or increments, frequency range, etc.
- Frequency stability. Amount of frequency deviation allowed variation with temperature, supply voltage, load, etc.
- Frequency determining device or network: Resistor-capacitor (RC), inductor-capacitor (LC), crystal, etc.
- Operating supply voltage and current draw.
- Packaging.

Sine Wave Oscillators

An oscillator is basically an amplifier with a feedback network that sends some output back to the input at a specific frequency. This positive feedback establishes oscillation. The amplifier supplies the gain to offset the attenuation in the feedback network. The overall loop gain is 1 to sustain oscillation. Figure 8.1 shows the concept. The frequency determining network can be RC, LC, or a crystal. This network usually attenuates the signal, and the amplifier supplies the gain to offset that attenuation. If the overall circuit gain is 1, oscillation occurs.

There are two kinds of sine wave oscillators, ones that use an inductor and capacitor (LC) resonant circuit to establish the frequency of operation and another that uses resistor-capacitor (RC) feedback networks to set the frequency. The LC oscillators tend to be used more at the higher frequencies (>1 MHz), while RC oscillators are used mostly at audio frequencies (20 Hz to 20 kHz) and up to about 1 MHz.

Wien Bridge Oscillator

Perhaps the most widely used RC oscillator is the Wien bridge shown in Fig. 8.2. It uses an

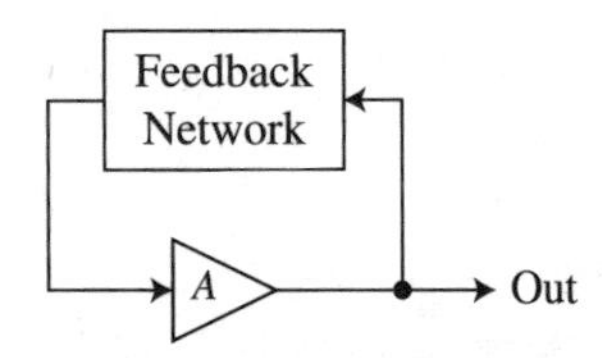

Figure 8.1 The basic concept of an oscillator. Output voltage is fed back to the amplifier input through a frequency-determining network.

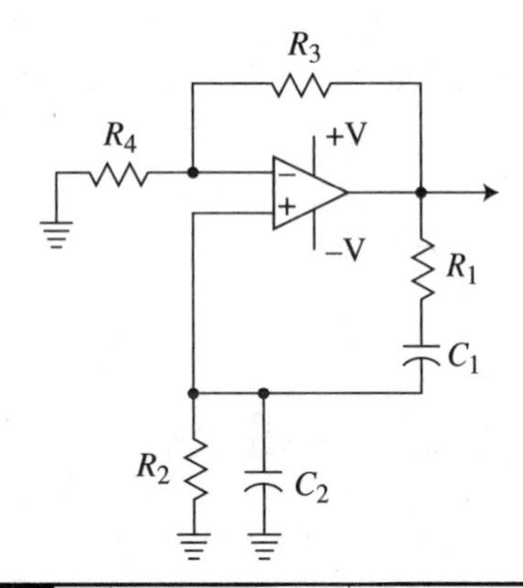

Figure 8.2 The popular Wien bridge oscillator.

op amp with positive feedback through the RC phase shift network made up of R_1, C_1, R_2, and C_2. At one frequency, the phase shift from op amp output to the + input is 0 degrees. Positive feedback is sending the output to the input. The RC network has a voltage divider loss of 3 to 1, so setting the op amp gain to 3 establishes the conditions for oscillation.

The frequency of operation is

$$f = 1/2\pi RC$$

$$C = C_1 = C_2$$

$$R = R_1 = R_2$$

The amplifier gain is set by R_3 and R_4.

$$A = 1 + R_3/R_4 = 3$$

The circuit generates a very clean, low-distortion sine wave. Its main problem is that drift in the components or supply voltage eventually causes the circuit to stop oscillating or causes the output to increase, clip, and become distorted. Several techniques are used to control this problem. See Fig. 8.3*a*. Using a light bulb

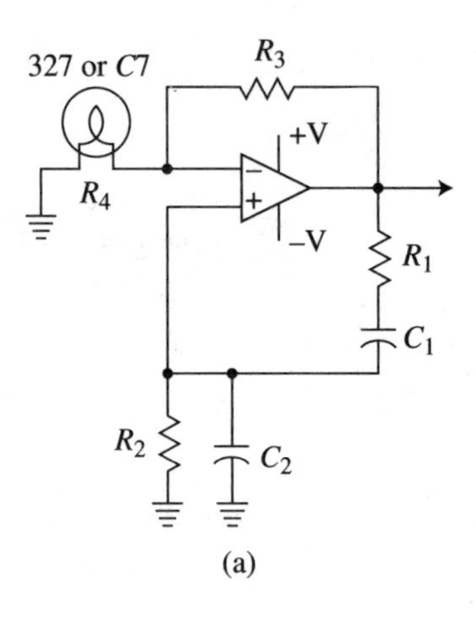

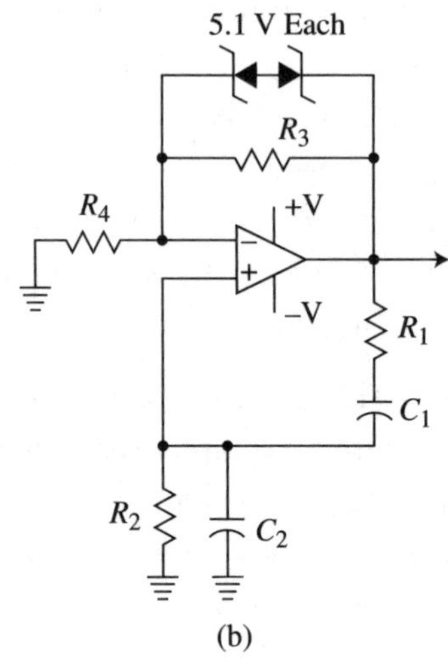

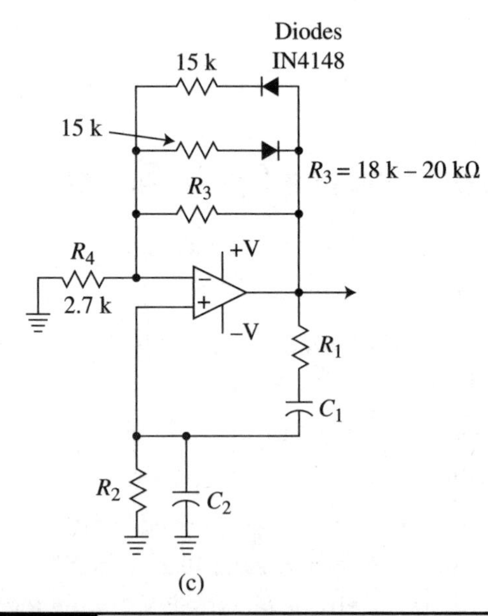

Figure 8.3 The Wien bridge oscillator with an (a) incandescent bulb to stabilize the output, (b) with zener diode stabilization, and (c) with diode output regulation.

for one of the op amp resistors, R_4 works to provide an automatic gain control (AGC) for the circuit. As the output increases, the current in the bulb increases, and its resistance increases. As a result, the gain decreases, the output decreases, and the distortion is eliminated.

The bulbs I have seen used are the 327, 1819, and the old C7 Christmas bulb. Bulbs with a 28-V, 40-mA rating seem to work. These are used as R_4 in the circuit. You will need to play around with the value of R_3 to get the gain set properly. If you use the 327, the feedback resistor R_3 should be in the 300- to 500-Ω range. With the 1819 bulb, a 1 k value for R_3 should work. Using the C7 bulb, R_3 will be in the 1000- to 1500-Ω range. In either case, connect a pot in the feedback path replacing R_3 then vary it until the circuit oscillates. You can leave the pot in there, but you also can take it out, measure its resistance value, and substitute R_3 with the closest standard resistor value.

Incidentally, I found all three bulbs with an internet search. There are multiple sources. You will need a socket to put them in; otherwise, I have been successful in soldering short wire leads on them. Just be careful.

The other technique to stabilize the output uses back-to-back diodes. One such arrangement is shown in Fig. 8.3*b*.

Some distortion occurs, but the output is usually satisfactory for most applications. Back- to-back zener diodes of 5.1 V will limit the output to about 10.2 V peak-to-peak. There are multiple other feedback arrangements that will help stabilize the output voltage. Figure 8.3*c* shows another. The feedback resistor R_3 needs to be in the 18- to 20-k range. Experiment until the circuit stabilizes. Put a variable resistor (pot) in place of R_3 and vary it until stability occurs. Incidentally, the circuits of Fig. 8.3*b* and 8.3*c* will produce some distortion of the output sine wave. The resistor values in Fig. 8.3*c* are approximate, some experimentation is required.

To vary the frequency of the Wien bridge oscillator, you need to change the capacitor values or resistor values. It is easier to change the resistors. One approach is to use two ganged pots whose values are the same and change together as you rotate the shaft.

The Wien bridge is very useful, and you should build the one in Fig. 8.3 on a small breadboarding socket to use as a signal source in other projects. A good useable test frequency is 1 kHz.

The following is the design procedure:

1. State the frequency of operation.
2. State the range of output voltage swing. It will be restricted to a value less than the power supply voltages. You can use the dual ± supplies or a single supply.
3. Select an op amp. Your choice. Almost any one will work.
4. Select a value for *C*. For frequencies below 5 kHz, choose 0.1 μF. For frequencies above 5 kHz, choose something in the 0.001-μF and 100-pF range.
5. Using the formula for frequency, rearrange it and calculate *R*.

   $f = 1/2\pi RC$, therefore, solving for *R*, we get:

   $R = 1/2\pi fC$
6. Select a value for *R*. If you need precision of frequency use 1 percent resistors.
7. Consider a stabilization method with the light bulbs or zeners. The popular zener diodes with designations 1N47xx or 1N52xx will work. The back-to-back zeners will keep the output voltage within the range of $\pm V_Z$, the zener voltage.

   Or use the stabilization circuit of Fig. 8.3*c*.

Colpitts Oscillator

One of the most popular sine wave oscillators is the Colpitts shown in Fig. 8.4. It uses a single transistor amplifier to supply the gain and an

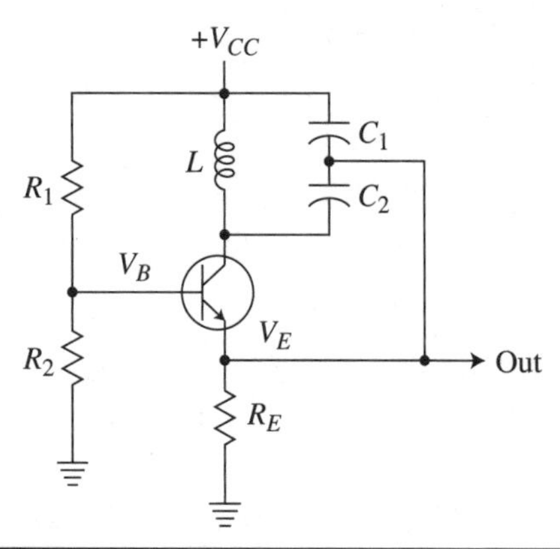

Figure 8.4 A Colpitts LC oscillator.

LC network (L, C_1, C_2) to provide a 180° phase shift at the oscillating frequency. C_1 and C_2 form a voltage divider that sets the amount of feedback to sustain oscillation. The frequency of oscillation is

$$f = 1/2\pi\sqrt{RC_T}$$

Here C_T is the value of C_1 and C_2 in series.

$$C_T = C_1C_2/(C_1 + C_2)$$

The design process is biasing the transistor and determining C_1 and C_2. If you are interested, here are the details and an example.

1. State the desired operating frequency. This oscillator generates a sine wave and can operate at frequencies up to about 30 MHz. You may want to build one to use in other projects. Assume an operating frequency of 10 MHz.

2. Select a transistor. Most will operate at this frequency with sufficient gain. Any of the general-purpose NPN devices will work. Let's select the 2N3904.

3. Designate the supply voltage. Select V_{CC} as 12 V.

4. Designate a collector current (I_C). This is a basic class A amplifier with feedback so is mainly a voltage amplifier and not a major power source. Almost any low current in the 1- to 20-mA range will work. Let's choose 5 mA.

5. Determine the emitter resistor (R_E) value. Since the emitter current is basically the same as the collector current, we can say that $I_E = I_C = 5$ mA. Now we need to know the voltage across R_E. As a rule of thumb, it has been determined that V_E should be about 20 percent of V_{CC}.

   So

   $V_E = 0.2(12) = 2.4$ V.

   Therefore

   $R_E = V_E/I_E = 2.4/0.005 = 480\ \Omega$.

   A standard value of 470 or 560 Ω would work. Choose 560 Ω.

6. Select a value of bias resistor R_2. Experience indicates it should be 2 to 20 times R_E. For this example, choose 10 times or $R_2 = 5600\ \Omega$.

7. Calculate R_1. Since R_1 and R_2 form a voltage divider with a minimal base current load, we can determine the value of R_1. Remember the base voltage is

   $V_B = V_E + V_{BE}$, where V_{BE} is emitter-base voltage of about 0.7 V.

   $V_B = 2.4 + 0.7 = 3.1$ V Round it off to 3 V

   $R_1 = (R_2)(V_{CC} - V_B)/V_B = 5600(12 - 3)/3$

   $= 16{,}800\ \Omega$

8. Select standard resistor values. $R_E = 560$ and $R_2 = 5600\ \Omega$ are standard values. R_1 is 16,000 Ω and a standard value.

   NOTE: This design procedure has been determined experimentally, so all values are approximations. It should produce a working circuit. You can fine-tune the values later. For example, you should adjust R_1 later to optimize the output voltage.

9. Design the tuned circuit. The tuned circuit is a parallel resonant circuit. At resonance, this circuit acts as a resistor. It is referred to as the equivalent impedance Z_{eq}. A rule of thumb is to make Z_{eq} between 5 and 20 times R_E with R_E equal to 560 Ω. Let's

make it 20 times or $Z = 560(20) = 11{,}200\ \Omega$. This impedance is effectively determined by the inductive reactance (X_L) of the inductor and its Q. The relationship is

$$Z = QX_L$$

Not knowing the Q, we can estimate. The Q needs to be 20 or more for the circuit to oscillate. Most inductors can achieve this.

Remember

$$Q = X_L/R$$

Here R is the winding resistance of the inductor. For this design, assume Q is 20, and hope the actual value is 20 or more.

10. Now find the inductance L.

$$X_L = Z/Q = 11{,}200/20 = 560\ \Omega$$

Remember too that $X_L = 2\pi fL$. Rearranging the formula, we can find L.

$$L = X_L/2\pi f = 560/2\pi(10 \times 10^6) = 8.9 \times 10^{-6}$$
or about 9 μH

The closest standard value would be 10 μH.

11. Calculate the capacitor values.

$$C_T = C_1C_2/(C_1 + C_2)$$

The resonant frequency is

$$f = 1/2\pi\sqrt{LC_T}$$

Rearrange this to find C_T. Use the 10 μH value for L.

$$C_T = 1/(2\pi f)^2L = 1/(6.28 \times 10 \times 10^6)^2(8.9 \times 10^{-6})$$
$$= 28.5 \times 10^{-12} = 28.5\ \text{pF}$$

As a general rule of thumb, ratio of C_1/C_2 should be in the 1 to 10 range. Assume $C_1/C_2 = 5$

Therefore $C_1 = 5C_2$

Substituting $5C_2$ for C_1 in the main formula

$$C_T = C_1C_2/(C_1 + C_2)$$

The result is $C_T = 5C_2/6$.

C_2 then is $6C_T/5 = 6(28.5)/5 = 34.2$ pF

$$C_1 = C_T(C_2)/(C_2 - C_T)$$
$$= 28.5(34.2)/(34.2 - 28.5) = 171\ \text{pF}$$

Using standard values, $C_1 = 180$ pF and $C_2 = 33$ pF.

Using these standard values will give us a frequency close to the 10 MHz we want but still different from the target frequency. That frequency difference can now be calculated. The new C_T with standard value capacitors is

$$C_T = C_1C_2/(C_1 + C_2)$$
$$= (180)(33)/(180 + 33) = 27.9\ \text{pF}$$

The frequency of operation where $L = 10$ μH and $\pi = 6.28$ is now

$$f = 1/2\pi\sqrt{LC_T}$$
$$= 1/6.28\sqrt{(10 \times 10^{-6})(27.9 \times 10^{-12})}$$
$$= 9.533\ \text{MHz}$$

That is close to 10 MHz, but there is some error. You can then experiment with the capacitor values to get closer to the 10 MHz target frequency.

12. Build the circuit and test it. Adjust component values as needed.

Crystal Oscillator

A crystal is a thin piece of quartz that vibrates at a very precise frequency with exceptional stability. Crystals are available for fixed frequencies from about 100 kHz to over 100 MHz. The crystal is used as the feedback network with a transistor amplifier to make the oscillator. The crystal equivalent circuit is a high Q series/parallel LC network.

Figure. 8.5 shows the oscillator circuit. The transistor is a popular 2N3904, but almost any common small signal NPN will work. Capacitors C_1 and C_2 set the amount of feedback as in the Colpitts. The output is a very stable sine wave at the crystal frequency. This frequency is very precise, and it does

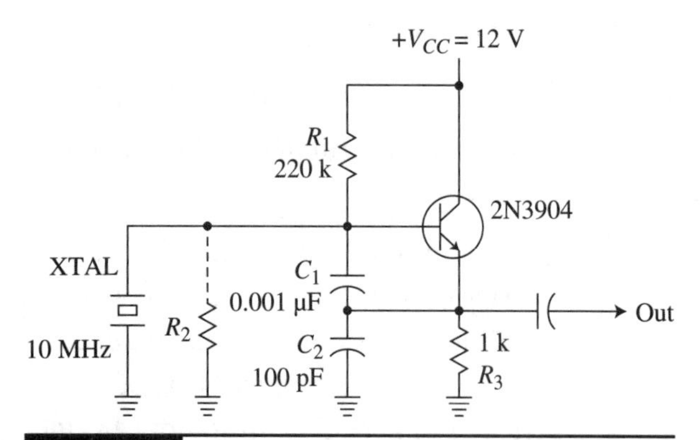

Figure 8.5 A simple crystal oscillator.

not drift much with temperature or other variables. Use this circuit when you need a very stable frequency. It works at frequencies up to about 20 MHz. Again, you may want to build one to use in other projects. A recommended crystal frequency is 10 MHz. Any crystal in the range between 5 and 20 MHz will usually work.

The circuit is basically a follower. You may need to tinker with the values to make it work. Another variation uses an extra bias resistor R_2 (see Fig. 8.5) and different capacitor values. Some other values that seem to work at 10 MHz are $R_1 = R_2 = 33$ k, $R_3 = 2.2$ k, $C_1 = 150$ pF, and $C_2 = 100$ pF.

To minimize the effects of loading on frequency stability, it is recommended that you add a buffer amplifier to the output. Also be sure that the supply voltage is regulated to prevent frequency variation.

Clock Oscillators

A clock oscillator is a circuit that generates a rectangular wave that is used to operate digital circuits. There are a variety of choices for all applications.

A Commercial Option

One popular choice is a crystal oscillator that generates a rectangular wave. You can buy these in the form of a component. Most are packaged in a metal can and operate from 3.3 or 5 V dc. The output is compatible with TTL and CMOS logic circuitry. Many popular frequencies are available from about 1 to 50 MHz. These are inexpensive and work great, and they offer crystal frequency stability in the ±100 ppm (parts per million) range. Plus, these devices will save you from having to design and build one of your own. The most common application is in digital circuit design.

Design Your Own IC Clock Circuits

However, if you wish to make your own, there are a few proven circuits that seem to work just fine. Two of these are shown in Fig. 8.6. One uses TTL logic inverters and the other uses CMOS logic inverters. The crystal provides the feedback and sets the frequency. They operate at frequencies up to about 20 MHz. You should build yourself one of these or get one of the prepackaged metal can units to use with digital logic projects later in this book.

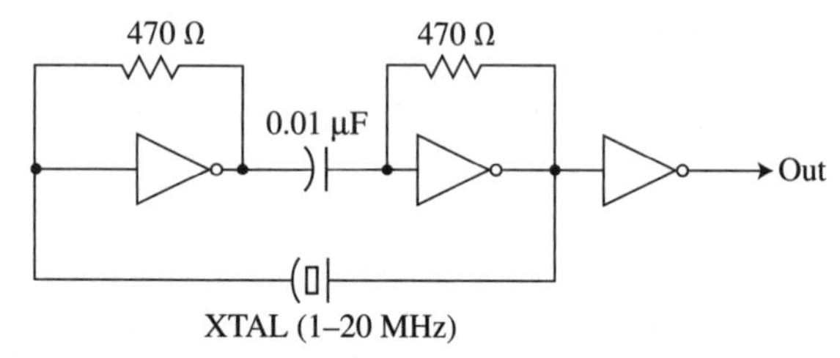

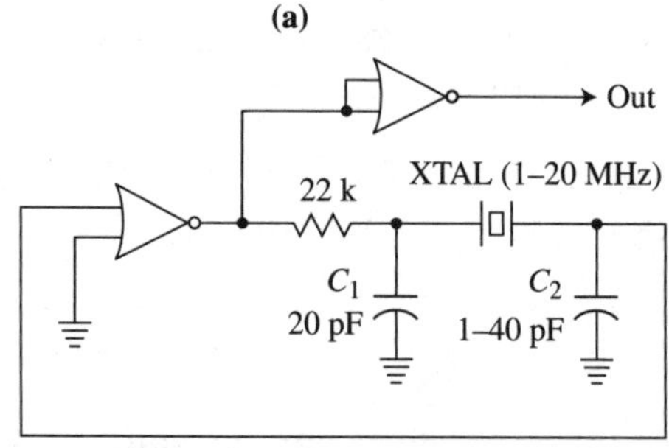

Figure 8.6 (a) A clock crystal oscillator using a TTL IC and (b) a clock crystal oscillator using a CMOS IC.

Multivibrators

Multivibrator is the old traditional name for a circuit that generates a rectangular wave. There are multiple circuits to choose from.

Schmitt Trigger Multivibrator

Figure 8.7 shows one of the simplest. It uses a TTL or CMOS logic Schmitt trigger inverter IC like the 74LS14 TTL or CD40106 CMOS. The charging and discharging times of the RC network along with the Schmitt trigger's upper and lower switching thresholds determine the frequency.

The frequency can be approximated with the following expressions:

$f = 0.8/RC$ for TTL and LS TTL

$f = 1.2/RC$ for HC and CMOS

Frequency range is wide from 0.1 Hz to about 20 MHz. The suggested feedback resistor values are fairly critical, 330 Ω for the TTL circuit and 1200 Ω for the CMOS circuit. Stay close to those resistor values. Experiment with capacitor values to get the frequency you want. This circuit is good for operating digital logic circuits and not much else.

The design procedure follows:

1. Select either TTL or CMOS.
2. State the desired frequency (f).

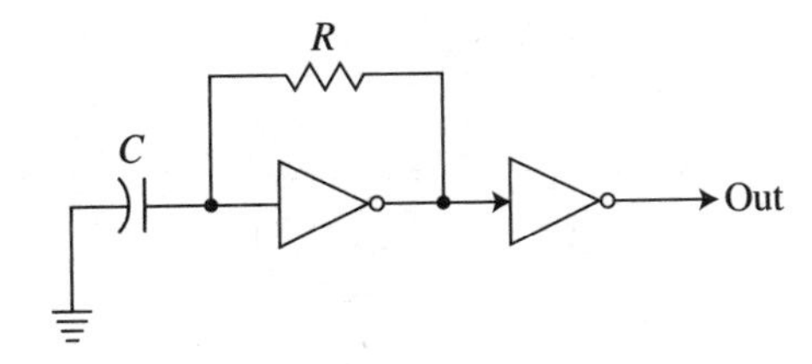

IC is 74LS14 TTL or
CD40106 CMOS, Schmitt triggers.
For 74LS14 R = 330 Ω
For CD40106 CMOS, R = 1200 Ω

Figure 8.7 A simple rectangular oscillator using a Schmitt trigger logic circuit, either TTL or CMOS.

3. Calculate C using frequency and the recommended resistor values.
   a. $C = 0.8/fR$ TTL
   b. $C = 1.2/fR$ CMOS

These formulas are only approximations. You will need to tinker with the C value to get the frequency you want. For R, try to stay in the range suggested.

555

The undisputed king of the multivibrator ICs is the ubiquitous 555. It is referred to as a timer, but it is most commonly connected as an astable multivibrator to generate a rectangular wave. This IC has been around for decades (introduced in 1972!) but is still widely used. You should get familiar with it and build one to use in future projects.

The popular astable multivibrator circuit is given in Fig. 8.8. This diagram shows the internal circuits of the 555, but you do not usually have to worry about this detail. The 555 is typically in an 8-pin DIP IC with the pin numbers labeled in Fig. 8.8.

The frequency of the rectangular wave output is set by the charging and discharging of C_1 through R_A and R_B.

$$f = 1.443/(R_A + 2R_B)C$$

The frequency range is enormous, from a fraction of a hertz up to about 100 kHz or so.

The circuit operates from a dc supply of 5 to 15 V. The supply voltage sets the output pulse amplitude. A pull-up resistor is needed at the output as shown at pin 3. Its value is not critical. 1 k to 10 KΩ is usually OK. The original 555 was made with bipolar devices and is still used. A newer CMOS version operates with a supply down to 3 V. Otherwise, everything else is essentially the same.

One issue with the 555 is that its rectangular output wave has a duty cycle greater than 50 percent. Some applications call for a 50 percent duty cycle square wave with equal on and off

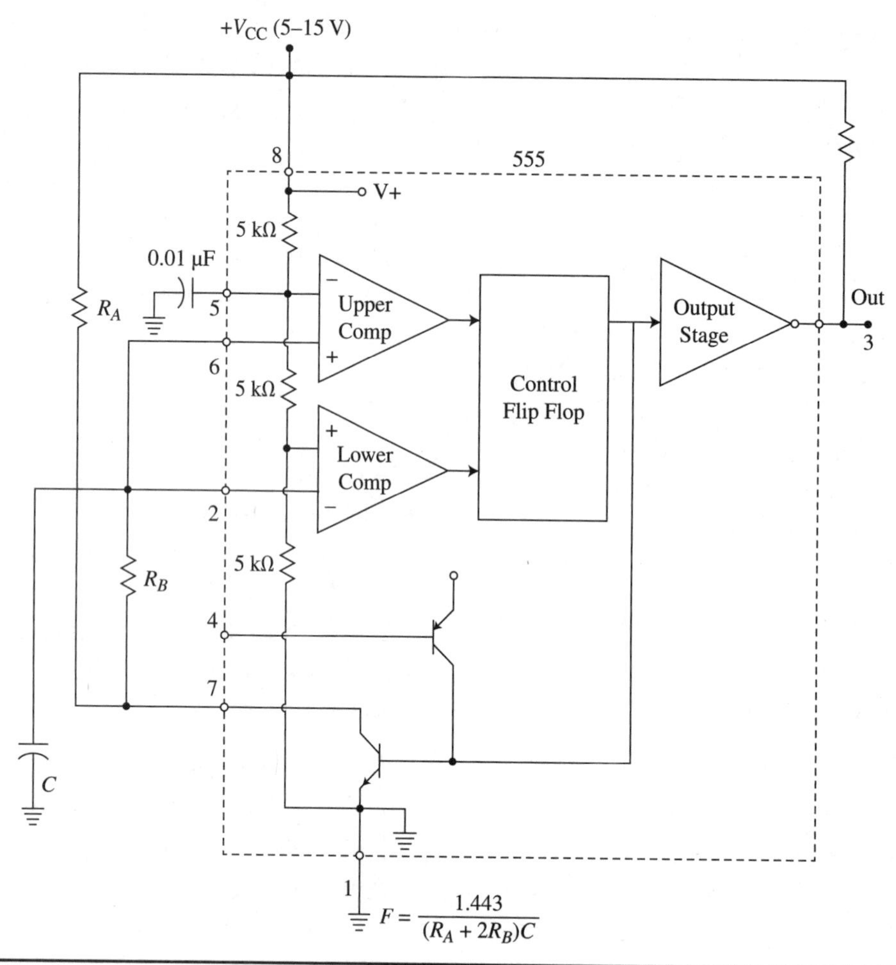

Figure 8.8 Circuit details of a 555 timer IC astable multivibrator.

times. You can achieve this with the modified 555 circuit shown in Fig. 8.9.

Another way to get a perfect 50 percent duty cycle is to send the 555 output to a type D or JK flip flop (FF) as Fig. 8.10 shows. Note the different connections for the JK and D FF. The FF will toggle or complement (change state) at the period set by the 555. This means that the

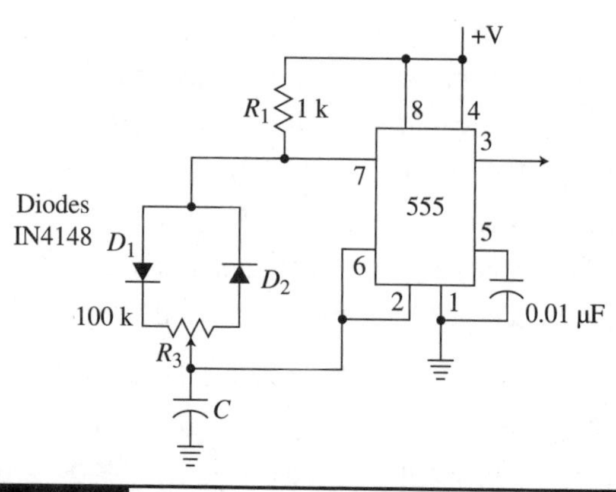

Figure 8.9 The 555 timer oscillator connected for variable duty cycle output.

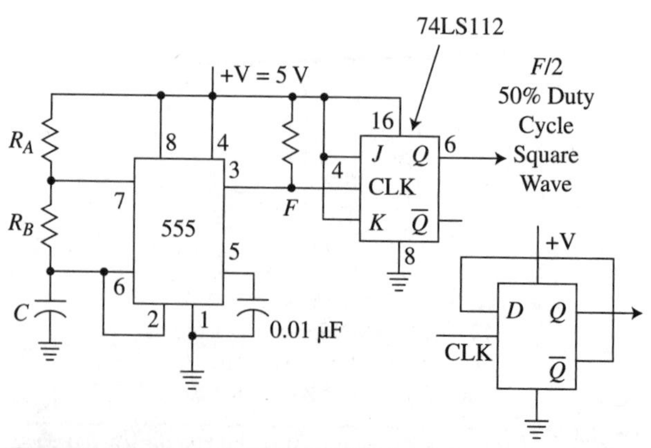

Figure 8.10 Another way to get 50 percent duty cycle output with a 555 timer IC and a flip flop.

FF output frequency will be one-half the 555 frequency.

If you use this method, design your 555 frequency twice the desired frequency. You will learn more about D and JK FF in Chap. 11, but here are a few choices to consider until then: TTL 74LS74, 74LS107, 74LS109, and 74LS112 and CMOS: CD4013.

The design procedure is pretty simple:

1. Select a supply voltage between 5 and 18 V. Common values are 5 and 12 V. The output voltage will switch between 0 V or near ground to the supply voltage.
2. Choose an output frequency. Here is the formula. $f = 1.443/(R_A + 2R_B)C$
3. Select values for R_A and R_B. These are very flexible. R_A values are usually in the 1-k to 1-MΩ range. Values for R_B can be in the 1-k to 10-MΩ range. As a general guide, use smaller-value resistors (1 to 10 kΩ) for the higher frequencies (>10 kΩ) and larger values (10 k to 1 MΩ) for the lower (<10 kΩ) frequencies. These values affect the frequency and duty cycle. You will probably need to experiment to get what you want. Choose some values and start experimenting. Try $R_A = 1$ k and $R_B = 100$ k as a starting point and adjust from there.
4. Consider duty cycle. You can get close to 50 percent but not quite. The 51 to 55 percent range is doable. Here is how the resistor values affect duty cycle D.

   $$D = (R_A + R_B)/(R_A + 2R_B)$$

5. Calculate C.

   $f = 1.443/(R_A + 2R_B)C$

   Rearrange to solve for C.

   $C = 1.443/(R_A + 2R_B)f$

6. Select a pull-up resistor. Anything from 1 to 10 k is OK.
7. Build the circuit and test it. Tinker with the values until you get what you need.

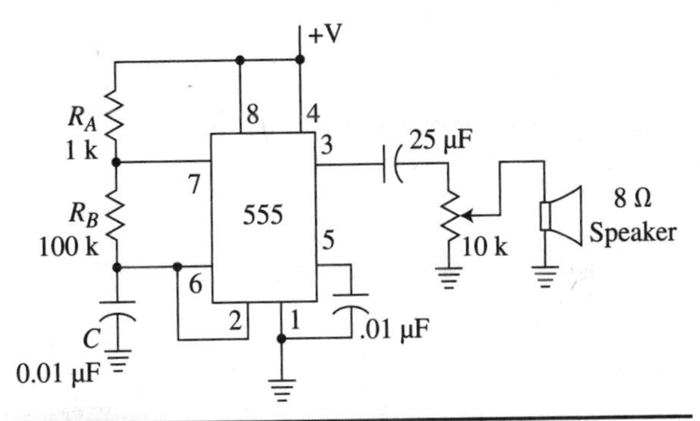

Figure 8.11 Using the 555 timer IC as an audio oscillator with volume control.

Here are a couple of other 555 circuits that may be useful.

Figure 8.11 shows a 555 connected as an audio oscillator. The output is connected to a speaker. With values shown, frequency should be around 700 Hz. The 10-k pot is the volume control.

Figure 8.12 shows the 555 flashing an LED. Large values of C are needed to slow down the oscillation to a visible rate.

Finally, the 555 may also be used as a monostable multivibrator or one-shot. This is a circuit that generates one output pulse when triggered at the input by a negative-going logic signal or trailing edge pulse. See Fig. 8.13. The 10-k resistor and pushbutton will generate the trigger pulse. The duration (t) of the output pulse is set by the RC values.

$$t = 1.1R_AC$$

Only one output pulse is generated for each input trigger. This circuit makes a great timer.

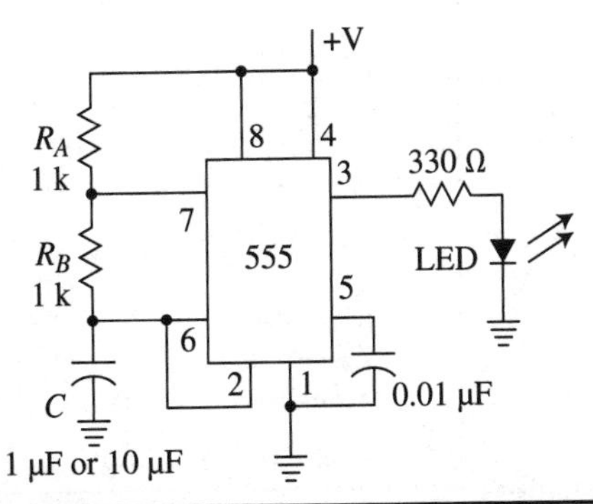

Figure 8.12 A 555 timer IC used to flash an LED.

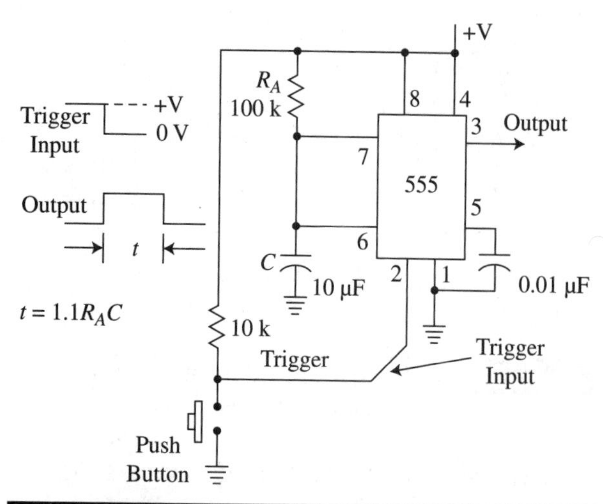

Figure 8.13 A 555 timer IC used as a one-shot or monostable multivibrator.

Try the circuit shown to flash an LED on one time for every button push.

There are other ICs for generating signals. Here are a few others that are popular. They can save you time by not having to design an equivalent circuit.

XR2206

This IC is called a function generator because it generates sine, square, and triangular waves simultaneously. The frequency range is from a fraction of a Hz up to 1 MHz. The frequency is set by an external RC network.

$$f = 1/RC$$

The sine output is very low distortion (<0.5 percent).

The 2206 also contains modulators so you can demonstrate amplitude and frequency (AM and FM) modulation, including FSK. The circuit can also perform voltage to frequency conversion. A great chip for general experimenting.

Figure 8.14 shows a complete function generator circuit you can build and use for testing and prototyping. Its maximum frequency is 100 kHz. This is an old but great chip—very useful. Another function generator chip to consider is the Maxim Integrated MAX038. I did not have time to explore it for this book, but you may want to take a look.

LTC6900

This is a series of IC oscillators from Analog Devices (previously Linear Technology) that generates square waves from 1 kHz to 20 MHz. The frequency is set by one external resistor.

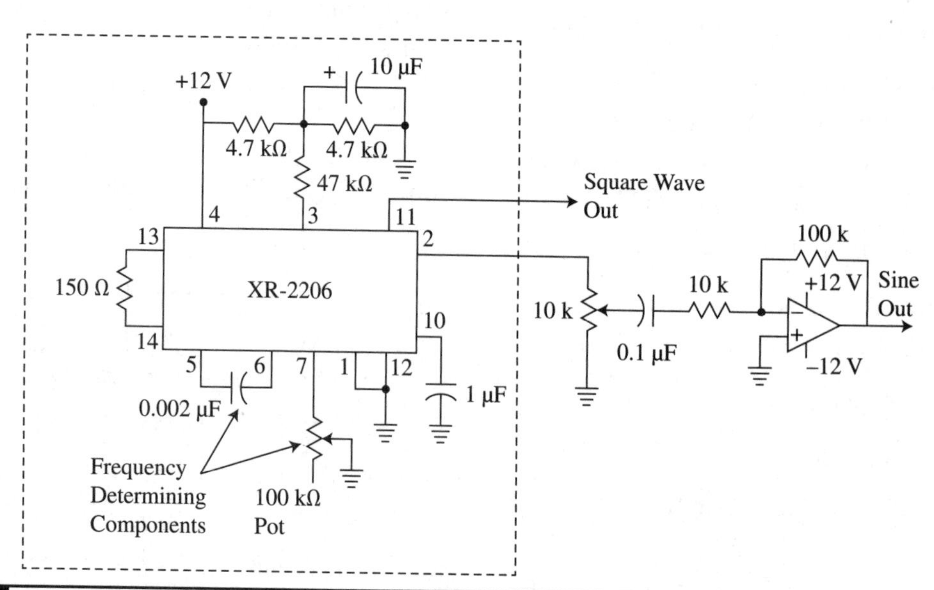

Figure 8.14 The XR-2206 IC connected as a general-purpose function generator.

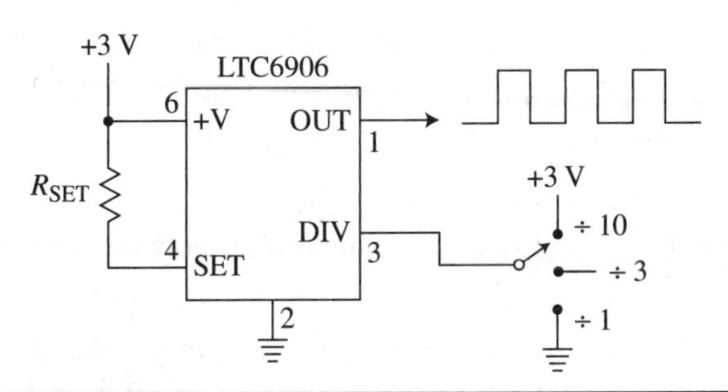

Figure 8.15 The Linear/Analog Devices all-purpose LTC6906 IC oscillator.

Another version, the LTC6906, covers the 10 kHz to 1 MHz range. See Fig. 8.15.

The frequency for the LTC6906 is determined by the resistor value R_{SET} and a frequency divider ratio N.

$$f = 1 \text{ MHz } (100 \text{ k}/NR_{SET})$$

N can be 1, 3, or 10.

$$R_{SET} = 1 \text{ MHz } (100 \text{ k/Nf})$$
$$= 1 \text{ MHz } (100 \text{ k})/10(12{,}000{,}000)$$

Another version the LTC1799 covers the 1-kHz to 33-MHz range.

These oscillators generate 50 percent duty cycle waves from a 2.7- to 5.5-V dc supply. A useful feature is that the oscillators can be used as voltage-controlled oscillators, where a dc input varies the output frequency. The package is an SOT-23 surface mount. With these devices, you never have to design your own.

Before you use any of these devices, be sure to get the data sheets with all the application details.

Frequency Synthesizers

Frequency synthesizers are the most sophisticated signal sources available. These are essentially mixed-signal integrated circuits that generate sine or rectangular outputs that are useful in making signal or function generators, carrier sources for radio transmitters, or local oscillators in radio receivers. They are complex but available in IC form, which makes them relatively easy to use. Their benefits are that they generate very stable and precise signals at frequencies that are easily set digitally in increments. There are two popular synthesizer types: phase-locked loop (PLL) and direct digital synthesis (DDS). The basic operation of these is described in the following sections.

Phase-Locked Loop

A basic PLL frequency synthesizer is shown in Fig. 8.16.

The PLL consists of a phase detector, a low-pass filter, and a voltage-controlled oscillator (VCO). The input to the phase detector is a crystal-controlled reference oscillator that provides high-frequency stability. The frequency of the reference oscillator establishes the incremental change in VCO operating frequency. The VCO output is connected to a frequency divider to provide an output frequency in some integer submultiple of the input frequency. A divide-by-100 frequency divider produces an output frequency that is one-hundred times the input or reference frequency. The synthesizer VCO output may be a sine wave or rectangular wave.

In the PLL in Fig. 8.16, the reference oscillator is set to 1 MHz. Assume that the frequency divider is initially set for a division (N) of 100. For a PLL to become locked or

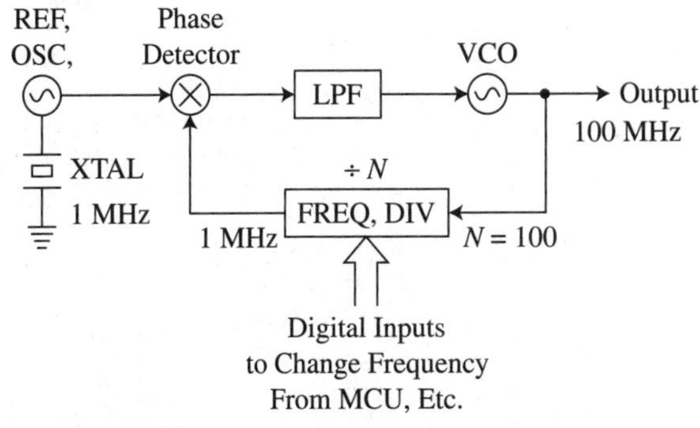

Figure 8.16 The concept of a phase-locked loop (PLL) frequency synthesizer.

synchronized, the second input to the phase detector must be equal in frequency to the reference frequency; for this PLL to be locked, the frequency divider output must be 1 MHz. The VCO output has to be 100 times higher than this, or 100 MHz. The VCO frequency is set to 100 MHz so that when it is divided, it will provide the 1 MHz input signal required by the phase detector for the locked condition. The result is a 100-MHz signal source. Because the PLL is locked to the crystal reference source, the VCO output frequency has the same stability as that of the crystal oscillator. The PLL will track any frequency variations and correct for them.

To vary the output frequency, the frequency division ratio is changed using special counters and various digital switching techniques. A microcontroller can also provide the correct frequency division ratio based on software inputs.

Varying the frequency division ratio changes the output frequency. For example, in the circuit in Fig. 8.16, if the frequency division ratio is changed from 100 to 99, the VCO output frequency must change to 99 MHz. Each incremental change in the frequency division ratio produces an output frequency change of 1 MHz.

A basic PLL synthesizer is the CMOS device CD4046 or the 74HC4046. This single IC contains a VCO and a selection of two-phase detectors. It requires a crystal reference oscillator, an RC low-pass filter (LPF), and a frequency divider. Maximum operating frequency is 1.4 MHz. This a relatively old part, but it is still available if you wish to experiment with it. Get the data sheet and any app notes for reference and design details.

An even older PLL is the LM565. It has mostly been phased out, but you can still find some at distributors and online. It is easy to use and makes a good demonstration of PLL action. Its maximum frequency range is 500 kHz. Download the data sheet for pin-out, applications, and circuits.

Direct Digital Synthesis

A direct digital synthesis (DDS) synthesizer generates a sine wave output digitally. The output frequency can be varied in increments, depending upon a binary value supplied to the synthesizer by a counter, a register, or an embedded microcontroller.

Conceptually, the DDS synthesizer is illustrated in Fig. 8.17. A read-only memory (ROM) is programmed with the binary representation of a sine wave. These binary values are fed to a digital-to-analog converter

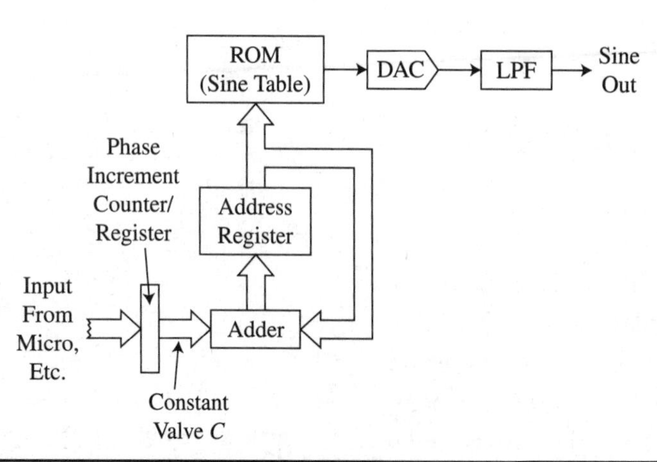

Figure 8.17 The concept of a direct digital synthesizer (DDS).

(DAC) that translates them into a stepped approximation of the sine wave. A low-pass filter (LPF) removes the harmonics and smoothes the output into a nearly perfect sine wave.

A binary counter/register is used to supply the address word to the ROM. A clock signal steps the counter that supplies an incrementally increasing address to ROM. The clock frequency determines the frequency of the sine wave.

The counter starts coming at zero and the sine values are sequentially accessed from ROM and fed to the DAC that produces a stepped approximation of the sine wave. If the clock continues to count, the counter will recycle, and the sine wave output cycle will be repeated.

An important point to note is that this frequency synthesizer produces one complete sine wave cycle for every 2^N clock pulses. With a 10-bit counter and its 1024 states, the sine values would be spaced every 360/1024 = 0.35°, giving a highly accurate representation of the sine wave. Because of this relationship, the output frequency of the sine wave $f_o = f_{clk}/2^N$, where N is equal to the number of address bits in ROM.

With a clock frequency of 10 MHz used with a 10-bit counter, the sine wave output frequency would be

$$f_o = 10{,}000{,}000/1024 = 9765.625 \text{ Hz}$$

The only way to change the frequency in this synthesizer is to change the clock frequency. This arrangement is not practical in view of the fact that we want the synthesizer output to have crystal oscillator precision and stability. To achieve this, the clock oscillator must be crystal controlled.

The most commonly used method to vary the synthesizer output frequency is to replace the address counter with a register whose content will be used as the ROM address but also one that can be readily changed. This register is used in conjunction with a binary adder, as shown in Fig. 8.17. The output of the address register is applied to the adder along with a constant binary input value (C). This constant value can also be changed. The output of the adder is fed back into the address register. This circuit is arranged so that upon the occurrence of each clock pulse, a constant C is added to the previous value of the register content, and the sum is restored in the address register. The constant value comes from the phase increment register, which in turn gets it from an embedded microcontroller or other source.

Assume that we set the constant value to 1. Each time a clock pulse occurs, a 1 is added to the content of the register. With the register initially set to zero, the first clock pulse will cause the register to increment to 1. On the next clock pulse, the register will increment to 2, and so on. As a result, this arrangement acts just as a binary counter.

Now assume that the constant value is 2. This means that for each clock pulse, the register value will be incremented by 2. Starting at 0000, the register contents would be 0, 2, 4, 6, and so on. Every other sine value is now being sent to the DAC, so the sine wave is generated at a more rapid rate. Instead of having 1024 amplitude values, the DAC gets only 512. The output has fewer steps and is a cruder representation. With an adequate low-pass filter, the output will still be a sine wave whose frequency is twice that generated by the circuit with a constant input of 1.

The frequency of the sine wave is adjusted by changing the constant value added to the accumulator. With this arrangement, we can now express the output sine wave frequency with the formula

$$f_o = Cf_{clk}/2N$$

The higher the constant value C, the fewer the samples used to reconstruct the output sine wave. You can change C until you get only two samples per cycle, which is the lowest number that can be used and still generate an accurate output frequency. Remember the Nyquist criterion, which says that to adequately reproduce a sine wave, it must be sampled a minimum of two times per cycle to reproduce it accurately in a DAC.

To make the DDS effective, then, the total number of sine samples stored in ROM must be a very large value. Practical circuits use a minimum of 12 address bits, giving 4096 sine samples.

The DDS synthesizer offers some advantages over a PLL synthesizer. First, if a sufficient number of bits of resolution in ROM word size and the accumulator size are provided, the frequency can be varied in very fine increments. Because the clock is crystal controlled, the resulting sine wave output will have the accuracy and precision of the crystal clock.

The second advantage is that the frequency of the DDS synthesizer can be changed much faster than that of a PLL synthesizer. To change the PLL synthesizer frequency, a new frequency-division factor must be entered into the frequency divider. Once this is done, it takes a finite amount of time for the phase detector to recognize the error and settle into the new locked condition. The storage time of the loop low-pass filter considerably delays the frequency change. This is not a problem in the DDS synthesizer, which can change frequencies within nanoseconds.

A downside of the DDS synthesizer is that it is difficult to make one with very high output frequencies. The output frequency is limited by the speed of the available DAC and digital logic circuitry. With today's components, it is possible to produce a DDS synthesizer with an output frequency as high as several hundred MHz. For applications requiring higher frequencies, the PLL is still the best alternative.

Single-chip DDS synthesizers are available from several IC companies. The complete DDS circuitry is contained on a chip. The clock circuit is usually contained within the chip, and its frequency is set by an external crystal. Parallel binary input lines are provided to set the constant value required to change the frequency. A 12-bit DAC is typical. A commercial DDS example is the AD98xx line of DDS chips from Analog Devices.

If you are looking for a chip to experiment with, seek out an Analog Devices AD9833. It is a single-chip DDS that can generate sine, triangle, and square waves with an upper frequency of 12.5 MHz. It is designed to be controlled by an external microcontroller by way of the serial peripheral interface (SPI) port available on most micros. Analog Devices makes an evaluation board to demonstrate its capabilities. You can even get an AD9833 "shield" for popular single-board computers like the Arduino.

Design Project 8.1

Design a Wien bridge audio oscillator circuit for a frequency of 1 kHz without feedback stabilization. Then use one of the methods described to stabilize the output.

Design Project 8.2

Design an oscillator like that in Fig. 8.7 for a frequency of 5 MHz. Use a CMOS Schmitt trigger IC. Build the circuit and test it. Get the data sheet for connections. Use a +5-V supply.

Design Project 8.3

Design a 555 astable similar to Fig. 8.12 for a frequency of 0.5 Hz. Flash an LED with it. Use R_A of 10 k and R_B of 100 kΩ. Build the circuit and test.

Design Project 8.4

Build the circuit in Fig. 8.9 that varies the duty cycle. Add the LED and 330-Ω resistor to the output. Demonstrate how changing the duty cycle by varying the 100 k pot varies the LED brightness.

Design Project 8.5

Build the 555 circuit in Fig. 8.11 that will produce an audio tone in the 700 Hz range. Connect the speaker to the output. Adjust the volume as necessary.

Design Project 8.6

Purchase one of the metal can TTL clock crystal oscillators mentioned in the text. Select a frequency that can be useful to you later. Frequencies from 1 to 50 MHz can be found. Something around 10 MHz is useful. Breadboard the oscillator and observe its output on your oscilloscope. It makes a good signal source for testing logic circuits. You can also use a frequency divider on the output to get lower frequencies.

Solutions are shown in App. B.

CHAPTER 9

Filter Design

A filter is a circuit that passes signals of some frequencies but blocks or attenuates those of other frequencies. The most common filters are RC or LC networks. Active filters using op amps and RC networks are also popular. High-frequency filters are made with LC networks, quartz crystals, ceramics, surface acoustic wave (SAW) devices, and other components. In this chapter, you will learn how to design basic RC and LC filters, as well as active filters with op amps. These filters are the ones you will most likely need or encounter. Online filter design calculators are available to simplify and speed up your design.

The primary characteristic of a filter is its selectivity. Selectivity is the sharpness of response between the passband where the desired signals reside and the reject band of undesired frequencies. The steepness of the response or sharpness of the transition between the passband and the reject band is a measure of the selectivity.

Types of Filters

There are four basic types of filters: low-pass, high-pass, bandpass, and band reject filters. Figure 9.1 shows the symbols commonly used to represent each filter type in block diagrams. The waves designate high, mid-, and low frequencies. Note the subtle marks across some of the waves that indicate which signals do **not** get through.

Figure 9.2 shows the ideal and realistic frequency response curves of each. The realistic curves on the right are referred to as Bode plots. They indicate the attenuation at different frequencies. The passband and a reject or stop band are indicated.

Low-Pass Filters

This filter passes signals below a specific cutoff point and attenuates signals above the cutoff point. The response below the cutoff is called the passband. The region above the cutoff is designated the stop band. See Fig. 9.2*a*. Signals, including dc, pass relatively unattenuated below the cutoff frequency (f_{CO}). The cutoff frequency is that point where the output drops to 70.7 percent (0.707) of the output at a lower frequency. If the filter attenuation is expressed in decibels (dB), the cutoff, f_{CO}, occurs at the point where the output drops to −3 dB from a lower frequency. The total passband or bandwidth is from dc to f_{CO}.

Above the cutoff, the attenuation increases linearly with frequency. The term used to describe this is roll-off at a constant rate. The steepness of this roll-off is a measure of the selectivity. In a basic RC filter, the roll-off rate is 6 dB/octave or 20 dB/decade.

Figure 9.2*a* shows the actual response of a low-pass filter. The cutoff frequency is 600 Hz. If the output is 5 V at lower frequencies, the output at cutoff is 0.707 (5) = 3.535 V. This is

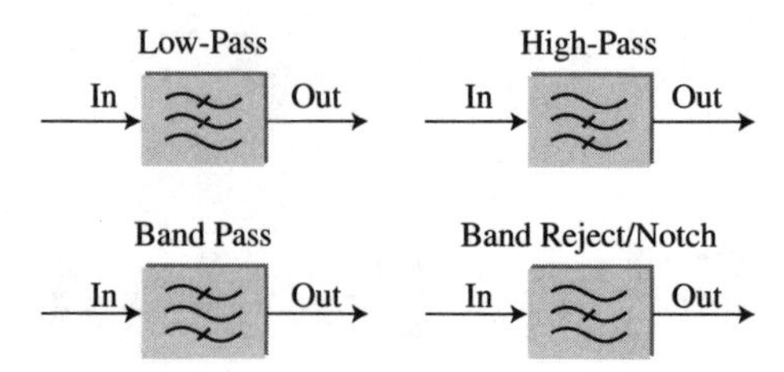

Figure 9.1 The four basic filter types, ideal and true responses.

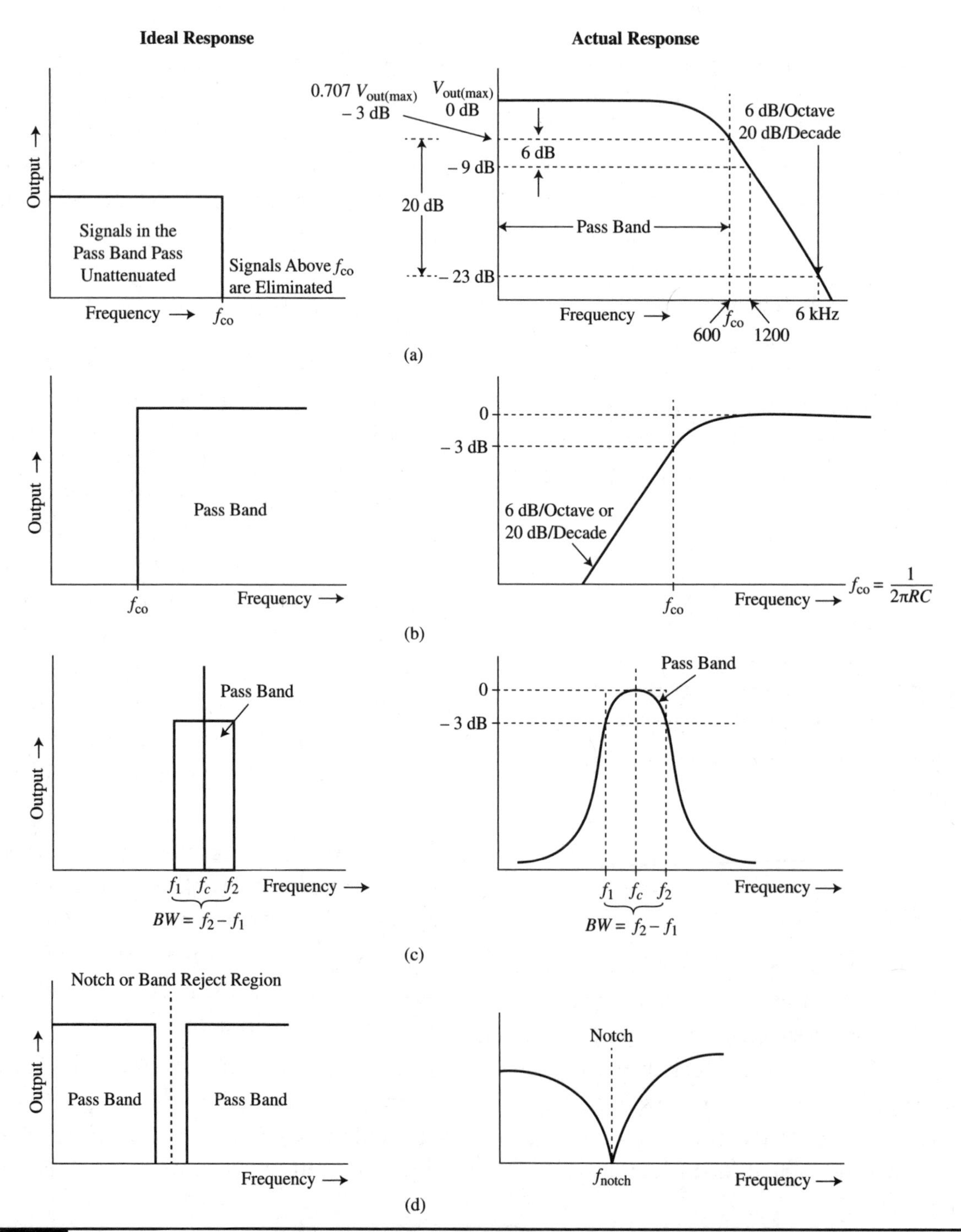

Figure 9.2 Block diagram symbols for the four filter types: (a) low-pass, (b) high-pass, (c) band pass, and (d) notch.

also the −3 dB point. The attenuation levels for one octave later (1200 Hz) and one decade later (6 kHz) are indicated.

High-Pass Filter

This filter lets frequencies above the cutoff pass with little attenuation, while frequencies below the cutoff are gradually attenuated. See Fig. 9.2*b*. The response below the cutoff is called the stop band. The region above the cutoff is designated the passband.

Bandpass Filter

Bandpass filters allow frequencies in a specific range known as the bandwidth to pass while attenuating frequencies above and below that range. The bandwidth is defined by two cutoff frequencies, f_1 and f_2, as illustrated in Fig. 9.2*c*. The bandwidth BW is

$$BW = f_2 - f_1$$

The filter also has a center frequency that is the mean between the cutoff frequencies.

$$f_C = \sqrt{(f_1 f_2)}$$

Band Reject Filter

A fourth filter type is the opposite of the bandpass filter. It greatly attenuates a narrow range of frequencies but lets others above and below a center frequency pass unattenuated. Refer to Fig. 9.2*d*. A better description of this filter is that it is a notch filter designed to filter out signals at the one specific frequency while others above and below pass with minimal attenuation.

Most filters are made with resistors and capacitors (RC) or inductors and capacitors (LC). Their function, of course, is dependent upon how the reactance of capacitors or inductors varies with frequency. These components are arranged mainly in voltage divider configurations to achieve their function. In this chapter, you will learn how to design the most common RC and LC filters.

Filter Specifications

The following are the basic characteristics to use in selecting a commercial filter:

Insertion loss. This is the attenuation of the signal in the passband. It is typically expressed in decibels (dB).

Attenuation. This is the loss in the filter because of the losses in the components.

Impedance. LC filters have a characteristic impedance. They are designed for a specific resistive load. For example, 50 Ω is a common filter impedance at radio frequencies (RF).

Group or envelope delay. The transit time of a signal through a device versus frequency and the effect that it has on signals with an envelope such as pulses and modulated signals. It results in a distortion because of the phase shift change with frequency.

Ripple. Variations in the filter amplitude response in the passband or stop band. Ripple means that the response is not flat or constant over the designated frequency range.

Order or number of poles. This refers to the number of RC or LC sections used in the filter. Discrete component filters tend to use multiple cascaded sections to get the desired selectivity. One pole is one RC section. In an LC filter, a pole is the number of reactive components per section. A single LC section is a two-pole filter. The more poles you use, or the higher the order, the sharper the response, and the greater the selectivity.

Cutoff or center frequency. This defines the filter frequency range.

Filter Design Guidelines

Here are a few rules of thumb that will help you with your filter needs and designs:

- Low-pass filters seem to dominate the filter world. You will encounter more and design more of these than any other kind. These filters are needed to filter out noise, harmonics, interfering signals, ripple, and other undesirable signals. A widespread example is a single capacitor connected at one point in a circuit that acts as a low-pass filter to provide some filtering out of the higher frequencies. This is called a bypass or decoupling capacitor.
- For frequencies below about 1 MHz, use a RC filter or an active RC filter. Inductors are larger and more expensive at the low frequencies. RC filters and active filters with op amps can achieve excellent results.
- Use an LC filter at the higher frequencies. Above 3 MHz or so, inductor sizes and prices are more reasonable. LC filters predominate at the radio frequencies from 3 to 300 MHz.
- Remember, you can probably buy just about any filter you want. There are many sources, especially for the higher frequencies (>10 MHz). Check out what is available before you design your own because it will save time but not money. Buying versus making is one of your design decisions. An example is an ac power line filter that is used in just about every ac power supply. You don't want to have to design and build this. This is a low-pass filter that works both ways: It keeps power supply noise off the ac power line and keeps high-frequency noise and pulses on the ac line out of the power supply.
- Keep in mind that there are other types of filters besides the traditional RC and LC types. Most of these are made for the higher frequencies. Some of your choices are filters using crystals, ceramic resonators, surface acoustic wave (SAW), bulk acoustic wave (BAW), mechanical, switched capacitor, and stripline.
- A major filter trend over the years from roughly the 1980s to the present is a digital signal processing (DSP) filter. This filter takes the analog signal to be filtered and digitizes it in an analog-to-digital converter (ADC) and stores the resulting samples. The digital samples are then fed to a digital signal processor (DSP) or FPGA, where special algorithms perform the digital equivalent of the desired filtering process. This is an expensive and complex approach, but the resulting filters give superb selectivity. Most if not all wireless devices used today (cell phones, Wi-Fi, etc.) incorporate DSP filters. Special DSP embedded controllers and processors are available to do the job.

Filter Response Options

The actual performance of a filter depends upon the way it is designed. There are four common filter behavior models to consider. These are Butterworth, Bessel, elliptical, and Chebyshev. These are compared in Fig. 9.3.

- **Butterworth.** The most common filter type is the Butterworth. The previously covered RC and active filters have a Butterworth response. Its output is very flat over the passband range of coverage. It then rolls off at a leisurely, constant rate. The Butterworth is not the most selective filter because the gradual roll-off characteristic does not adequately filter some inputs. Also, the phase response is nonlinear, causing it to distort some signals. Your first consideration for a filter is its frequency response. As the reactances in the filter change with frequency, a phase shift is also taking place. This may or may not be relevant to the signals you are covering.

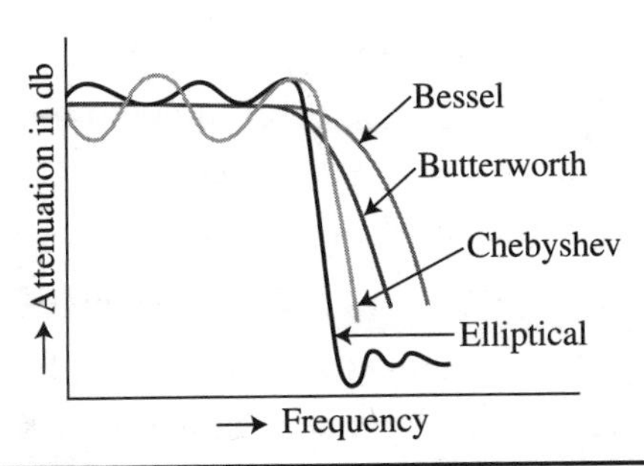

Figure 9.3 Common filter response curves.

- **Bessel.** A Bessel response has a somewhat slower roll-off than the Butterworth but also has a flat response in the passband. The key characteristic is its linear phase response. Other filter types like Butterworth and Chebyshev have nonlinear phase responses that cause distortion to some types of signals, such as modulated signals and pulses. The Bessel filter is a good alternative because it does not distort and it does provide a minimum of frequency selectivity.
- **Chebyshev.** This filter has better selectivity because of its rapid roll-off above or below the cutoff. One potential downside is that it has ripple in the passband. This response variation may or may not be a problem.
- **Elliptical.** The elliptical filter has the best selectivity of all of the types of filters. Its roll-off is rapid at cutoff. Its potential disadvantage is ripple in the passband and stop band.

RC Filter Design

The simplest RC low-pass filter is shown in Fig. 9.4.

The resistor and capacitor form a voltage divider. At low frequencies, the capacitive reactance (X_C) is high, so most of the input signal appears across the capacitor. At the higher frequencies, the capacitive reactance begins to decrease, so less voltage drop occurs across it. At

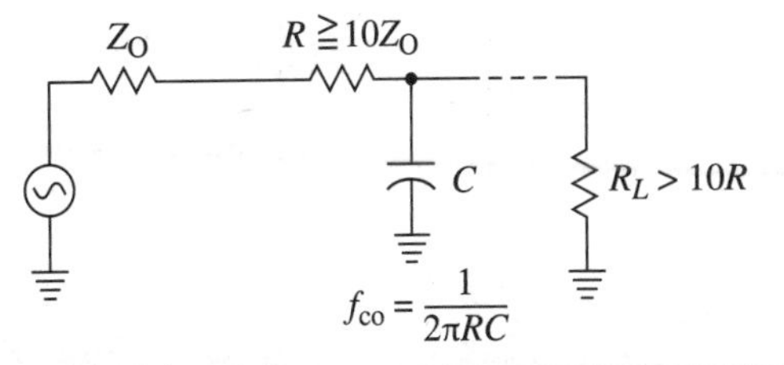

Figure 9.4 Simple one-section RC low-pass filter.

some frequency $X_C = R$, the output drops to 0.707 of the input. This is the cutoff frequency f_{CO}.

$$f_{CO} = 1/2\pi RC$$

If the output voltage (V_O) is 0.707 of the input voltage (V_I), this is the equivalent of −3 dB. Remember the formula for decibel (dB) measure.

$$\text{dB} = 20 \log(V_O/V_I) = 20 \log(0.707) = -3 \text{ dB}$$

Above the cutoff, the attenuation increases linearly at the rate of 6 dB/octave or 20 dB/decade. The frequency response plot in Fig. 9.2*a* shows this roll-off. Keep in mind that an octave means a ratio of 1 to 2 or 2 to 1. One octave above 500 Hz is 1000 Hz or 1 kHz. A decade is a 1 to 10 relationship. One decade higher than 400 Hz is 4000 Hz or 4 kHz. Note in Fig. 9.2 the cutoff is 600 Hz. One octave higher (1200 Hz), the attenuation drops another 6 dB for a total of 9 dB. One decade from 600 Hz is 6 kHz, and over that range the attenuation drops 20 dB more to 23 dB.

By the way, keep in mind that this RC circuit also produces a phase shift. The output frequency is shifted by 45° at the cutoff frequency. Overall, the output voltage lags the input by some amount from 0° to 90° depending upon the frequency.

The operation of a high-pass RC filter is similar. Reversing the positions of the resistor and capacitor as shown in Fig. 9.5 produces a high-pass filter.

The cutoff frequency is

$$f_{CO} = 1/2\pi RC$$

The attenuation at the cutoff is also 3 dB. The roll-off rate is the same as for the low-pass filter.

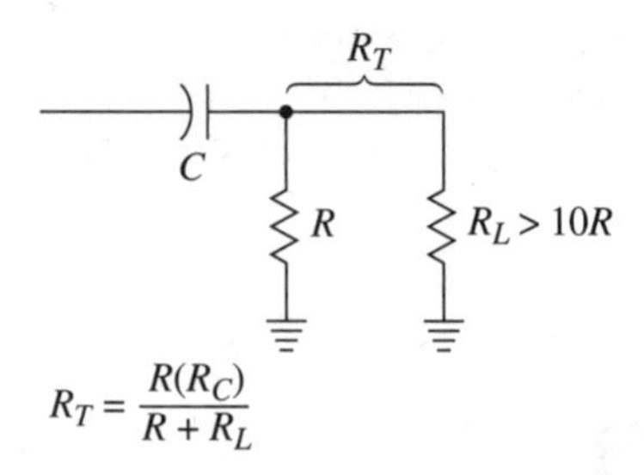

Figure 9.5 Simple one-section RC high-pass filter.

Step-by-Step RC Filter Calculations

Use the following steps to design either the low-pass or high-pass RC filter. An example is included.

- Specify low- or high-pass.

  Low-pass

- State the cutoff frequency.

  4 kHz

- Specify the output impedance Z_O of the circuit driving the filter. This is needed so that the filter will not "load" the driving source, producing additional attenuation. Remember, low output impedance is the most desirable.

  Assume Z_O is 1 kΩ.

- Determine the filter resistor value. Ideally, it should be at least 10 times or more of Z_O. The higher the better. The exact value is not generally critical.

  $R = 10(Z_O) = 10(1\text{ k}) = 10\text{ k}\Omega$ minimum. Let's choose $R = 50$ k. Standard values of 47 k or 51 k will work. We will use 47 k.

- Calculate the capacitor value. Knowing the cutoff frequency and the value of R, you can rearrange the cutoff formula to solve for C.

  $f_{CO} = 1/2\pi RC$

  $C = 1/2\pi f_{CO}R = 1/6.28(4000)47{,}000 = 847\text{ pF}$

  Standard values are 820 and 910 pF. Either will work with a small shift in the cutoff frequency.

- Notice that no load on the filter was specified. So any load should be minimal. Any resistive load should be at least 10 times or more the value of R to minimize the deviation from the design performance.

Use this same procedure to design a high-pass filter. The main difference is that any resistive load on the output is going to have a significant effect on the cutoff frequency. Figure 9.5 shows that a load on an RC high-pass filter puts the load directly in parallel with the filter R. The total parallel equivalent of the two resistors is the value to use in the design.

$$R_T = RR_L/R + R_L$$

If the load on the filter is minimal, use the design value for R. Minimal means that R_L is at least 10 or more times the filter R value. The higher the better.

Cascading RC Filters

There are design instances where the roll-off rate is not fast enough. Figure 9.6 illustrates the problem. The desired signal f_1 passes successfully. Some of the undesired signal f_2 still gets through. The more gradual response lets in more of f_2. Increasing the selectivity and roll-off rate gets rid of f_2.

One solution is to increase the roll-off rate by cascading RC sections as shown in Fig. 9.7*a*.

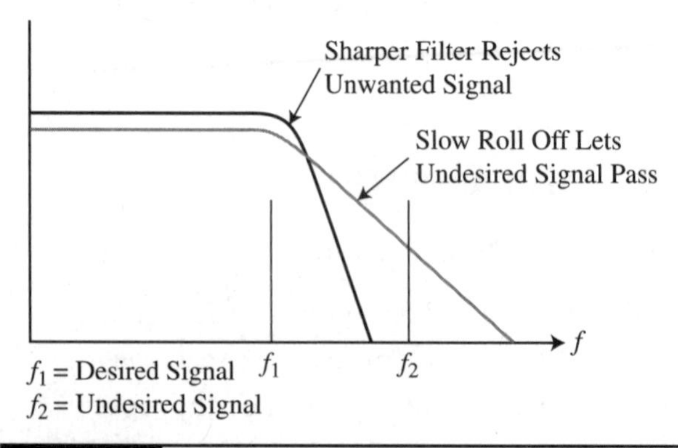

Figure 9.6 Response of low-pass filters with different roll-off rates and their success in minimizing unwanted signals.

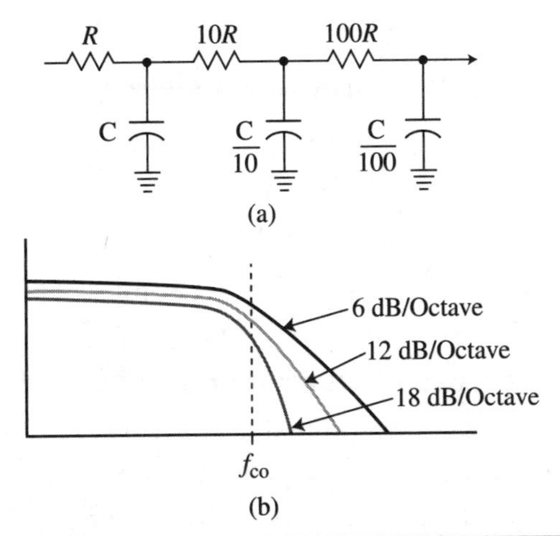

Figure 9.7 A three-section RC low-pass filter used to improve selectivity.

Each RC section attenuates at a rate of 6 dB/octave, so the combined attenuation rate for a three-section filter is 18 dB/octave or 60 dB/decade. The curves in Fig. 9.7*b* are exaggerated to show the concept.

The problem with this arrangement is that each section loads the previous section. This loading effect shifts the cutoff frequency, making the filter something different than what you may need. To correct this problem, make each section resistor value 10 times or more the value of the resistor in the previous section. The most effective approach is to make all RC sections the same, but put an isolation amplifier between sections as shown in Fig. 9.8. Here an op amp follower with its unity gain, high input impedance, and low output impedance effectively eliminates the loading problem.

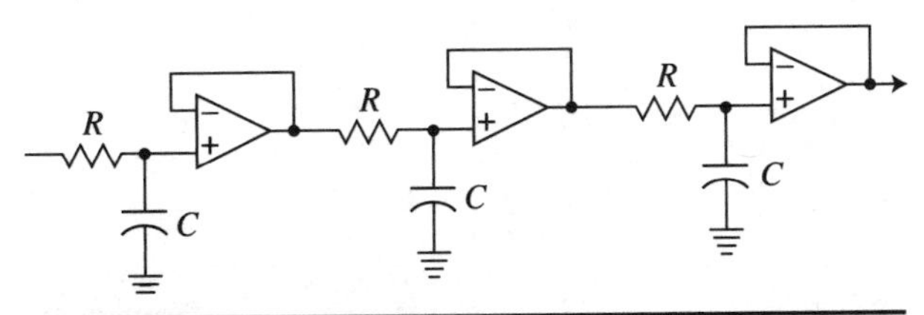

Figure 9.8 A three-section RC low-pass filter with interstage isolation with op amps.

Bandpass LC Filters

Bandpass RC filters are rarely used. However, you can make BPFs with simple resonant LC circuits as Fig. 9.9 shows.

Here is the procedure and an example.

The center frequency of the filter is the resonant formula. Assume $f = 7$ kHz.

$$f_C = 1/2\pi\sqrt{LC}$$

A good starting point is to select an available inductor with a known value. Select a large value like 100 mH for low frequencies (<10 kHz).

Next, knowing the center frequency and the inductance, we can rearrange the resonance formula to solve for C.

$$C = 1/(2\pi f)^2 L = 1/(6.28 \times 7000)^2(100 \times 10^{-3})$$

$C = 0.005\ \mu$F A standard value, but 0.0047 should also work.

Recalculating f with that later value of C gives 7345 Hz.

One solution is to put another capacitor in parallel with the 0.0047 μF to get 0.005 μF. We need

$$0.005 - 0.0047 = 0.0003\ \mu\text{F or } 300\ \text{pF}$$

There is no 300 pF, but there is 330 pF or 0.00033 μF.

Putting the two capacitors in parallel gives

$$C = 0.0047 + 0.0033 = 0.00503\ \mu\text{F}$$

Now recalculating f, we get

$$f = 7100\ \text{Hz}$$

closer to the desired 7000 Hz.

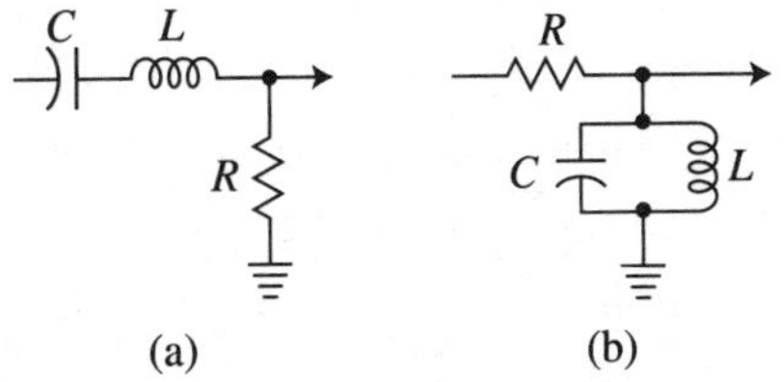

Figure 9.9 (a) Series and (b) parallel LC bandpass filters.

The selectivity is determined mainly by the Q of the circuit. Recall, that Q is

$$Q = X_L/R$$

X_L is the inductive reactance that is computed with

$$X_L = 2\pi fL = 6.28(7100 \times 100 \times 10^{-3}) = 4459\ \Omega$$

Of course, f is the center frequency, and L is the inductance in henrys.

R is the total circuit resistance in a series filter like Fig. 9.9*a*. It is made up of the output resistor R plus the winding resistance of the inductor, R_W. The Q of the circuit is

$$Q = X_L/(R + R_W)$$

Remember that in a series resonant circuit the inductive and capacitive reactances are equal, so they cancel one another. That leaves only the coil resistance R_W. At the resonant frequency the whole equivalent circuit is a voltage divider made up of R_W and R. R is made much larger than R_W, so the attenuation is minimal.

To find R_W, measure the inductor resistance with your DMM. A data sheet or catalog will also have that value. Let's assume it is 20 Ω. We want the output resistor at least 10 times that or 200 Ω. Make it 2000 Ω.

$$Q = X_L/(R + R_W) = 4459/(2000 + 20)$$
$$= 4459/2020 = 2.2$$

As for bandwidth, it is calculated with the expression

$$BW = f/Q = 7100/2.2 = 3227\ \text{Hz}$$

$$Q = X_L/R + R_W$$

Remember that bandwidth is

$$BW = (f_2 - f_1)$$

This filter is not very selective with a 3227 Hz bandwidth. One solution is to make the output resistor less to increase Q. We can use the previous estimate of 10 times R_W or 200 Ω. The new Q value is

$$Q = X_L/R + R_W = 4459/(200 + 20) = 20.27$$

The new bandwidth is

$$BW = 7100/20.27 = 350.3\ \text{Hz}$$

This is better but not great. Another type of filter may be needed to get a narrower bandwidth.

As for the parallel resonant circuit of Fig. 9.9*b*, most of the formulas are the same as the series circuit. The new information is that at resonance the parallel LC circuit is equal to a pure resistance, since the inductive and capacitive reactances cancel. That equivalent resistance is

$$R_W(Q^2 + 1)$$

Q is still the same.

$$Q = X_L/R_W$$

At resonance, the equivalent circuit is a voltage divider made up of R_{EQ} and the input resistor R. R_{EQ} is much higher than the input resistor R, so most of the input reaches the output.

Band Reject Filters

Twin-T Filter

One of the best RC filters is the notch filter shown in Fig. 9.10. It is called a twin-T or parallel-T. If you look carefully at the filter, you can see that it is a low-pass filter and a high-pass filter in parallel. The passbands overlap and define an attenuation notch. The twin-T is designed to null out one particular frequency

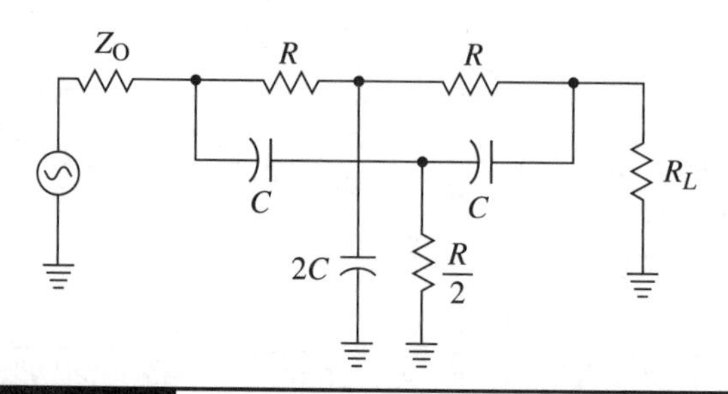

Figure 9.10 The popular twin-T RC notch filter.

and pass those above and below the center frequency. It has been widely used to notch out 60 Hz ac power line interference.

The notch or center frequency is determined by the familiar expression

$$f_C = 1/2\pi RC$$

Like other RC filters, its effectiveness is greatly affected by the output impedance of the driving circuit and the load. As usual, the driving impedance should be as low as possible and the load as high as possible. Here is the basic design procedure.

- State the notch frequency.

  Let's use 360 Hz.

- Select a value for R. It should be at least 10 times higher than the Z_O of the driving stage and lower than a factor of 10 or more of the load R_L.

  If Z_O is 200 Ω and the load is 100 kΩ, a value of 10 k for R would work.

- Calculate C. Rearranging the formula to solve for C,

  $C = 1/2\pi f_{CO}R = 1/6.28(360)10{,}000 = 0.044\ \mu\text{F}$

  A standard 0.047 µF capacitor would work but would shift the notch a bit to

  $f_C = 1/2\pi RC = 1/6.28(10{,}000)(0.047 \times 10^{-6})$

  $= 339\ \text{Hz}$

  This is usually not suitable, so a capacitor value closer to the calculated value must be used. You can experiment with putting capacitors in series or parallel to get closer to the desired value.

- Calculate $2C$ and $R/2$. These values are required as Fig. 9.10 indicates.

  Using the previous design results,

  $2C = 2(0.044) = 0.088\ \mu\text{F}$

  $R/2 = 10\ k/2 = 5\ \text{k}\Omega$

This filter design is very sensitive to component values. They must be right on the desired calculated value to achieve good attenuation at the center frequency. It is possible to achieve an attenuation of up to about 40 dB at the notch frequency, but only if the actual calculated values are used. Typically, you must resort to 1 percent resistors and capacitors to get close enough to the desired values.

Another approach is to select the capacitor first. Choose a 2 percent or 5 percent capacitor, then using that value, calculate the resistor value. Use the closest 1 percent value. Since there are more closely spaced resistor values, you are more likely to get close to the ideal.

One helpful modification is to use a variable resistor for the $R/2$ value. This will let you vary the notch frequency a small amount so you can optimize the attenuation of the undesirable signal.

A more selective twin-T can be created by using it in conjunction with an op amp. See Fig. 9.11.

This circuit uses two op amp followers to provide feedback that significantly sharpens the selectivity. The degree of selectivity is determined by the amount feedback provided by the voltage divider R_1 and R_2. A good way to implement this circuit is to use a pot in place of the two resistors. This way you can adjust the response to fit the application.

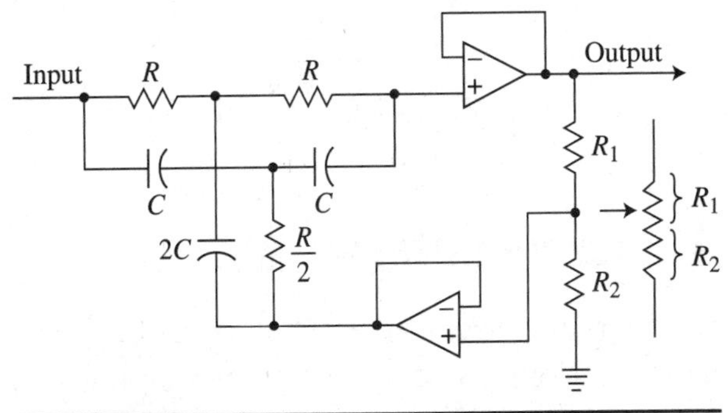

Figure 9.11 The twin-T RC notch filter with feedback to improve selectivity.

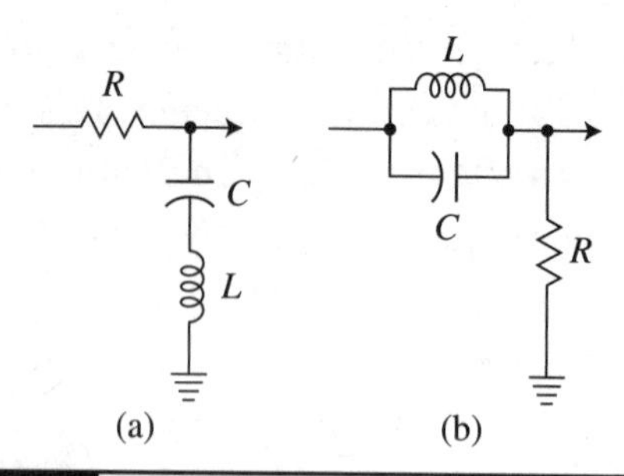

Figure 9.12 (a) Series and (b) parallel LC notch filters.

LC Notch Filters

You can also make some simple band reject filters with LC tuned circuits. See Fig. 9.12. The center or notch frequency is the resonant frequency of the LC circuit:

$$f_C = 1/2\pi\sqrt{LC}$$

All of the formulas and procedures used on the bandpass filters apply to these band reject filters. Keep in mind that at resonance, the inductive and capacitive reactances cancel one another out, leaving only resistance. In the series circuit of Fig. 9.12*a*, the series resistance is just the inductor winding or coil resistance. In the parallel resonant circuit of Fig. 9.12*b*, the equivalent parallel resistance can be computed with any of the following formulas:

$$R_p = Q(X_L) \text{ or}$$
$$R_p = R_w(Q^2 + 1)$$
$$R_p = L/C(R_w)$$

Other relevant formulas are

$$Q = X_L/R_w$$
$$X_L = 2\pi fL$$

R_w is the winding resistance of the inductor.

RC Active Filters

RC filters are cheap and easy to design, but they have limits. The main one is that they do not work well above about 1 MHz. At the higher frequencies, stray capacitances and inductances affect the design outcome. Second, they are not very selective. A better choice for higher frequencies is to use LC filters. However, there is another class of RC filter that is widely used. It is the active filter. The term *active* means that the RC network is enhanced by an amplifier that provides feedback and isolation. The frequency range is also extended above 1 MHz. Here are the design procedures for the basic forms of active filters. These design instructions are taken from Texas Instruments' (TI) Application Report SLOA093. That document is still available from the TI Web site. Otherwise, the procedures presented here are simplified.

The active filters described here use op amps. Many different types are candidates. Even the older IC op amps work well. You will need the ± power supplies for a full output swing, but single-supply versions are also useable. Here are the preliminary design steps.

- Determine the output voltage needed. A peak-to-peak specification is best. This will tell you what power supply voltages are needed. For older bipolar op amps, make the power supply voltages 1 to 3 V greater than the peak-to-peak output desired.
- Select the filter type. You should be able to determine what frequencies will pass and which will be attenuated based upon your application.
- Establish the cutoff frequency. For low- and high-pass filters, you will need the cutoff frequency. For bandpass and band reject filters, you need to know the center frequency and/or the bandwidth.
- Choose a capacitor value. It is not too critical, but at least you can specify standard available values. As a start, select 0.1 μF for low frequencies (below about 10 kHz) and 100 pF for the higher frequencies (above 100 kHz).
- Calculate the resistor values.

- If the calculated resistor values are too low or high, select another capacitor and start over.
- Despite the use of an op amp, most of these circuits do not produce any gain or loss but do offer good isolation between input and output with low output impedance.

Low-Pass Filter

Figure 9.13 shows the circuit.

Here is an example.

- Output voltage is less than 10 V peak-to-peak. Choose ±9 V or ±12 V.
- Choose filter type. Low-pass.
- State cutoff frequency f_{CO} is 5 kHz.
- Choose C_1. Try 0.1 μF for frequencies less than about 10 kHz.
- Calculate C_2, where $C_2 = 2C_1 = 0.2$ μF.
- Calculate R_1 and R_2. $R_1 = R_2 = 1/2\sqrt{2}(\pi C_1 f_{CO})$ = 225 Ω. Use a 220-Ω resistors.
- The single-supply low-pass circuit is given in Fig. 9.14. You probably want to block the dc from the input and output. The coupling capacitors C_{IN} and C_{OUT} should be about 100 to 1000 times C_1. This would be a value of 1 μf to 10 μF.

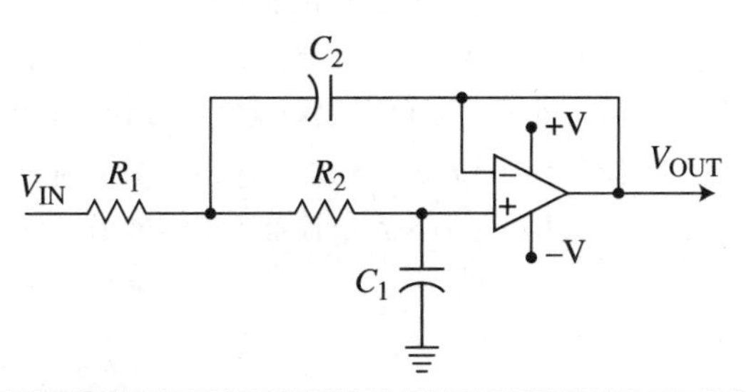

Figure 9.13 Active RC low-pass filter.

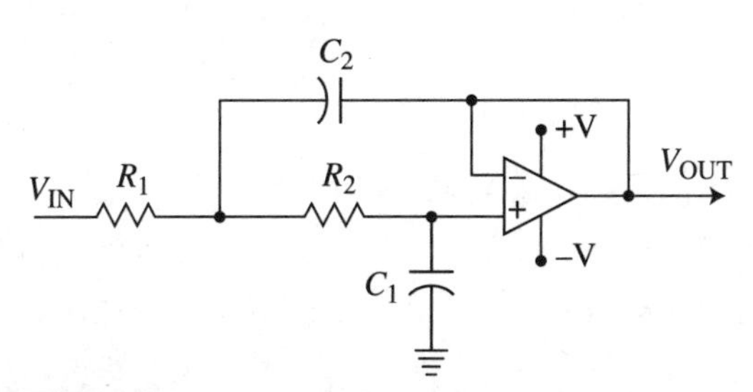

Figure 9.14 Active RC low-pass filter using a single-supply op amp.

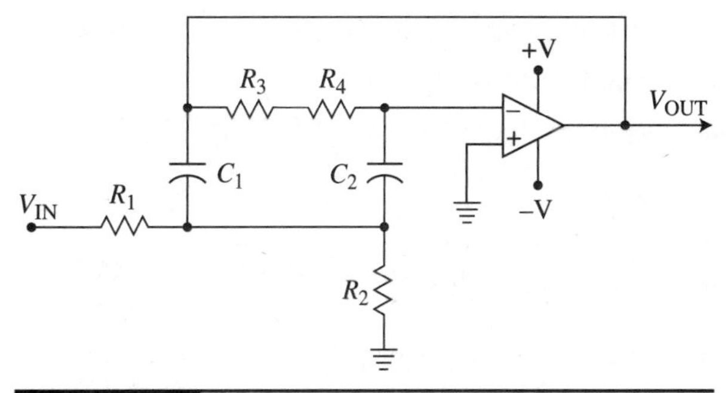

Figure 9.15 Active RC bandpass filter.

Bandpass Filter

This is the procedure for a narrow bandpass filter. The circuit is shown in Fig. 9.15. This filter also produces a gain of 10 (20 dB) at the center frequency.

The following is a typical design:

- Determine the desired center signal frequency (f_C). Assume 7000 Hz.
- Select $C_1 = C_2$ = Let's choose 0.01 μF.
- Calculate $R_1 = R_4 = 1/2\pi C_1 f_C = 1/6.28(0.01 \times 10^{-6} \times 7000) = 2275$ Ω
- Calculate $R_3 = 19R_1 = 19(2275) = 43{,}221$ Ω
- Calculate $R_2 = R_1/19 = 2275/19 = 120$ Ω
- Use the closest standard resistor values. If necessary, use 1 percent resistors for precision in setting the center frequency.

For the design details of the other filter types, download the Texas Instruments Application Report SLOA093 as recommended earlier.

LC Filter Design

Because of the complexity of designing LC filters, I suggest you do as I have done, and use an online calculator. There are several good filter design tools you can access for free. Here is an example to show you how.

1. https://www-users.cs.york.ac.uk/~fisher/lcfilter/
2. http://www.rfwireless-world.com/calculators/RF-filter-calculator.html
3. rf-tools.com/lcfilter
4. www.wa4dsy.net/filter/filterdesign.html
5. https://www.coilcraft.com/apps/lc_filter_designer/lc_filter_designer.cfm

Here is a design example using tool #1 above.

A radio transmitter operating at 30 MHz is producing a distorted output and is generating harmonics. The second and third harmonics are 60 and 90 MHz. We need to get rid of the harmonics so they will not interfere with signals near the harmonics. We need to design a low-pass filter that will have a cutoff of 30 MHz. Assume a filter impedance of 50 Ω. Design for a Butterworth response with five reactive components or order.

Now repeat that same filter design with the procedure using tools #3 and #4. The result for all three is the same and is shown in Fig. 9.16.

Note that the generator output impedance is 50 Ω as well as the load. Here are the odd, nonstandard C and L values.

$$L_1 = L_2 = 0.4292\ \mu\text{H} = 429.2\ \text{nH}$$

$$C_1 = C_3 = 65.57\ \text{pF}$$

$$C_2 = 212.21\ \text{pF}$$

The design process is pretty painless using these software design tools. Implementing the filter is a challenge. You want to get components with values close enough to those calculated so the filter will work. That is easier said than done. Standard capacitor values in 5 percent tolerance units are usually not close enough. Inductors are even more difficult. You may need to build your own. For standard inductors, search the distributors first. Then try the inductor manufacturers like Coilcraft and Bourns (J.W. Miller).

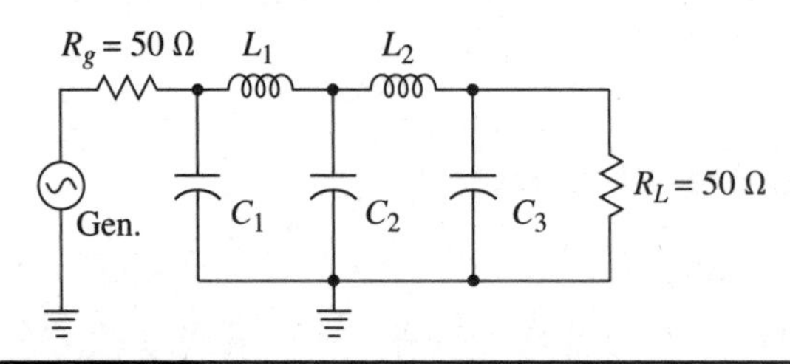

Figure 9.16 Low-pass LC filter with five poles.

Switched Capacitive Filters

An interesting and different type of filter is the switched capacitor filter (SCF). It uses a network made up of multiple capacitors, each with a MOSFET switch. The network of capacitors each with its on-off switch is connected to the input of an op amp. The capacitors are switched in and out of the circuit by a clock signal that drives the MOSFET's gates. The result is a bandpass or low-pass filter response. Most SCFs are available as ICs and are programmable, so you can select low, high, or bandpass devices with any of the four responses (Butterworth, etc.). The clock frequency determines the center frequency or cutoff of the filter. Most SCFs have an upper frequency limit of 100 kHz or so. For that reason, applications are limited to these lower frequencies. With up to eighth-order variants, selectivity is superb. If you need a small IC filter with excellent selectivity and low cost, SCFs are a good choice.

SCFs are available from several IC manufacturers, including Analog Devices, Maxim Integrated, and Texas Instruments.

Making a Sine Wave Out of a Square Wave

A good example of a filter application is using a filter to convert a square wave into a sine wave. It is based upon the Fourier theory that says that any nonsinusoidal like a square wave is actually a composite of multiple sine waves at harmonic frequencies added together. A square wave (50 percent duty cycle) is made up of a fundamental frequency sine wave and all the odd harmonics. Remember that a harmonic is a sine wave that is some multiple of the fundamental square wave frequency. For example, the third harmonic of 2500 Hz is 7500 Hz.

Figure 9.17 shows a frequency domain plot similar to what you would see if you connected the square wave to a spectrum analyzer. Each vertical line represents the amplitude of a sine wave. The fundamental, third harmonic, fifth harmonic, etc. Note the diminishing amplitudes of the harmonics with frequency.

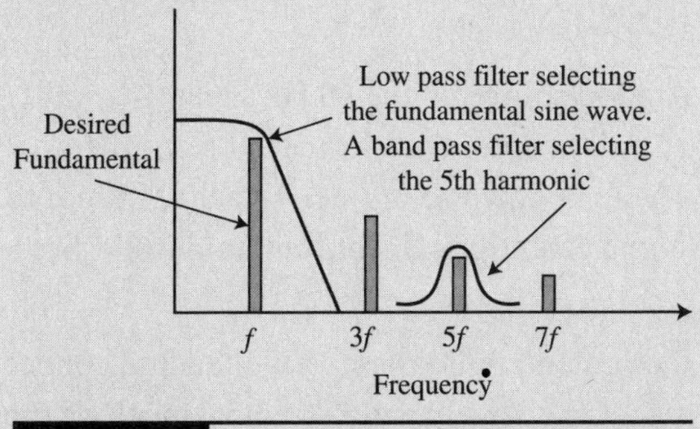

Figure 9.17 Using low-pass and/or bandpass filters to select the fundamental or harmonic components of a rectangular wave to generate a sine wave.

To produce a sine wave, you can design a bandpass filter that will select out one of the desired sine wave signals. The selectivity of the filter needs to be good so that it passes only the desired sine wave/harmonic. Figure 9.17 shows the response curve of a bandpass filter selecting the fifth harmonic. Another possibility is to select the fundamental sine wave with a low-pass filter as Fig. 9.17 shows.

This technique works well and is an alternative to making a sine wave oscillator that is harder to design than a square wave generator.

DSP Filters

DSP means digital signal processing. It is a technique that substitutes digital mathematical calculations to substitute for conventional analog or linear signal processing. The idea is to digitize the analog signal to be processed and store the resulting data. The digitized signal is a set of samples representing the signal at short intervals. With the signal in this form, it can be subjected to any one of multiple DSP programs that will perform the same function as an analog circuit.

The most common application is filtering. A processing algorithm can perform any type of filtering. A DSP IC or any micro capable of DSP functions can duplicate the effect of any filter type, for example. A DAC at the output of the DSP converts the filtered signal back to its analog equivalent—or any other filter type for that matter. Replacing a handful of inductors and capacitors or resistors and capacitors with a DSP IC or some microcontroller does not seem practical, but it is. Furthermore, the digital filter performs better than the equivalent analog circuit. It can have lower loss and much better selectivity than any LC or RC filter.

The truth is, a DSP filter is really software. The filter algorithms are programmed on a standard microcontroller or on a special DSP IC. They are also implemented on programmable logic devices (PLDs) like field programmable gate arrays (FPGAs).

A DSP device can implement a wide range of other linear circuits. DSPs can perform modulation, demodulation, equalization, mixing, error correction, spectrum analysis, and many other analog operations. A variety of math techniques are used. For example, in filtering, the overly simplified explanation of the process is multiplying the signal samples by coefficients and adding them together. DSP is a complex topic beyond this book, but it is an option you may not have known you have.

Design Project 9.1

Design an RC low-pass filter with a cutoff frequency of 8 kHz. The output impedance of the driving circuit is 300 Ω. Simulate the filter in software if that is available to you. Use the sweep frequency capability, and create a Bode plot. Build the circuit, and test it.

Design Project 9.2

Design an LC low-pass filter using the online tools suggested. The cutoff frequency is to be 10 MHz with an impedance of 150 Ω. Design for 7 poles or order 7.

Design Project 9.3

Design an RC active bandpass filter to select the third harmonic out of a 1-kHz square wave input to produce a sine wave.

Design Project 9.4

Design a twin-T notch filter that will remove the 60 Hz interference from a 1-kHz sine wave. Figure 9.18 shows the circuit. An op amp summer linearly mixes a 1 kHz sine wave (the desired signal) and a 60 Hz signal (the undesired or interfering signal). The signals can be derived from function generators or oscillators that you build. Put pots on each input so you can adjust the amplitudes. Suggested input levels are 2 V peak-to-peak each.

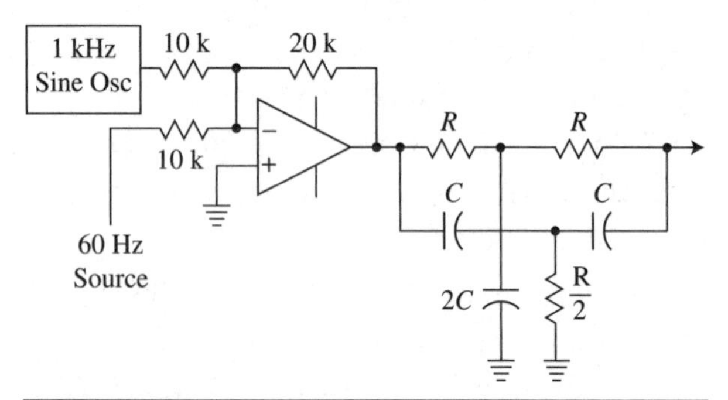

Figure 9.18 A circuit demonstrating how a twin-T filter can remove 60 Hz of noise from a 1 kHz signal.

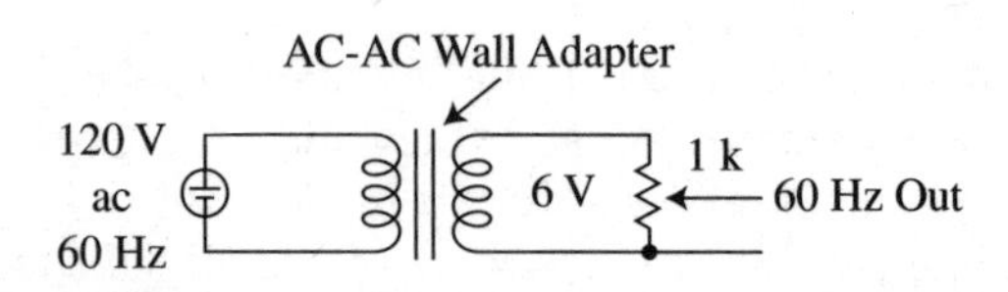

Figure 9.19 A simple and useful source of low voltage 60 Hz.

A good source of the 60 Hz signal is a wall wart/wall adapter transformer. It steps the 120-V ac in down to 6 V ac. Attach a 1 k pot to it so you can adjust the output amplitude. See Fig. 9.19.

Design and build the notch filter, and connect it to the op amp output to see how much of the 60 Hz is removed. You may want to put a pot as *R*/2 so you can adjust the center frequency to improve the attenuation of the 60 Hz. Measure the peak-to-peak 60-Hz voltage at the output of the op amp and then at the output of the filter. Calculate the dB attenuation.

See solutions in App. B.

CHAPTER 10

Electromechanical Design

Some of your designs may use electromechanical devices. This chapter covers the most common devices that include switches, relays, solenoids, motors, and servos. Most of the design effort involves understanding how the devices operate and paying attention to the specifications on each device's data sheet. The circuits are simple, and the more common configurations are covered here.

Switches

You already know what a switch is, so I won't elaborate other than to review that a switch is a set of manually operated mechanical contacts designed to carry electric current. There are dozens of different configurations and types. Only the most common operations and types are covered here.

Switch Contacts

There are three basic switch contact arrangements you will encounter. These are summarized in Fig. 10.1. Single pole single throw (SPST) switches (Fig. 10.1*a*) have one set of contacts that are either open or closed and two mechanical positions. These are typically used for on-off switches. A variation is a double pole single throw (DPST) that has two sets of contacts (Fig. 10.1*b*) that are mechanically ganged together and switch simultaneously. It lets you operate two separate circuits at the same time. Connections are provided to which wires can be soldered or that can be soldered to a printed circuit board (PCB).

Another configuration (Fig. 10.1*c*) is a single pole double throw (SPDT). The configuration lets you switch a single contact between two other contacts. A popular use of the SPDT is the use of two switches separated by a distance both of which can operate a device like a light bulb. See Fig. 10.2. Home lighting wiring uses this method. Here either switch can turn the bulb off or on. Check for yourself.

A fourth configuration (Fig. 10.1*d*) is a dual contact arrangement. Two circuits can be controlled with one push. When the button is not pressed, contacts A and B are closed. Contacts C and D are open. When the button is pressed, contacts A and B open, and contacts C and D close momentarily. If you connect contacts A and C together, the switch functions as a SPDT.

Switch Types

The type of switch refers mainly to its physical appearance or arrangement of the contacts and mechanical switching mechanism. A short list of the most common types you will encounter follows:

- Slide switch. The manual lever slides the contacts open or closed. Simple and inexpensive. Mostly PCB mounted.
- Toggle switch. Old style with a metal handle and mounted in a single hole on a panel.

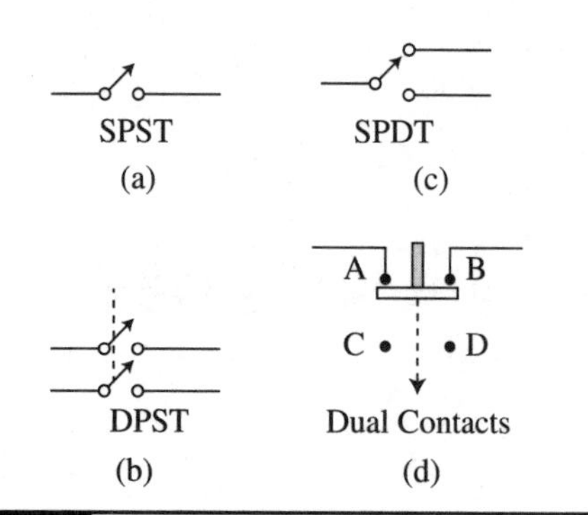

Figure 10.1 Standard switch contact configurations: (*a*) single pole single throw (SPST), (*b*) double pole single throw (DPST), (*c*) single pole double throw (SPDT), and (*d*) dual contacts.

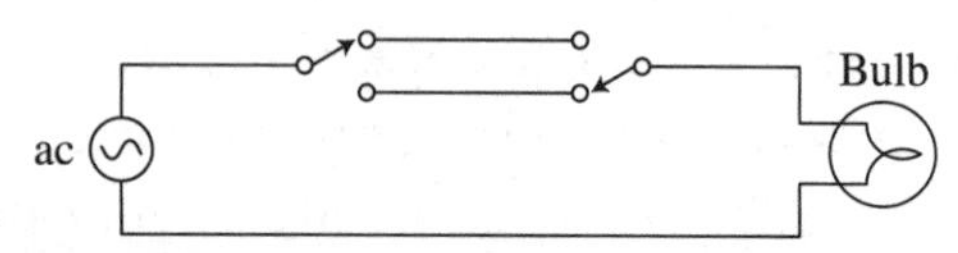

Figure 10.2 Two SPDT switches can be used to operate a bulb or other device, one near the bulb and another more remote connected by cable. Either switch can turn the bulb off or on.

- Rocker switch. The mechanical actuator is a rocker or seesaw assembly with a distinctive click and feel.
- Push button switch. Operation is self evident. Most are of the momentary contact type where the contacts close when the switch is pushed but spring back open when the finger is removed.
- Limit switches. Also called snap action switches. These are fitted with a flexible mechanical lever or handle designed to be operated by something other than a finger or hand. The lever usually comes into contact with some other moving device (machine tool, door, etc.) that will trigger a switch closure or opening. Also used as a limit switch to signal that some moving object has reached the limit of its movement.
- Rotary switches. A shaft rotates a contact between multiple contacts arranged in a circular manner. Not as common as they once were but available when you need to switch between multiple circuits or components.
- DIP switches. Dual in-line package switches have a physical configuration similar to the common DIP integrated circuit. They include multiple (4, 8, 10, etc.) miniature slide or rocker switches and are used to enter binary codes into digital circuits.

Switch Ratings

These are the specifications regarding how much current and voltage the switch contacts can withstand. The most common specification is simply a statement of the maximum current the contacts can handle and the maximum voltages that can be used. A common specification is 3 A at 120/240 V ac. The switches are rated to handle ac line voltages but can handle dc voltages also.

Relays

A relay is an electromechanical switch. A relay is made up of a set of mechanical switch contacts that are operated by an electromagnet. The electromagnet is a coil of wire on a core. When you apply voltage to the coil, current flows and a strong magnetic field is produced that causes the switch contacts to close or open. The contacts remain closed or open as long as there is current flowing in the coil. Removing the voltage from the coil causes the magnetic field to collapse and the contacts to revert to their unpowered state.

Fig. 10.3*a* shows one common form of relay with the key components identified. This one has SPDT contacts. The upper contact (A) is normally closed (NC) with the arm (B) when no voltage is applied to the coil. The lower contact (C) is normally open (NO) with the arm (B) when no voltage is applied to the coil. Contacts B and C close when power is applied.

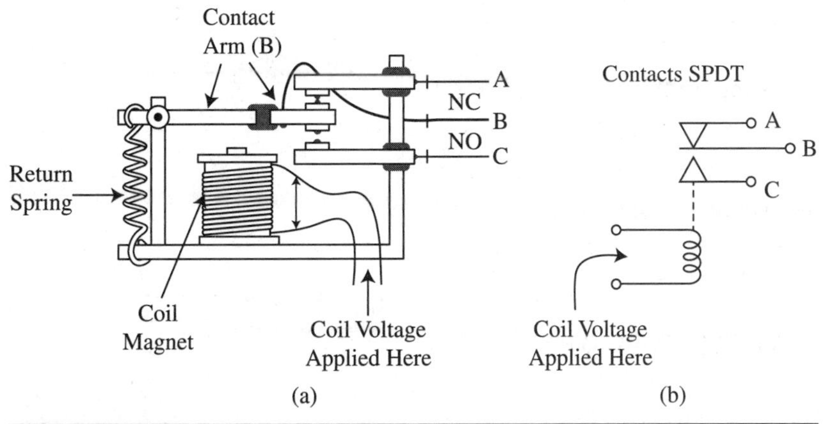

Figure 10.3 (*a*) The mechanical details of one type of open frame relay and (*b*) the electrical equivalent.

Such relays are usually packaged in a clear plastic housing and designed to plug into a socket of some kind for easy replacement. Note the schematic symbol in Fig. 10.3*b*.

Common Application

Most relays are used in circuits where a small switch with limited current-handling capability can operate a relay with contacts that can handle much higher current. Another use case is where a small, simple switch can control a larger relay from some distance. It may be inconvenient to locate the control switch near the load to be switched.

A good example is the starter switch in a car. The starter motor is a heavy-duty motor that takes 100 A or more to start the engine. It is not convenient to run heavy, high-current cables into the passenger area and install a high-current switch. A small ignition switch or button can operate a remote heavy-duty relay called a contactor. Pressing the starter button engages the relay, whose high-current contacts close and connect the battery to the starter motor.

Relay Terminology

When a relay is activated, contacts either close or open. Contacts that close are said to "make." An older slang term for a relay that has been activated is "picked." Contacts that open are said to "break."

Relay Contacts

Like switch contacts, relay contacts come in many forms, such as SPST, SPDT, DPDT, etc. However, you will also see relay contacts referred to as shown in Fig. 10.4. Form A is the same as a SPST with normally open (NO) contacts deenergized. Form B is also SPST but with normally closed (NC) contacts with coil deenergized. Form C is similar to SPDT contacts. Many variations are available. DPDT is very common, as is 4PDT or what amounts to as four SPDT contacts.

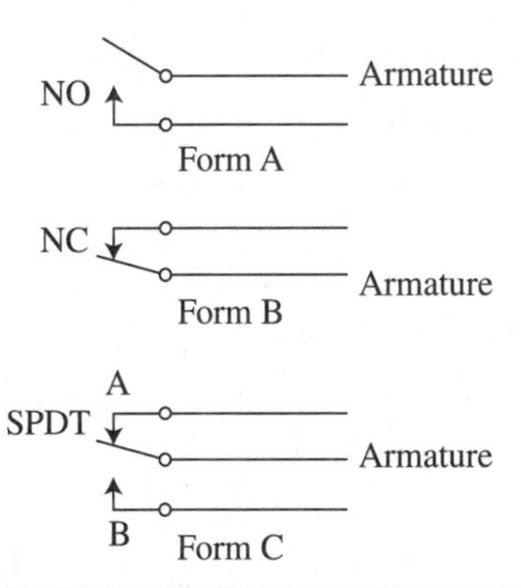

Figure 10.4 Standard relay contact designations.

Relay Specifications

These are similar to switch contact specifications. Other specifications relate to the coil. Most relays are designed for dc operation, although ac voltage relays are available. Only the dc relays are covered in this book.

- Contact rating. The maximum amount of current the contacts can handle, typically 1 A or more.
- Coil voltage. The amount of voltage you have to apply to the coil to get the relay to pull in or engage the contacts. The most common voltage ratings are 5, 6, 12, and 24 V. AC relay coils are usually activated by power line voltage of 120 V.
- Coil resistance. The dc resistance of the coil—with this value and the voltage you can, of course, calculate the current.
- Holding voltage or current. This is the amount of voltage or current needed to keep the relay activated. It is less than the normal "make" voltage or current. Any lower current will cause the relay to drop out.
- Switching time. This is the time it takes for the contacts to close (or open) once voltage is applied to the coil. Being mechanical devices, they are slow compared to a transistor switch. Switching times are usually many milliseconds (ms).

Reed Switches and Relays

A reed switch is made up of a pair of thin mechanical reeds that have contacts on the ends that overlap. The reeds are housed in a clear glass tube. See Fig. 10.5. Other housings are common.

The reeds can be easily magnetized by an external magnetic field. That magnetic field can come from a permanent magnet or an electromagnetic coil that is usually wrapped around the reed switch as the figure shows. When a magnetic field is applied, the two reeds become

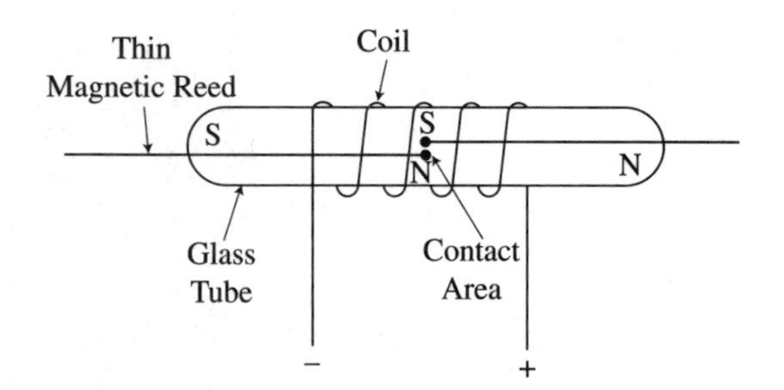

Figure 10.5 The basic structure of a magnetic reed relay.

magnetized with the north (N) and south (S) polarities shown. Since opposite magnetic poles attract, the two contacts pull together and make the connection.

A common use of reed switches is as position or proximity sensors. These are the sensors used with a bar permanent magnet to detect open windows or doors in a security alarm system. When the window or door is closed, the magnet and reed switch line up, and the contacts are activated or closed. Opening the window or door moves the magnet away and the contacts break, signaling the circuit to set off an alarm.

Reed relays are also available in DIP housings for PCB mounting. A coil around the glass housing provides the magnetic field. Operating voltages are usually either 5 or 12 V. Reed relays also operate faster than other forms of relays with switching times of a few hundred microseconds (μs) compared to milliseconds for more conventional relays.

Latching Relays

One special form of relay you may encounter or need in a design is a latching relay. This is a relay whose contacts remain closed even when the coil voltage is removed. Some latching relays use a mechanical latching mechanism to hold the contacts closed when power is removed from the coil.

You can also use an electrical latching approach as illustrated in Fig. 10.6. Note that

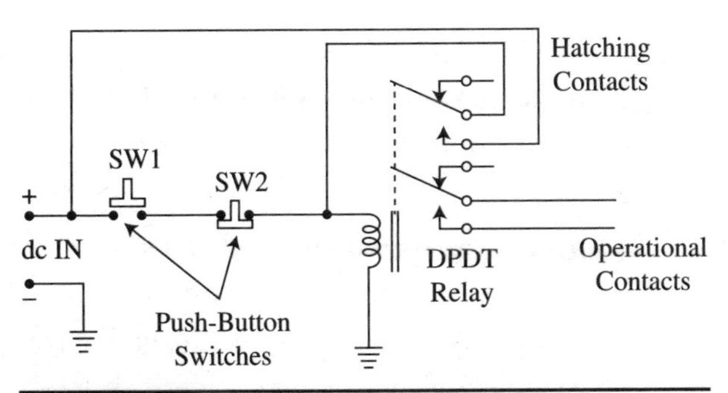

Figure 10.6 A relay connected to latch or maintain its position if it is energized momentarily. A push button (SW1) is pressed to open the circuit to the relay coil, and the relay returns to its normal off state.

an extra set of contacts on this DPDT relay is used to implement the latching function. When the NO push button, SW1, is pressed, voltage is applied to the coil, and the relay pulls in or makes. Releasing the push button opens its contacts and disconnects the voltage. However, the upper set of relay contacts close, keeping the coil energized. To turn the relay off, a separate NC push button, SW2, must be pressed to open the coil circuit.

Solid State Relays

There is such a thing as a solid state relay (SSR) that uses all electronic components rather than mechanical components. The switching contacts are replaced by a semiconductor component called a triac or a metal oxide semiconductor field effect transistor (MOSFET). Each is an electronic switch that turns off or on according to an input signal to its gate. The triac conducts in either direction as long as a gate signal is applied. For that reason, it can switch ac voltages as well as dc.

Most SSRs switch ac loads. Different models can switch 120- or 240-V single-phase circuits or 208-, 240-, 480-, or 600-V three-phase circuits. Typical maximum current level is usually in the 1- to 10-A range.

Another key specification is control or drive signal to gate. This drive voltage is typically in the 1.5- to 30-V range. Special input options are available. A popular one is a signal that switches at or very near the zero crossing points of an ac signal that is being switched.

The primary benefit of an SSR over a mechanical relay is that it is faster. Switching times are in the low millisecond range.

Other characteristics to consider in selecting an SSR are the number of poles needed and the package type and mounting. Most SSRs are SPST or form A types. Other arrangements are possible. There are multiple mounting options. Be sure to gather as much information as possible from the manufacturers before selecting and designing with SSRs.

Solenoids

A solenoid is another electromechanical device. It consists of a coil to produce a magnetic field and a magnetic plunger or bar inside the coil. Figure 10.7 shows the basic construction of a solenoid. When voltage is applied to the coil, a magnetic field is produced. That field interacts with the magnetic field produced by the internal bar or plunger. This causes the plunger to move linearly either into or out of the coil assembly. The bar is physically constrained to stay inside the coil. The plunger produces considerable force that can be used to initiate some external mechanical action. An example is a door latch. Any action that requires a push or pull can be

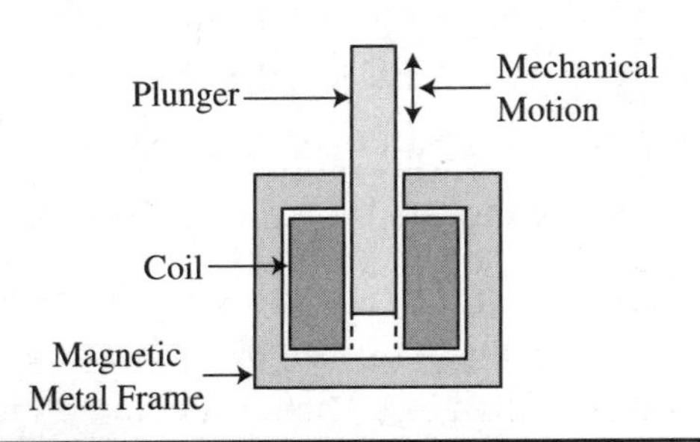

Figure 10.7 One form of solenoid.

operated by the solenoid. The linear movement of the plunger is not great, usually an inch or less, depending upon the size of the solenoid.

Most solenoids operate from 12 or 24 V. Typical coil resistances are low, necessitating high currents in the 300-mA to 1-A range.

Solenoids can be operated by just a mechanical switch or by a transistor driven by some circuit. The BJT and MOSFET switches covered in Chap. 5 can be used. Be sure to put a diode across the solenoid coil as described for relays to protect the driving transistor.

Motors

There are dozens of types of motors both dc and ac. For many projects, a basic dc motor is adequate. They are plentiful and cheap, and they are simple to control. This chapter only covers the permanent magnet (PM) brush motors. They operate from voltages in the 1.5- to 24-volt range. Speeds range from a low of about 1500 rpm to almost 20,000 rpm. The speed can be controlled by controlling the applied voltage. Lower speeds can be achieved by using external gears or belts. However, gear motors are also available. These motors contain an internal gear set that reduces the speed downward to a few rpm or several hundred rpm.

The basic specifications for design are dc operating voltage, current, speed, and torque. Torque is a mechanical term defining the mechanical turning force of the motor. Torque is expressed in grams per centimeter (g-cm). The driveshaft size is also a consideration. Most shafts are pretty short, usually less than an inch. Shaft diameter also varies from 0.07 up to 0.25 in.

In selecting motors for a project, you can use this procedure:

1. Identify your power supply voltage and current capability. The most common motor voltage ratings are 3, 6, and 12 V, although you can find special motors that will work down to 1.5 V. Be sure your power supply can furnish the amount of current needed by the motor.
2. Determine the speed you need. This can vary widely, depending upon the application. A specific need will usually give you a hint. You can guess or experiment. Specify a gear motor if your speed needs are very low.
3. Determine your torque needs. Again this could be any number, depending upon the application. You may have to experiment unless you have some clue to your needs.
4. Define any special needs, such as shaft rotation reversal, speed control, or shaft couplings.

Motor Control

The motor can be controlled by a simple switch to turn it off or on. It will run at its rated speed. Two things can be controlled, direction of rotation and speed.

Direction

One option you may need is to reverse the motor's direction of rotation. To do this, all you really need to do is reverse the polarity of the applied dc. This is easier said than done. One approach is to use a DPDT switch to reverse the polarity as Fig. 10.8 shows. A separate off-on switch is provided.

Another way to reverse the polarity is to use an H-bridge circuit as shown in Fig. 10.9.

The circuit uses MOSFET switches in a bridge configuration to perform the polarity reversal. Note the protection diodes across each MOSFET. In this circuit, the SPDT switch controls the direction. With the switch in the A position, Q_1 and Q_4 are biased on turning the

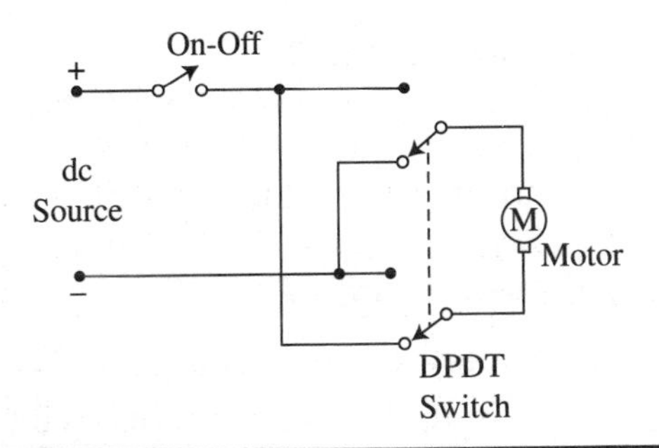

Figure 10.8 How a DPDT switch can be used to reverse the direction of rotation of a motor by reversing the polarity of the operating voltage.

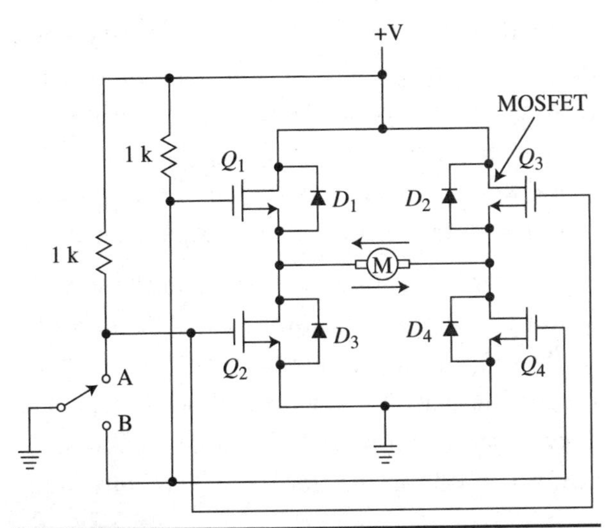

Figure 10.9 An H-bridge circuit made with MOSFETs used to electronically reverse the motor rotation direction.

motor on in one direction. When the switch is set to the B position, Q_2 and Q_3 conduct, reversing the polarity on the motor and reversing its rotation. You can make this circuit with discrete MOSFET switches. Use the design procedure for MOSFET switches described in Chap. 5. Alternately, you can buy an H-bridge IC.

Speed

The most common method of speed control is pulse width modulation (PWM). This is a technique that applies dc pulses to the motor rather than a steady dc. The speed is determined by the average voltage the motor sees. The width

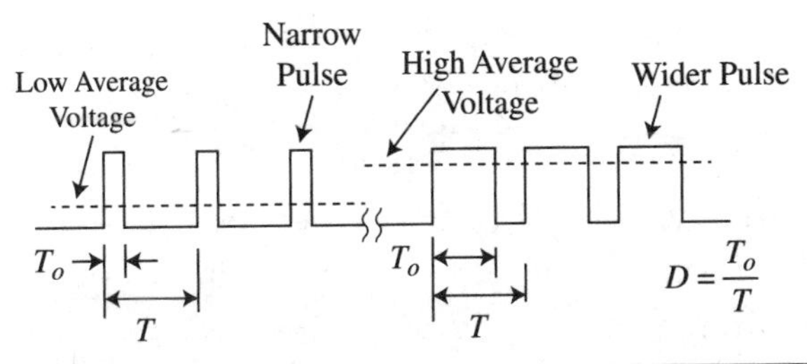

Figure 10.10 Pulse width modulation (PWM) uses variable duty cycle voltage to vary the speed of a permanent magnet brushed motor.

of the pulses determines the speed. Refer to Fig.10.10.

PWM is a technique of varying the duty cycle of the pulses. The duty cycle (D) is the ratio of the pulse on time (T_o) to the period of the pulses. The period T is the reciprocal of the frequency of the pulses (f).

$$T = 1/f$$

$$D = T_o/T$$

Duty cycle is a fraction, but it is sometimes multiplied by 100 to express it as a percentage.

The frequency of the pulses is kept constant. Because the on time varies, the average voltage as seen on a voltmeter or a motor varies.

With narrow, short on time pulses, the average dc voltage is low. A dc motor would turn at a low speed. Increasing the on time or pulse width increases the average voltage, so the motor will turn faster. Many of the microcontrollers have a way to generate programmable PWM pulses that can be used for speed control.

Special duty cycle controller ICs are available for motor control. An example is the SG3525 made by ST Microelectronics and ON Semiconductor. A PWM motor controller kit is also available from Velleman. Designated the K8004, this kit uses the SG3525 IC. Its frequency of operation can be from 100 Hz to 5 kHz. It will take an input voltage in the 8- to 35-V range, and the output can be adjusted over a 2.5- to 35-volt range with a current limit of 6.5 A. That will control a very large motor.

Many microcontrollers come with a PWM output feature. The MCU generates a pulse at the desired frequency. Instructions in the MCU let you program and control the pulse width and the motor speed.

Just So You Know

PWM can be used to control other things besides motor speed because it can vary the average voltage that a device receives. Another popular use of PWM is for brightness control of LEDs. PWM can also control a dc heating element to vary its temperature.

Servo Motors

Another type of motor you are likely to encounter is a servo motor. These are used in remote control applications, such as model airplanes, boats, cars, drones, and robots. The term *servo* is derived from the term *servomechanism*. A servomechanism is a type of feedback control system that incorporates a motor whose speed and direction can be controlled, usually by radio. Here is a quickie tutorial on servo motors, in case you need to use one.

Internal Components

Figure 10.11 shows the basic structure of a servo. The motor is a small permanent magnet brushed motor. Because of its natural high speed, a gearbox is added to slow the output down considerably. The output is taken from the gearbox output shaft that is connected to whatever mechanical part it is controlling.

The gear output shaft also drives a potentiometer. A dc voltage on the pot is varied by the gear set. This pot provides a feedback signal that tells the motor control circuitry where the shaft output is. That feedback signal is sent to some internal control circuitry that drives the motor. That control circuitry takes an input signal that adjusts the speed and direction of the motor and the position of the output shaft.

It is important to understand that the servo motor does not spin continuously. The position of the output shaft can only rotate over a 180° range. The shaft sits in a normal position at 90°. Then it can turn up to −90° counterclockwise (CCW) or +90° in the clockwise (CW) direction.

Controlling a Servo

You control a servo by applying a pulse width modulation (PWM) signal to the control input.

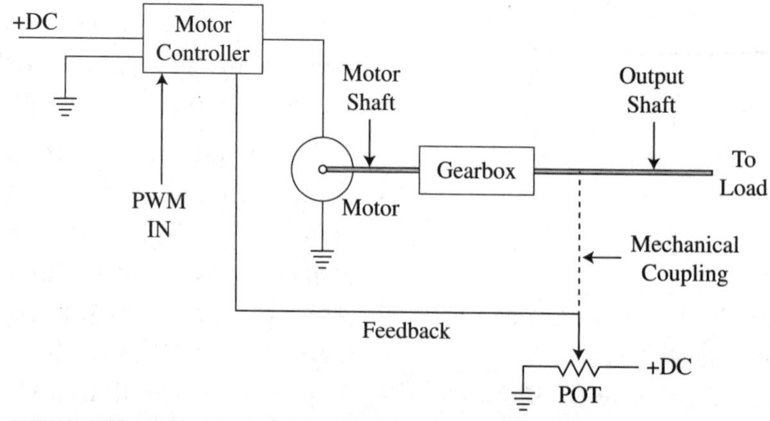

Figure 10.11 A simplified block diagram of a servo motor. It uses PWM to control the shaft position.

PWM was covered earlier in the previous section, but here is a review targeting servos.

PWM is a pulse wave train with a constant frequency. For servos, this signal is usually in the 50- to 60-Hz range. The modulation of this pulse train is variation of the signal duty cycle. The duty cycle is the ratio of the pulse on time to the signal period (T). See Fig. 10.10. The period is fixed because the frequency is constant.

$$T = 1/f$$

But the pulse width T_O changes.

The duty cycle is

$$D = T_O/T$$

For a very short duration pulse, D is very low. A wider or longer pulse gives a higher D.

Assume that the frequency of the pulses applied to the servo is 50 Hz. The period then is

$$T = 1/f = 1/50 = 0.02 \text{ s or } 20 \text{ ms.}$$

The standard pulse time is 1.5 ms, and that keeps the output shaft at the 90° neutral position. A 1 ms pulse will move the shaft CCW to the −90° position. A 2 ms pulse moves the shaft CW to +90°. See Fig. 10.12.

Small dc motors typically do not have high torque, so their load must be minimal. However, the gearbox greatly increases the torque, so it can operate stronger mechanisms. If the device connected to the output shaft tries to oppose the motion, the servo feedback senses this and corrects it, keeping the output shaft solidly in its position.

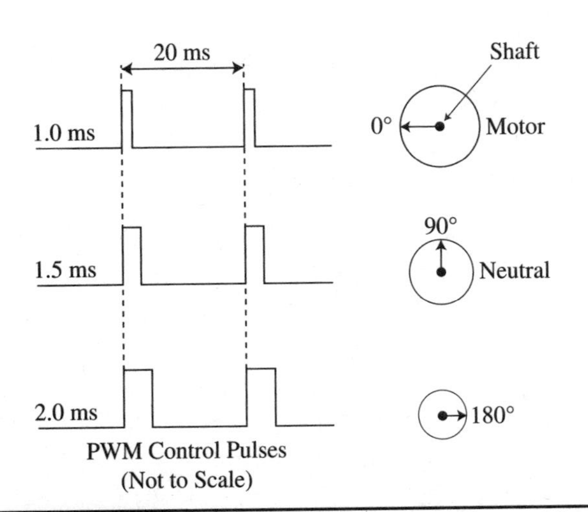

Figure 10.12 How pulse width affects the position of a basic servo motor. Most servos expect a 50-Hz PWM signal.

Design Project 10.1

Acquire a small dc motor that will operate with a voltage in the 1- to 12-V range. I found many motors at suppliers All Electronics and Jameco. 3-, 5-, and 6-V motors are common. Prices are only a few dollars at most. Be sure to get the electrical specifications, such as current draw at the rated voltage. A data sheet may be available, and if so, get it. A 3-, 5-, or 6-V motor will probably be best.

Design and build an H-bridge like that shown in Fig. 10.9. Use some heavy-duty MOSFETs like the IRF510—one of my favorites. Get the data sheet, too. Use 1N4001 protection diodes. Wire the circuit as shown. The SPDT switch selects the transistors that determine direction of rotation. When the switch is in the A position, the gates to Q_2 and Q_3 are at ground, so they do not conduct. The 1-k resistor on the B position of the switch applies voltage to the gates of Q_2 and Q_4, turning them on. Electrons flow through the motor from right to left.

Apply power to the circuit. Notice the direction of rotation. Set the switch to the other position. Again, note the direction of rotation, which should be the opposite of what you observed when the switch was in the other position.

Design Project 10.2

Design a motor speed controller using PWM. Use the circuit in Fig. 10.13 that shows the 555 timer circuit configured for PWM. Set the supply

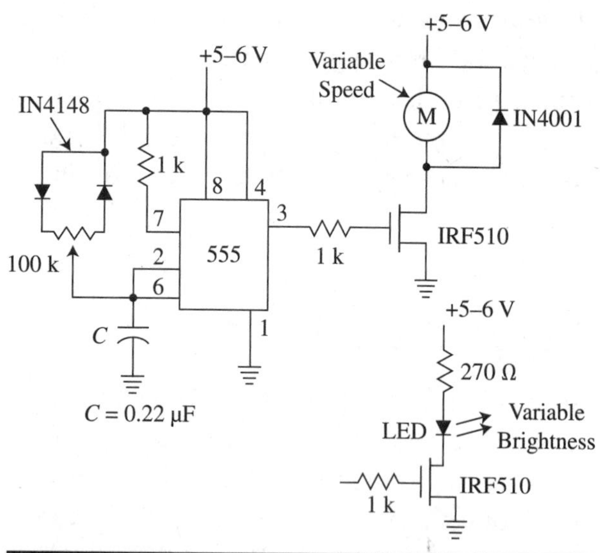

Figure 10.13 A variable duty cycle motor speed control circuit using a 555 timer IC connected to provide a 0 to 100 percent duty cycle output. A power MOSFET is used to operate the motor. The MOSFET can also operate an LED or group of LEDs and provide dimming with the PWM.

voltage to the motor voltage rating at 5 or 6 V. Varying the pot 100 k varies the duty cycle and the speed. The MOSFET is needed to handle the current of the motor. Monitor the 555 output at pin 3 on your oscilloscope. Vary the 100-k pot over its full range to see the duty cycle vary.

To test the circuit, apply power, and the motor should run. Adjust the pot over its full range, and note the apparent speed change in the motor.

To demonstrate how PWM controls the brightness of an LED, simply replace the motor with an LED and resistor as shown in Fig. 10.13. The transistor controls the LED brightness as the duty cycle is varied.

Design Project 10.3

Modify the circuit in Fig. 10.13 to set the frequency as close to 50 Hz as possible. Connect the servo to it. Research how to configure the circuit and wire the output of the PWM circuit to the yellow wire of the servo. Use your scope to zero in on the 50 Hz output. Show how varying the duty cycle makes the servo perform its function.

CHAPTER 11

Digital Design

Digital equipment dominates electronics today. Logic circuits, embedded microcontrollers, and computers are at the heart of most electronic gear these days. Analog and linear circuits have not gone away and will never disappear. They continue to play a support role to the digital circuitry. However, every engineer needs to know how to design basic digital circuits. There is lots to it, but this chapter attempts to boil it down to some basic procedures you can use to implement most new designs.

NOTE: This chapter assumes that you have some digital background, including knowledge of binary numbers, basic logic gates, flip flops, and the most common digital ICs. However, some of these fundamentals will be reviewed briefly as a refresher before we dive into the design process.

Three Design Approaches

There used to be only one way to create digital equipment. You had to build it from scratch with basic logic circuits. Today, the majority of digital circuit implementation is by way of programmable logic devices (PLDs) or microcontrollers. Here is a look at three options available to you today.

Basic Logic ICs

You can still design equipment using the older TTL or CMOS gates, flip flops, functional logic circuits, and memory. It is still a workable solution today since the ICs are still available. However, it is no longer practical or economical for new commercial products. Many ICs are usually needed for most designs, and they require lots of PC board space. This makes a product larger and more expensive. If you have to make changes, a costly new PC board layout is most likely needed. That's expensive and time consuming, yet for some small personal one-of-a-kind products, basic logic is still an OK choice. For big projects, you are probably better off going with a microcontroller.

Why Traditional Logic?

Why present this legacy, even retro, approach to digital design? Most digital projects are made with a microcontroller today. Some special products need a massive amount of logic and extra high speed, in which case an FPGA can be used. No one makes commercial products these days with conventional logic ICs unless they are small and simple products. I thought this traditional design procedure was valuable for the following reasons:

- It helps you to better learn and understand digital logic down at the gate and flip flop level.
- It is still taught in colleges. (Most textbooks still include it.)
- It is an inexpensive approach to small projects because most legacy ICs are still available and cheap.
- It is compatible with many logic software simulators.
- It is a good choice for many experimenter projects.
- It is fun and rewarding to design and build a digital product and see it work.
- For devices using PLDs like FPGAs, you still need to do the design before you enter the code

into the hardware description language (HDL) or enter the schematic diagram.

- With the use of so much design software, there is a growing abstraction that seems to insulate the designer from the actual hardware. Engineers designing with software often do not have a clue what the real logic hardware is like. By working with the nitty-gritty of the digital circuits, such as gates and flip flops, you really get to see what the circuits are doing. This leads to a better understanding of how the design works. It improves your perspective and your creativity in design.

Programmable Logic

Another approach is to use some form of programmable logic IC. Programmable logic devices (PLDs) are ICs containing multiple logic elements, such as gates and flip flops, that can be programmed from an external source to form the exact logic circuit you need, and all on a single chip. Programmable logic devices come in multiple forms, from a dozen gates or so to thousands or even a million logic elements on a chip. Once-popular simple programmable logic devices (SPLDs) that were popular from the 1970s through the 1990s, like the programmable logic array (PLA), the programmable array logic (PAL), and the generic array logic (GAL), have essentially faded away. Most of these parts have been declared obsolete and are no longer made. The only exception I found to this is from the company Microchip Technology/Atmel, which still makes a few of these SPLDs. Generally, I do not recommend these SPLDs for any new commercial product designs. They could disappear from the market at any time, leaving you with a redesign crisis.

On the other hand, for one-of-a-kind personal projects, they are often a good fit. They are a better match to the smaller designs than the larger complex programmable logic devices (CPLD) and the field programmable gate array (FPGA) chips so popular today. These older SPLDs are summarized in Chap. 12, and those parts still available are identified.

One type of PLD that is still available is the complex programmable logic device (CPLD). It is larger and more capable than the older PLA/PAL/GAL devices. It is similar to a big GAL and somewhat less complex than the FPGA. Some models contain up to 10,000 gates or equivalent. A popular product is Xilinx's CoolRunner. Maxim Integrated, Lattice Semiconductor, and Microchip Technology/Atmel also make CPLDs.

Most of the SPLDs have been replaced by the larger of these chips, the field programmable gate array (FPGA). The FPGA was once an expensive and complex device, but today prices are lower and manufacturers have made it easier to use thanks to support software. For larger more complex projects, the FPGA is an alternative.

FPGAs are complex devices with thousands, even millions, of gates and flip flops. These programmable devices are suitable primarily for large-scale projects or ones needing the speed advantage over most MCUs. In addition, you need to learn a special programming language, such as VHDL or Verilog, to develop your product.

Microcontrollers

Perhaps the most widely used approach today is digital design with a microcontroller, also known as an embedded controller or microcontroller unit (MCU). You can do almost any digital operation with an MCU by programming it. It will do basic logic functions like AND and OR, add and subtract, as well as practically any other operation. In most cases, these one-chip wonders replace dozens of other types of digital ICs. Many designs can be implemented by a single-chip MCU, making them small and cheap. If changes are required, you just reprogram the chip. The only downside to this approach is that when very high speeds are needed, the MCU may not be fast enough, so the design usually goes to a faster MCU, a CPLD, or an FPGA.

That said, keep in mind that virtually all—and I mean ALL—electronic products contain at least one embedded controller. For that reason, your first design choice should be an MCU. The main challenge is programming the device to do what you want. You will need to learn a programming language like C and become familiar with development software, algorithm development, subroutine libraries, and the like. Easier said than done. Nevertheless, there are some projects that can be done with logic ICs. The remainder of this chapter focuses on digital design with TTL or CMOS devices. It will help you learn and understand the digital design process. SPLDs, CPLDs, and FPGAs are covered in Chap. 12. You will learn more about designing with microcontrollers in Chap. 13.

Digital design is typically divided into two parts: combinational logic design and sequential logic design. The combinational logic is the static logic made up of gates. Sequential logic is the dynamic part made of flip flops, counters, and registers.

Preliminary Design Decisions

Before you launch into any new design, you need to ask some basic questions. Record your answers in your design notebook.

1. Is this a commercial design or one for personal use?
2. If it is a commercial venture, how many units would be manufactured?
3. What power sources are to be used? Battery or ac mains? Is low power consumption critical?
4. What are the physical design and size requirements?
5. What are the inputs? Other logic signals, analog signal requiring an analog-to-digital converter (ADC), sensors, keyboard, or what?
6. What loads or outputs will be required? LEDs, relays, LCD readout, motor, etc.
7. Knowing the inputs and outputs, draw the black box model.
8. Determine what interface circuits are needed for the inputs and outputs you defined.
9. What is the frequency range of the design? Clock speed?
10. Are there any requirements for meeting electromagnetic interference (EMI) requirements (FCC, CFR 47, Part 15)? None for personal products.
11. Will any testing and certifications be required? For commercial products, UL approval may be needed. None for DIY projects.
12. Are secondary sources for all parts needed? Commercial products only.
13. What is the timetable for the design, prototype, final design, and transfer to manufacturing?
14. Do you have the software (language) to write your code?
15. As an initial step, write out a sequence of steps you think will be needed to solve the problem, and translate that into a flowchart.
16. If you can, at this point, generate a truth table that lists all the possible input states and the corresponding outputs.

Combinational Logic Circuits

Most of you are probably already familiar with some of this material, but here is a quickie summary of basic logic functions as a refresher. Figure 11.1 covers gates and inverters in their various forms. Recall that the small circles at the outputs of some gates are inverters. For more

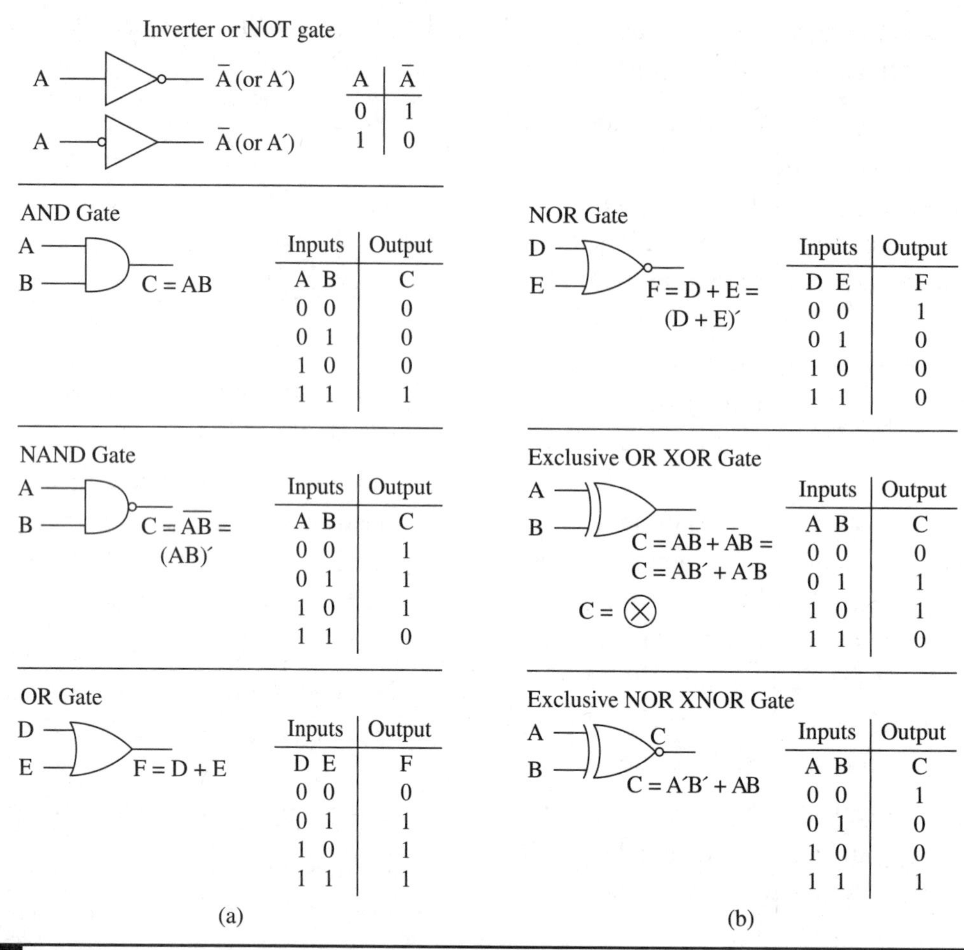

Figure 11.1 Summary of the basic logic operations and circuits.

details, refer to any basic digital book available. See App. A for some suggestions.

Remember that the function of a gate or other logic circuit is expressed in terms of boolean algebra. It uses a format like standard mathematical algebra. Each AND gate has an output that is said to be a *product* of its inputs. The input variables are written together as you would in multiplying two or more variables as in standard algebra. A AND B = AB. You also do A(B). An apostrophe after the term means that that term has been inverted or is the complement of the term. We say A NOT for A′. The truth table associated with each gate in Fig. 11.1 tells what the output is for all possible sets of inputs. Gates can also have more than two inputs. Three and four input gates are common. The OR function is designated with a + sign because it is like a summation of AND terms.

An example circuit is given in Fig. 11.2. It shows how to write the boolean expression for a logic circuit. The AND gate outputs are sent to an OR gate to create the sum-of-products (SOP) expression that is the output. In this case F = AB′ + A′BC′ + D. The ANDed variables are the products, and the + represents the OR function or sum.

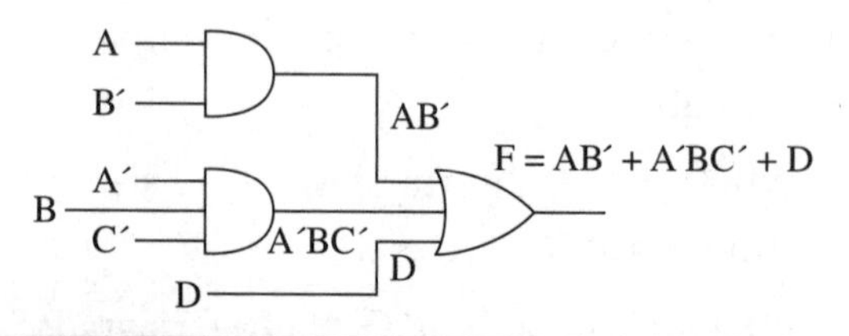

Figure 11.2 Writing the boolean equation from the logic circuit.

DeMorgan's Theorem

DeMorgan's theorem is expressed in two boolean equations that you need to know. They say that a NAND gate can be used as a NOR gate and that a NOR gate can be used as a NAND gate. The first expression

$$(AB)' = A' + B'$$

The left side of the equal sign is the boolean expression for a NAND gate. The right side of the equal sign is an OR expression with the terms inverted. It is called a negative NOR. Check back with Fig. 11.1 to confirm if you need to.

The other version of DeMorgan's is

$$(D + E)' = D'E'$$

The left side of the equation is the standard NOR expression. The equal sign says that it can also perform a NAND function. It is a negative NAND.

Why is this useful?

It is useful because one type of gate can perform two different operations. This means you don't need special gates for most operations. You can do everything with a NAND gate. It can do NAND as well as AND, NOR, OR, and NOT. Figure 11.3*a* shows the details. Keep in mind that the circle or bubble at the input or output of a gate is an inverter or NOT circuit.

You can also implement all the basic logic functions with NOR gates. Figure 11.3*b* shows the details. However, using NOR gates usually makes the circuit larger and more complicated.

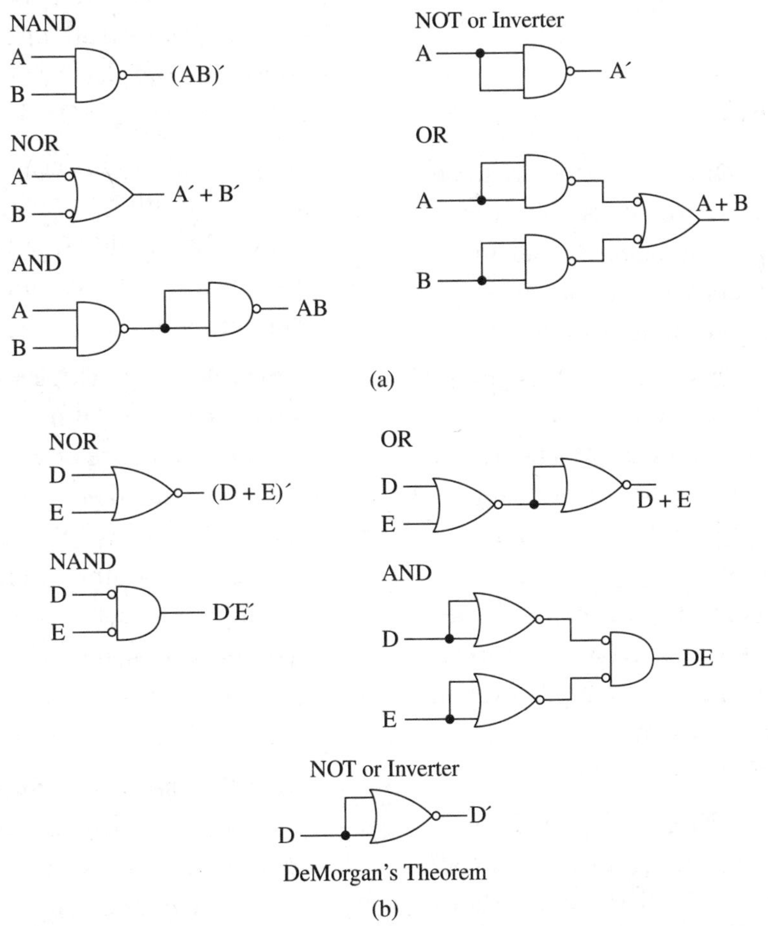

Figure 11.3 DeMorgan's theorem shows how (*a*) NAND and (*b*) NOR gates can be used to implement almost any logic circuit.

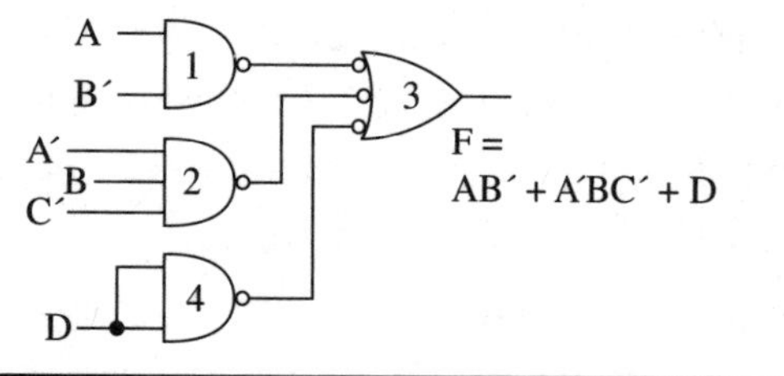

Figure 11.4 How the example in Fig. 11.3 is implemented using NAND gates.

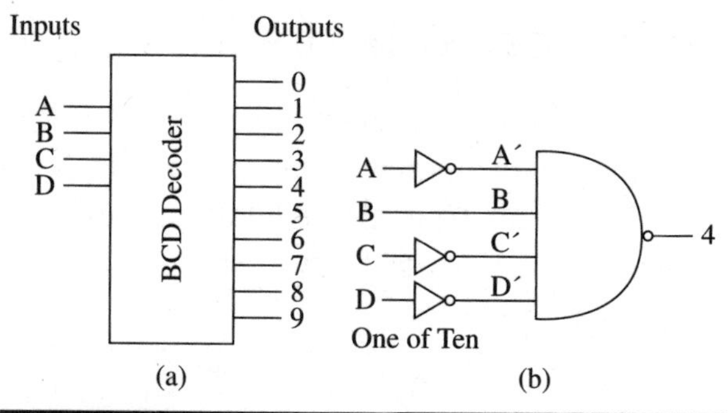

Figure 11.5 A decoder functional IC.

As an example, consider the expression used earlier in this chapter (Fig. 11.2).

$$F = AB' + A'BC' + D.$$

Figure 11.4 shows an all NAND solution, including gate 3 that is being used as a NOR and gate 4 used as an inverter. Note the circles or bubbles. Two bubbles connected to one another cancel one another. In a boolean expression, A NOT inverted is just A.

$$(A')' = A$$

Functional ICs

These are complete circuits, not just gates to be connected. They perform some commonly used digital functions that occur again and again in practice like decoders and multiplexers. A simple block diagram illustrates its function. Keep in mind that many common combinational logic circuits have already been manufactured as integrated circuits. Many designs can be implemented using just one or more of these ICs. Many special designs can use available ICs with some external modifications or enhancements. **The secret to designing with these ICs is to know what is available and to understand what each does.** The following examples explain how they work. Typical commercial TTL and CMOS chips are identified.

Decoders (binary or BCD)—Figure 11.5*a* shows a decoder that detects one binary code or multiple related codes. The decoder shown has a 4-bit BCD input, so it can decode or recognize 10 codes from 0000 to 1001 or 0 through 9. A BCD decoder has 4 inputs but 10 outputs. For example, if the BCD code 0100 appeared at the input (A′BC′D′), output 4 would be binary 1 and all other outputs would be binary 0. Figure 11.5*b* shows one of the 10 decoder gates implementing this example. This chip can also be used to realize many sum-of-product boolean expressions derived from a truth table. It can also be used as a demultiplexer. Typical devices: TTL-74LS42, CMOS-4028.

You can also implement all the basic logic functions with NOR gates. Figure 11.3*b* shows the details. However, using NOR gates usually makes the circuit larger and more complicated.

Data selectors (multiplexers)—Figure 11.6*a* shows the circuit. It has 8 inputs and 1 output. A 3-bit input code or address selects one of the inputs that then appears at the output. If the AB′C input is 101, the input 5 would be passed to the output. Figure 11.6*b* shows how one input is selected. This IC can also be used to implement many sum-of-product boolean expressions derived from a truth table. Typical devices: TTL-74LS151, CMOS-4512.

Data distributors (demultiplexers)—Figure 11.7. This device distributes one input to one of multiple outputs, such as 1 to 4 or 1 to 16. A 2-bit address is used with the 1 to 4 and a 4-bit address is used with the 1 to 16. If the address input is 1001 (AB′C′D), the input

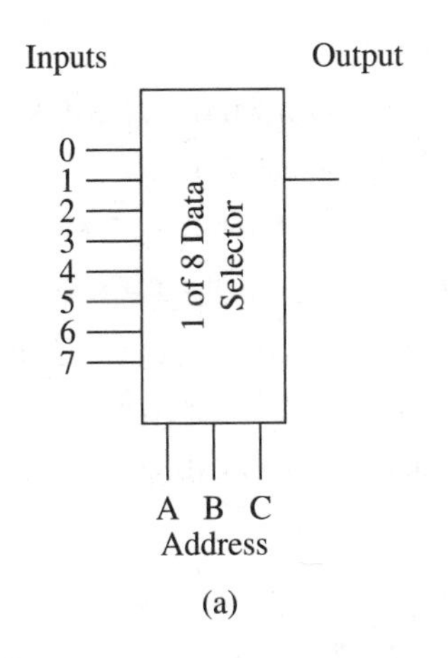

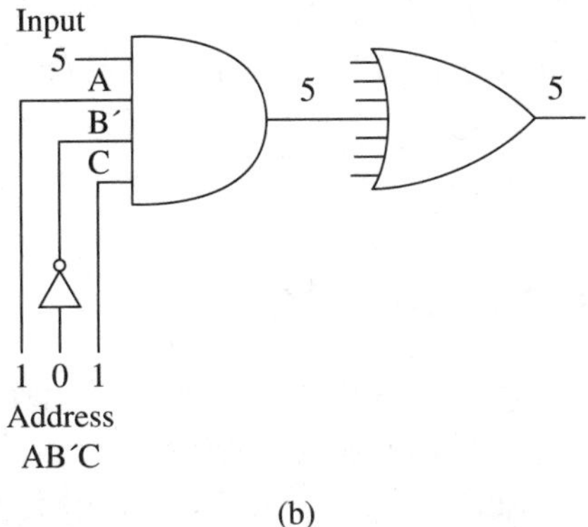

Figure 11.6 (*a*) 1 of 8 data selector functional IC and (*b*) an example of how it selects a specific input.

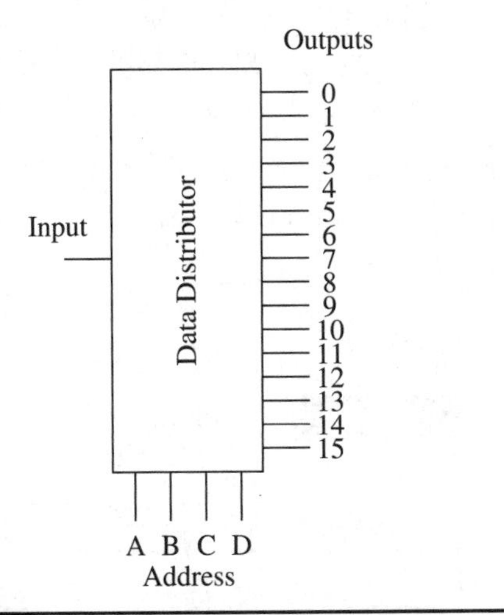

Figure 11.7 A data distributor functional IC.

will be routed to output 9. Typical devices: TT-74LS154, CMOS-4512.

Comparators—Figure 11.8 shows how this IC compares two 4-bit binary words and provides an output that indicates whether

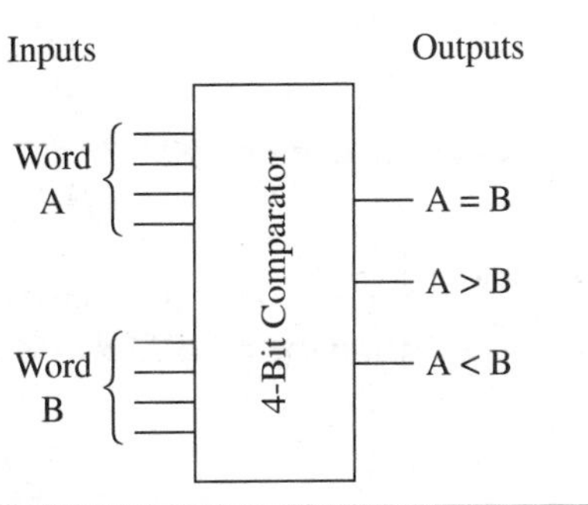

Figure 11.8 A 4-bit comparator functional IC.

input A is equal to, more than, or less than input B. Typical device: TTL-74LS85, CMO-4063. If all you need is a 1-bit comparator, use a XNOR gate. The is an exclusive OR with an inverted output. See Fig. 11.1.

Adder—See Fig. 11.9. This IC fully implements the addition of two 4-bit numbers A and B. Typical devices: TTL-74LS283, CMOS-4032.

Practical Digital Design Procedures

Here is a step-by-step process you can use to create almost any digital circuit. It is based upon the black box concept that assumes you can view any project as a black box of logic that accepts inputs from one or more sources. These inputs

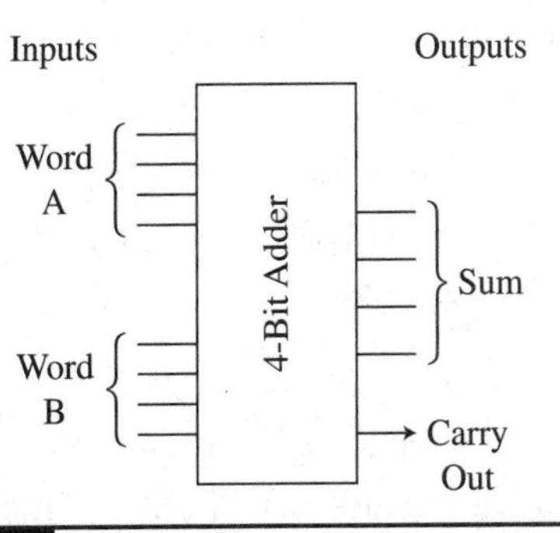

Figure 11.9 A 4-bit adder functional IC.

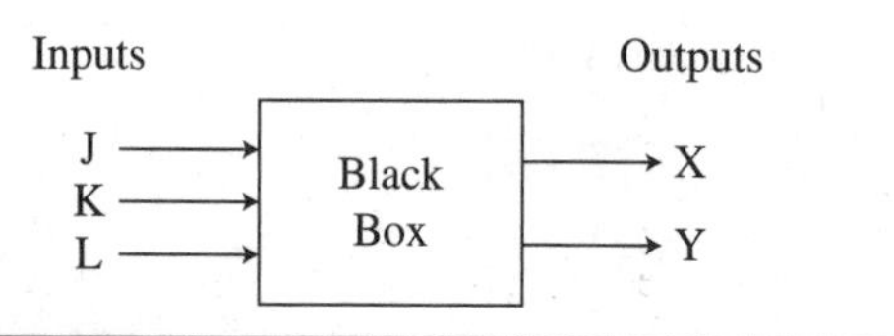

Figure 11.10 Summarizing an initial digital design using the black box method that shows all inputs and outputs.

are then processed by the circuit in the box to produce the desired outputs. See Fig. 11.10.

The approach described here works for most projects. An example follows. You can try it yourself and write the information in your notebook.

1. Write out a brief description of what the circuit is supposed to do.
2. State and name the desired logic inputs and outputs. You do not have to use individual letters or the alphabet. Instead use short designations that say or imply what the signal is. Examples: EN for enable, CLK for clock, RST for reset, or Q′ for flip flop output 1.
3. Using these inputs and outputs, create a truth table. The number of possible input combinations is given by 2^N where N is the number of inputs. For 4 inputs there $2^4 = 16$ possible input combinations for instance. These are 0000 through 1111.
4. Next, write the boolean logic expression for each output. You can write the boolean expression for this circuit from the truth table. Use the sum-of-products (SOP) form, that is, OR-ing together each AND expression based on the inputs.
5. Once you have the boolean equations, you can implement them directly with gates or some larger IC.
6. Before you create the circuit, you can use boolean algebra or Karnaugh maps to minimize the number of gates. This is just an option. At one time, this was standard procedure in design. If you are interested in more combinational logic design, go to Design Projects 11.1 and 11.2. Refer to App. D where the design process is further explained. The solutions use Karnaugh maps.
7. Now, draw the logic circuit. Convert the equations you derived from the truth table or into logic gates connected to implement the logic.
8. Simulate the circuit using one of the simulation software packages named in Chap. 3.
9. Select either TTL or CMOS logic type.
10. Build the circuit and test it.

Design Example

Design a logic circuit that will give a binary 1 output if any 3-bit binary number containing a majority (2 or 3) of binary 1s appears at the input. The inputs are named A, B, and C, and the output is F. With three inputs there are $2^N = 2^3 =$ 8 possible binary input combinations. These are 000 through 111 or 0 through 7. The truth table describing this is shown in Table 11.1. There are four places where F is 1. These define the circuit.

In the left column is the decimal value, while the middle column is the standard binary count sequence, 0 to 7 (111).

Table 11.1 Truth Table for Majority of 1s Detector

Decimal	A B C	F
0	000	
1	001	
2	010	
3	011	1
4	100	
5	101	1
6	110	1
7	111	1

You can write the boolean expression for this circuit from the truth table. Use the sum-of-products (SOP) form, that is, OR-ing together each AND expression based on the inputs. You can write the boolean expression for this circuit from the truth table. There will be four terms, one for each of the inputs sets that deliver a binary 1 output. For each output where there is a binary 1 output, write an AND expression using the A, B, and C inputs. Use the inverted or complement of the letter where a 0 is present. An example is a 3-bit binary word with bits named A, B, and C. The value 101 would be written as AB′C. The value 001 would be A′B′C.

Here are the boolean equations from the truth table.

$$F = A'BC + AB'C + ABC' + ABC$$

Once you have the boolean equations, you can implement them directly with available NAND/NOR logic ICs as shown in Fig. 11.11. The three inputs ABC are each inverted to provide the complements A′ B′ C′. Each state requires a 3-input NAND gate. These product expressions are then ORed together in a NOR gate. It takes three chips as shown.

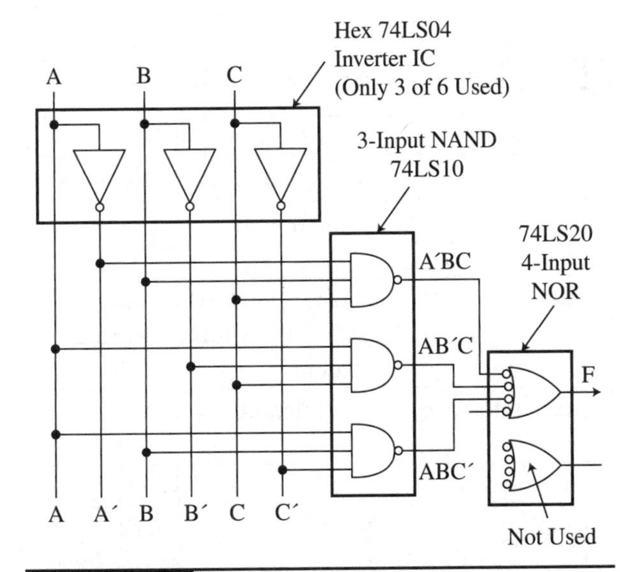

Figure 11.11 How the majority detector is implemented with TTL circuits.

Designing with a Programmable ROM

As indicated earlier, you do have the option of using programmable logic chips to simplify your combinational logic circuits. The older PLAs, PALs, and GALs may no longer be available, but two other options are available. The programmable ROM and the FPGA are good choices for more complex logic. The FPGA is covered in Chap. 12, so let's take a look at the programmable ROM.

When using the programmable read only memory (PROM), you use the address lines for your logic inputs and the memory contents as the logic outputs. In this way, you can implement a truth table directly. Figure 11.12 illustrates the concept. This hypothetical PROM is a 4-bit device with only eight storage locations.

	Inputs (Address) A B C	Outputs (Gray Code) D E F
0	0 0 0	0 0 0
1	0 0 1	0 0 1
2	0 1 0	0 1 1
3	0 1 1	0 1 0
4	1 0 0	1 1 0
5	1 0 1	1 1 1
6	1 1 0	1 0 1
7	1 1 1	1 0 0

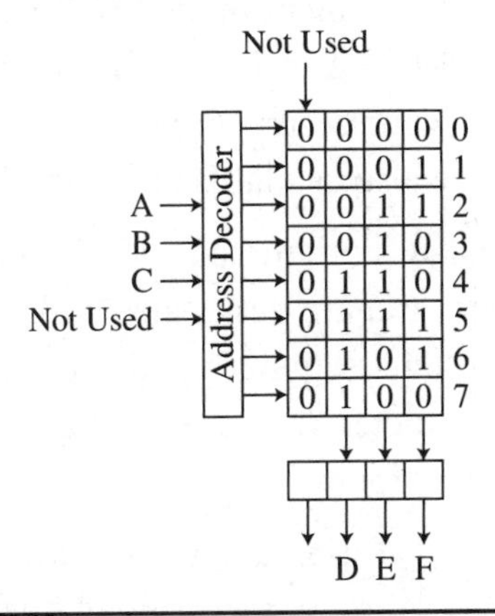

Figure 11.12 Using a programmable ROM to implement a logic function by storing the desired outputs at memory locations identified by an address formed by the inputs.

This PROM implements a binary to Gray code converter.

The A, B, C inputs representing the standard binary code are used for the address. At the corresponding memory location, you store the desired output. The D, E, and F outputs are the Gray code. This is a quick and easy way to implement a single-chip solution. The cost may be a little more than using standard ICs, but it uses less space and power. Some ROMs require a programming device to enter data into the chip. An electrical erasable ROM does not require the programmer. Most available ICs are overkill in terms of storage capacity for most designs. The popular 28C16A EEPROM, for example, contains up to 2 k (2048) bytes (8 bits). You will probably need only a fraction of that memory capacity, but it is an inexpensive quickie solution in some cases.

TTL vs. CMOS

A design choice you must make with simple combinational logic is between the two primary types of logic circuits: transistor-transistor logic (TTL) and complementary metal oxide semiconductor (CMOS). The TTL ICs are older and made with Schottky bipolar junction transistors (BJT). CMOS is another technology and uses enhancement mode MOSFETs. Both TTL and CMOS are viable, but one may be a better fit to your design. Some of the original TTL ICs are also available in a CMOS version.

The TTL ICs originated with Texas Instruments and are designated with a 74xxx number. A letter sequence indicates their speeds or other characteristics. A few of the various classifications follow:

7400. Original circuits—available but not recommended.

74LS. A low-power Schottky version of the original circuits still used for small projects.

74AS. Advanced or faster Schottky.

74ALS. Faster low-power Schottky.

NOTE: The term *Schottky* refers to the fast-switching Schottky diodes used in the 7400 series circuits to speed up the logic circuits.

74HC. A high-speed CMOS version of the original circuits. Very popular. Recommended for most new designs.

These are the most commonly used devices, but other specialized versions that are faster are available. These are variations of the original circuits to make them faster or use lower voltages. For most personal projects, the 74LS series is a good option because prices are low, power consumption is less, and there are more IC types. The 74HC CMOS series is also a good choice.

The original CMOS ICs (not TTL related) came from RCA and are referred to as the 4000 series. Most of these are still available. Table 11.2 shows the primary differences between the two.

A simplified decision-making process goes like this: For highest speed (>10 MHz), go with TTL. For largest list of available TTL circuits,

Table 11.2 TTL Compared to CMOS

Feature	TTL	CMOS
Supply voltage	1.8 V, 3.3 V, or 5 V	3 V to 18 V
Power consumption	Medium to high	Lowest
Speed	Rates to 350 MHz at the lower supply voltages	Rates to 10 MHz at 5 V
Noise tolerance	Less tolerant	More tolerant
ESD tolerance	Good	Poor
Number of functional ICs	Most	Less
Cost	Medium to high	Lowest

choose the 74LS series. For lowest power usage, as in battery-operated equipment, choose CMOS, either 74HC or the 4000 series. For noisy environments, select CMOS. Both types can operate from 5 V, the most common supply voltage.

Go to the Texas Instruments website (www.ti.com) and download a copy of their Logic Guide. Other sources of TTL and CMOS ICs follow:

http://www.ti.com/lit/sl/scyd013b/scyd013b.pdf Digital Logic Pocket Data Book

http://ecee.colorado.edu/~mcclurel/ON_Semiconductor_LSTTL_Data_DL121-D.pdf

https://en.wikipedia.org/wiki/List_of_7400-series_integrated_circuits

For the 4000 series CMOS chips, check out the following sites:

https://www.cmos4000.com/

https://en.wikipedia.org/wiki/List_of_4000-series_integrated_circuits

Remember, when you want to work with a specific IC, download the data sheet.

Sequential Logic Circuits

A sequential logic circuit is one that can exist in a fixed number of states. It transitions from state to state, creating output pulses that meet some need. The states are stored as a binary code in storage circuits called counters or registers made up of flip flops (FF). The circuit changes state based upon one or more inputs and the current stored state. A clock signal initiates the state change once the input conditions are settled.

Figure 11.13 shows a generalized diagram of a sequential circuit. A storage register or counter acts as a memory where the different states are stored. The state changes are initiated by the inputs from a combinational logic circuit, the current states, and the clock.

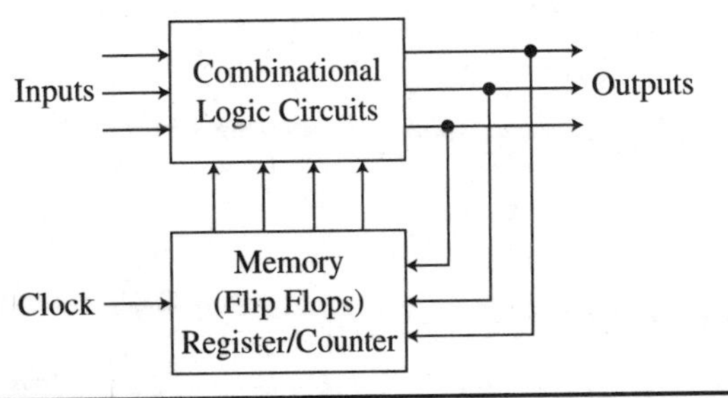

Figure 11.13 A simplified representation of a sequential logic circuit.

The basic building block of sequential circuits is the flip flop. These circuits are used to create a variety of sequential circuits called counters and registers. Here are the details.

Flip Flops

Another basic logic building block is the flip flop. It is the basic logic element in sequential logic circuits that will be covered later. As with combinational logic, sequential logic circuits can mostly be implemented with available ICs. The flip flop is the core of sequential logic circuits. First, let's get familiar with the most commonly used flip flops and how they work. See Fig. 11.14.

Figure 11.14*a* shows the set-reset FF or latch. It is made with NAND gates but is often drawn in a more simplified form as shown. A logic low (0) on the S input sets the FF. A set FF is said to be storing a binary 1. The Q output goes high (1) to indicate the set condition. A logic low on the R input resets FF. Q goes low (0). A reset FF is storing a binary 0. You observe the Q output to determine the state of the FF: 1 = set, 0 = reset. The Q NOT or Q′ FF output is the complement of the Q output.

Figure 11.14*b* shows the D-type FF. The D input receives a logic signal from some other source. It can be a 0 or 1. A clock signal is applied to the CLK input. The clock signal is a stable rectangular wave pulse train. When the clock signal goes positive, the FF assumes

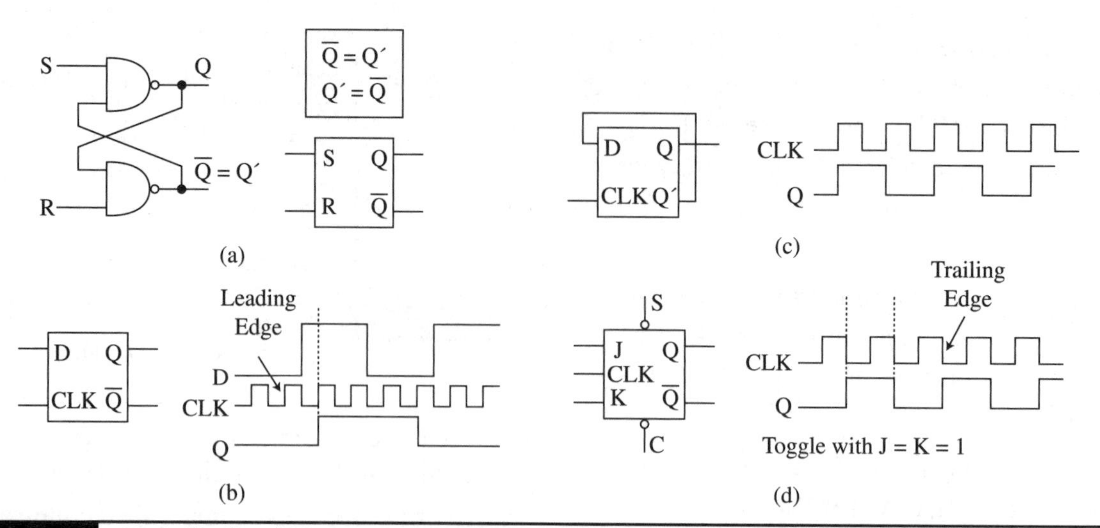

Figure 11.14 Basic flip flop (FF) circuits. (*a*) latch or RS FF, (*b*) D-type FF, (*c*) a D FF connected to toggle with each clock input, (*d*) JK FF.

the state of the D input. The FF is said to be positive edge triggered.

A common version of the D FF is shown in Fig. 11.14*c*. The complement output Q′ is connected back to the D input. That means that when the positive edge is input from the clock, the FF state will simply reverse. If the FF was set, the next clock pulse will reset it. The constant reversal of the FF state with each clock pulse is called toggling.

The JK FF is shown in Fig. 11.14*d*. It is driven by a clock signal applied to the CLK input (sometimes called the T input). This type of FF is often triggered by the negative-going or trailing edge of the clock. The state of the FF is determined by the J and K inputs. See Table 11.3. Q is the output. Q + 1 means the Q output after a clock pulse occurs with the inputs given. X means it could be either 0 or 1.

Table 11.3 JK FF Output

Input	Input	Output Q	Output Q + 1
J	K	Q	Q + 1
0	0	X	X
0	1	X	0
1	0	X	1
1	1	X	X′

A common connection is setting J = K = 1. With this input condition, and the clock applied, the Q output changes state or "toggles" for each negative transition of the clock pulse. Also note the two additional inputs S and C. These inputs are used to preset the FF to a desired state before additional operations occur. Bringing the S input low (0) sets the FF. Bringing the C input low (0) resets or clears the FF.

Counters and Registers

Counters and registers are the key circuits in sequential logic. There is a wide variety of IC counters and registers available to build sequential circuits. Now for the details.

Counters

Figure 11.15 shows how to make a basic binary counter with JK FF. The output of each FF is connected to the clock input of the next FF. The truth table generated by the counter is shown in Table 11.4. Also shown in Fig. 11.15 are what the outputs of each FF would look like on an oscilloscope. Note that each FF changes state on the negative-going edge of its input pulse. Look closely and observe each of the 16 states in both the truth table and waveform outputs.

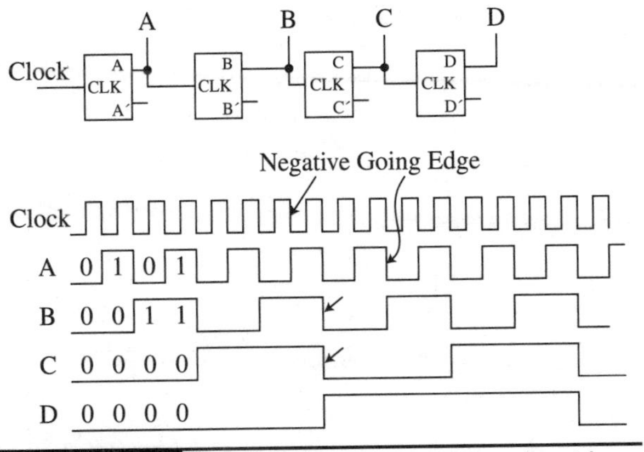

Figure 11.15 A 4-bit binary counter made with JK FF along with the inputs and outputs for all 16 states.

Table 11.4 Binary Counter Output Sequence

DCBA (Binary)	Decimal Equivalent
0000	0
0001	1
0010	2
0011	3
0100	4
0101	5
0110	6
0111	7
1000	8
1001	9
1010	10
1011	11
1100	12
1101	13
1110	14
1111	15

There are two things to remember. First, the number of bits or FFs determines the maximum count for the counter: 2^N where N is the number of FFs, $2^4 = 16$ states. These states are 0000 through 1111 or 0 through 15. Second, note in Fig. 11.14 that the output of each FF is one-half the frequency of its input frequency. This makes the counter a frequency divider as well.

In most cases, circuits like this counter are already available as a single IC. Some of these common sequential logic circuits are summarized next.

- Binary counters (up, up/down) 4 or 8 bit. Up counters count clock or input pulses up in the traditional binary sequence from 0000 to 1111 or down from 1111 to 0000. See Fig. 11.16*a*. For each clock (CLK) pulse or other input, the counter increments (or decrements) so that the number of input pulses that have occurred is stored as a binary value in the FF and is available at the FF outputs. The FF outputs are labeled A, B, C, and D. FF A is usually the LSB (least significant bit). Counters have a count or clock input. Most counters, binary or BCD, also have a clear input that sets all FFs to binary 0. The maximum count value of a counter K depends upon the number of FFs used, represented by the letter N. $K = 2^N - 1$. With 4 FFs, the maximum count value is $K = 2^4 - 1 = 16 - 1 = 15$ or 1111. Note: Most counters also have a clear or reset input that is used to put the counter in the 0000 state.
- BCD counters (up, up/down). Counts in the 10-state BCD code from 0000 to 1001 or down from 1001 to 0000. Again, see Fig. 11.16*a*.

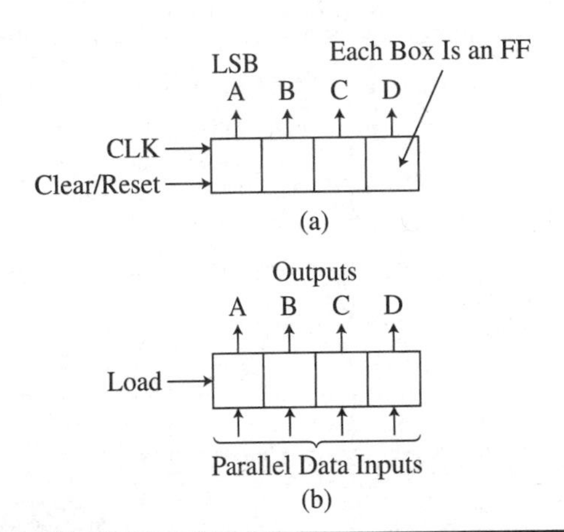

Figure 11.16 Simplified representation of common FF applications (*a*) binary counter and (*b*) storage register.

- Remember, too, that counters can also be used as frequency dividers. Each FF in a binary counter divides its input frequency by 2. Therefore, a 4-bit binary counter divides the clock input frequency by $2^N = 16$, where N is 4. A BCD counter divides its input by 10. Special dividers can be made for any integer (whole number) value. Cascading dividers increase the division factor by the product of the number of divisions used. A divide by 5 followed by a divide by 3 gives a division of 15. A 30-MHz input would produce an output frequency of 30 MHz/15 = 2 MHz.

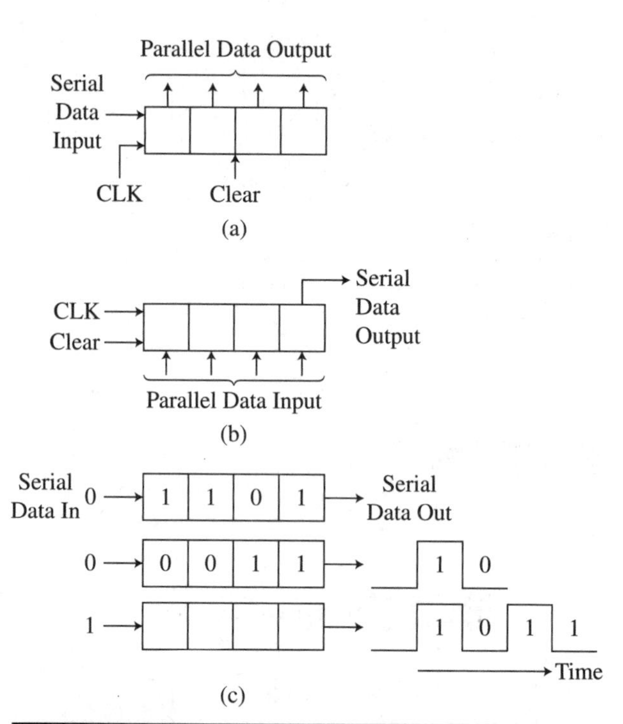

Figure 11.17 A summary of how shift registers work. (*a*) serial-to-parallel conversion, (*b*) parallel-to-serial conversion. (*c*) a serial-in-serial-out explanation.

Registers

There are basically two main kinds of registers: the storage register and the shift register. Storage registers are used to temporarily store data. Data is usually loaded into the register from a parallel bit source. A load input transfers the input data into the FF, then that data appears at the outputs. See Fig. 11.16*b*. IC registers come in 4- or 8-bit sizes.

Shift registers transfer the bits they contain from one FF to another internally or externally. Some common operations are shift left, shift right, serial in parallel out, parallel in series out. IC versions come in 4- and 8-bit sizes. A sequence of flip flops is used to shift bits to the right or left with each clock pulse. Shift registers are used for serial-to-parallel conversion (Fig. 11.17*a*), parallel-to-serial conversion (Fig. 11.17*b*), or some special function. Figure 11.17*c* illustrates a shift right operation. After two clock pulses, the two right-most bits are shifted out. At the same time, two 0 bits are shifted in.

State Machine Design

Of particular interest is a form of sequential circuit called a state machine. It is designed to produce a specific set of states for a given application. It generally cannot be implemented with commonly available IC counters or registers. The following is a guideline to designing a state machine for special occasions when they occur.

Here is the step-by-step design procedure.

1. Define the application. Write out a general description of the conditions you wish to create. Define in words the input and output signals.
2. Draw a state diagram. The state diagram generally consists of circles that represent each of the states. Figure 11.18 is an example. Lines with arrows show the transitions from one state to the next. The number of states determines the number of flip flops you are going to need to build the circuit.
3. Determine the number of FFs needed. Remember the number of states is

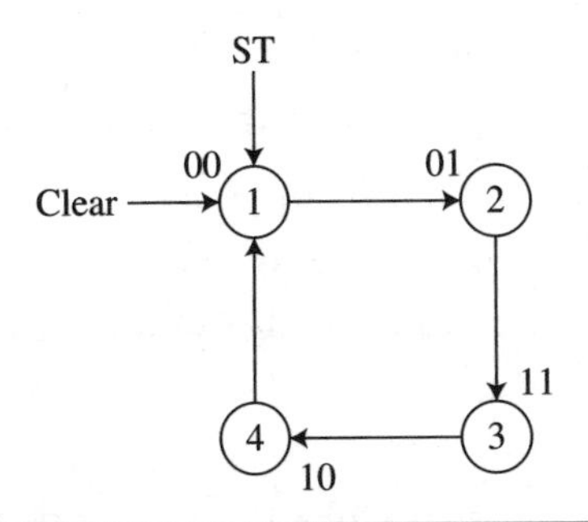

Figure 11.18 A state diagram showing the desired number of states to create the sequential logic circuit.

calculated with 2^N, where N is the number of FFs. Two FFs define four states. Three FFs define eight states. For example, if you need five states, you need three FFs. Two FFs only provide four states, so the next choice is three FFs with eight states. You would only need to use five FFs of the eight states; three others would just be unused.

4. Combinational logic circuits are needed to guide the FF from one state to the next, depending upon the external input conditions.
5. Generate a truth table for the circuit. Write down all the states possible with the number of bits you are using. Identify the states you will be using.
6. Select the type of flip flop. For most sequential circuits, you need flip flops that are triggered by a clock signal. Both D-type and JK flip flops can be used in this case.
7. Write the input conditions for the JK flip flops that will cause the states to change from one state to the next. This will be inputs to the J and K inputs on the flip flops. Write the boolean equation for each J and K input.
8. Draw the logic diagram showing all gates and flip flops.
9. Simulate the circuit if you have the software, and correct any problems.
10. Build and test the circuit.

Design Example

Here is a design problem to show the process.

1. Define the circuit. Assume that the assignment is to design a four-state machine or counter that counts in a Gray code sequence. Remember that the Gray code is one in which each code word changes by only one bit in transitioning from one word or state to the next.
2. Draw the state diagram. See Fig. 11.18. The states are numbered, and the related binary code is listed. Any special input conditions are also noted. A state machine is like a special counter and is made up of flip flops and gates but has specific states that occur in a unique sequence. The transitions from one state to another depend upon external inputs and logic. Each is specially designed for the application. The special conditions noted here are a start signal. ST is needed to begin the cycle. A reset or clear input is used to set the counter to its initial state of 00.
3. Determine the number of FFs needed. The number of states determines the number of flip flops needed. With four states, only two FFs are needed.
4. Make a truth table with the Gray code sequence. The FF outputs are designated A and B.

State Number	Gray code
	FF Outputs
	A B
1	0 0
2	0 1
3	1 1
4	1 0

5. Select the type of flip flop. For most sequential circuits you need flip flops that are triggered by a clock signal. Both D-type and JK flip flops can be used in this case. Let's select the JK FF.

6. Write the input conditions for the JK flip flops that will cause the states to change from one state to the next. This will be inputs to the J and K inputs on the flip flops. What input conditions for each J and K input must exist in getting from one state to the next? Write the boolean equations for each.

   For example, to get the B flip flop to change from zero (0) to one (1), the J input must be 1. It is developed by looking at the initial state 00 or A′B′. A start signal is used to initiate the sequence, so the complete expression for J_B is A′B′(ST). If that condition is met, the B FF will set when the clock pulse occurs. $J_B = A'B'(ST)$. The new AB state is 01.

   From the 01 state, the next state is 11, which requires the J input to the A flip flop to be a binary 1. The previous state sets up the conditions, therefore, $J_A = A'B$. With this set of input conditions on the J_A input when the clock pulse occurs, the A flip flop will be set, therefore, the new state 11 is achieved.

   Next, the B flip flop must be reset. This calls for a binary 1 at the K input of the B flip flop. The conditions are developed from the present state, and this is $K_B = AB$. When the clock pulse occurs, the state will change to 10.

   Finally, the A flip flop must now be reset. This means a binary 1 on the K_A input, and it is developed from the current state AB′. This causes the counter to cycle back to the beginning of 00, $K_A = AB'$.

7. The logic diagram can now be drawn given these logic conditions. Figure 11.19 shows the basic implementation. There are two JK flip flops and four AND gates that develop these logic signals. You can trace through the logic circuit to show how the various states set up the logic conditions on the J and K inputs.

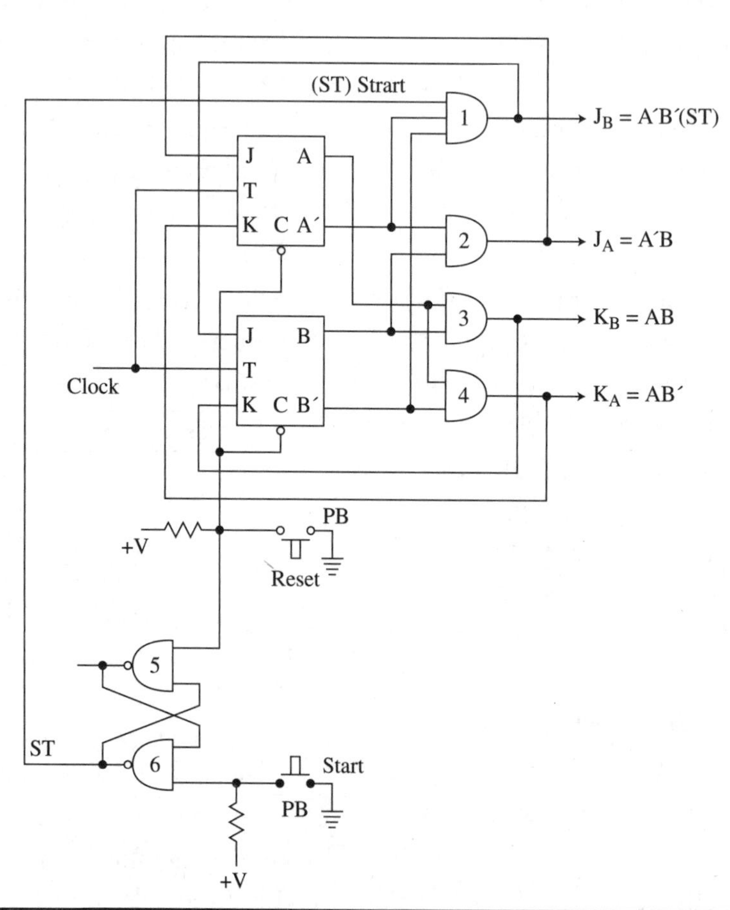

Figure 11.19 The complete circuit of the example state machine.

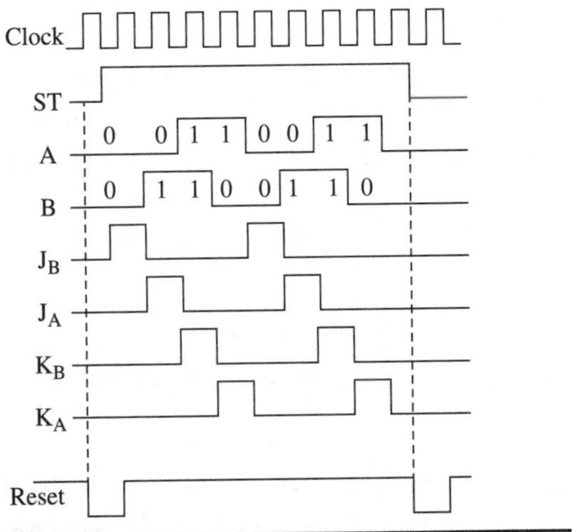

Figure 11.20 Input and output waveforms of the state machine.

Putting the two FFs in the 00 state, you press the reset button. This brings the C inputs to the FFs low, resetting them both. To start the count, press the start push button. This sets the latch made up of the two NAND gates 5 and 6. The ST line goes high, enabling gate 1. The clock pulses trigger the FFs, and they go through their sequence. Keep in mind that the state changes occur with the trailing or negative-going part of the clock pulse. The sequence will be repeated until the reset PB is pressed, returning the counter to the 00 state. The clock frequency determines the recycle rate. Figure 11.20 shows the pulse outputs from the various parts of the circuit. Look at the states as they occur in the A and B signals. Note that when the ST signal goes back to 0, the counter will complete its current cycle and then stop.

Data Conversion

Many if not most modern electronic equipment contains data conversion circuits. These circuits are often buried within other ICs like embedded microcontrollers, yet individual ICs are also available. Since data converters are so common, you need to know how to design with them.

There are two types of data conversion: analog-to-digital and digital-to-analog. An analog-to-digital converter (ADC) changes a linear or analog signal, such as voice or a temperature variation, into a series of binary words. Once the input is in digital form, it can be stored in a memory, processed by a computer, or otherwise manipulated with an MCU.

A digital-to-analog converter (DAC) does the opposite. It converts a sequence of binary words into an analog or linear signal. Examples are digitized music on a CD or from an online music site that is converted into the original analog version.

A/D

Figure 11.21 illustrates the A/D process. The analog signal is sampled (measured) by an ADC at regular intervals, and each sample produces a binary number that is proportional to the sampled voltage level. The key to this process is to sample the signal as many times as possible to produce a faithful digital representation. The minimum sample rate is at least two times the highest frequency in the analog signal. For example, to digitize a 20 kHz audio sine wave, you would need an ADC with a sampling rate of at least 40 kHz. That's why digital music sources use 41 kHz and 48 kHz rates and higher.

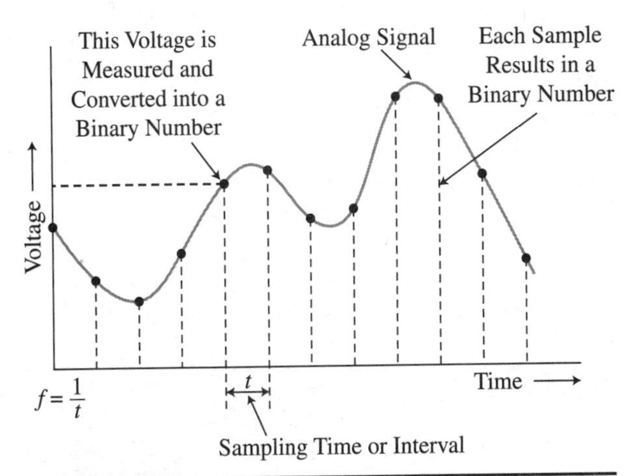

Figure 11.21 A continuous analog signal is being sampled or measured every *t* seconds.

The most significant characteristic of an ADC is its bit size. You can get ADCs with 8, 10, 12, 14, 16, 20, and 24 bits of output. The number of bits defines the conversion resolution. The resolution is the smallest increment of voltage that the ADC produces. For example, an 8-bit ADC can represent $2^8 = 256$ levels of resolution. An external dc voltage is used as a reference. With a 5-V reference, the resolution is $5/256 = 0.01953$ V or 19.53 mV. The binary output value can vary from 00000000 to 11111111 or 0 to 255. Each binary value has a resolution of 19.53 mV or 0.01953. The maximum output value then represents $255 \times 0.01953 = 4.98$ V.

D/A

Once the binary version is available, it may be stored in a digital memory or further processed. If you want to recover and reproduce this signal, you feed the binary samples to a DAC, and a stair-step version of the original signal is generated. See Fig. 11.22. This is only a crude reproduction of the original signal, but you can pass it through a low-pass filter to smooth it out. If you use more samples and increase the resolution on the ADC, you will get a signal with many more smaller steps that is closer to the original.

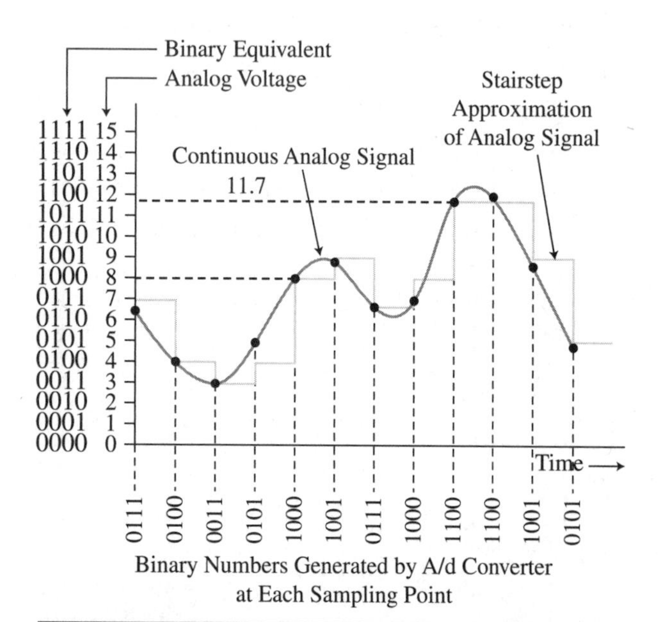

Figure 11.22 The stepped approximation of the original analog signal is produced by the DAC.

Data Conversion ICs

There are dozens of ADC and DAC ICs available from multiple manufacturers. The devices used in this discussion are older ICs but still for sale. They are inexpensive and easy to use. The ADC is the ADC 0804. The DAC is a DAC 0808. Be sure to go online to get the data sheets on these devices. You should look at several data sheets from multiple manufacturers because each has different information.

Figure 11.23 shows both the ADC and DAC connected together so that you can test them. The input V_{IN} is on the left and is set up to supply a 0- to 5-V dc signal for the ADC to convert. The ADC0804 operates from a 5-V supply. The sampling interval or rate is set by a clock signal. This ADC has a built-in clock whose frequency is set by an external resistor and capacitor on pins 4 and 19. The frequency of the clock is $f = 1/1.1RC$. This the values shown in Fig. 11.23. The clock frequency is 606 kHz. With that clock frequency, the conversion time is about 100 μs. That translates to a sampling rate of $t = 1/t = 1/100 \times 10^{-6} = 10$ kHz. That means that the highest frequency ac input that can be converted is one-half that or 5 kHz. NOTE: A START button is provided to reset and start the conversion process.

The parallel 8-bit output word is displayed on the eight LEDs and is sent as an input to the DAC. The DAC0808 and the op amp operate from ±15-V supplies. The DAC output is the

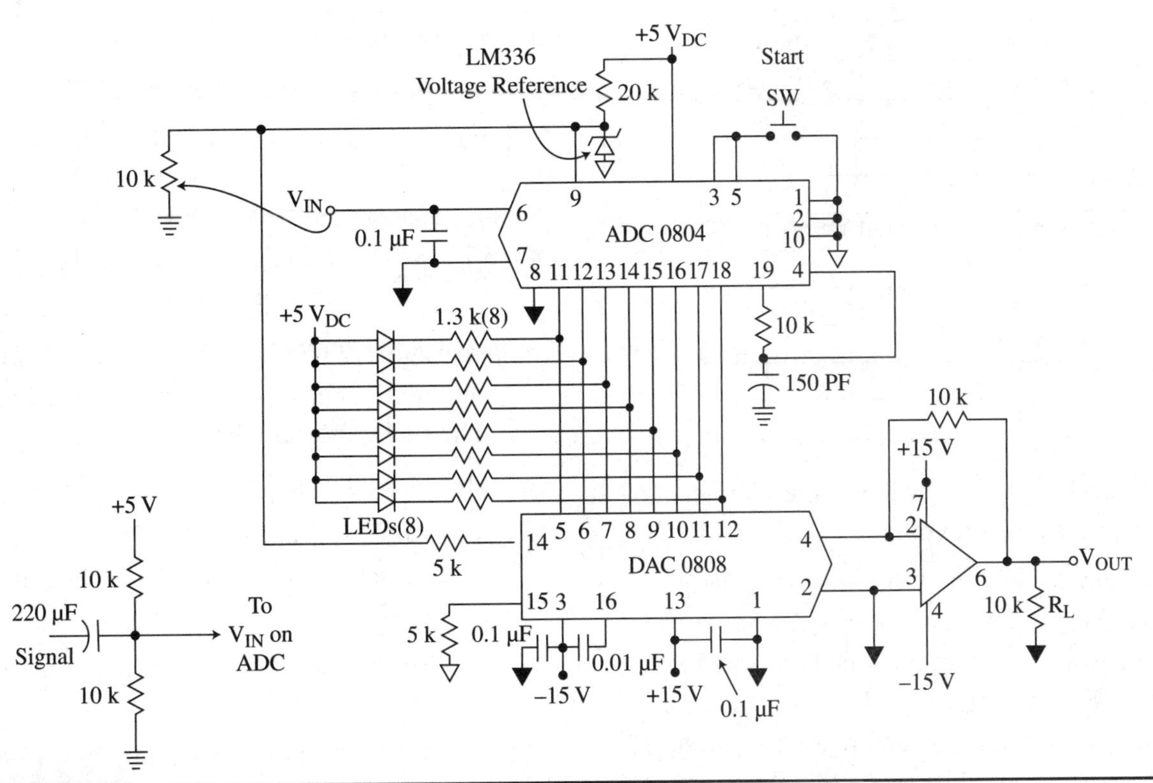

Figure 11.23 A complete ADC/DAC test circuit using the ADC0804 and the DAC0808.

analog signal that is sent to an op amp for buffering.

If you build or simulate the circuit, you can apply a dc at the ADC input by setting the 10 k pot. That input voltage should be reproduced at the op amp output.

Design Project 11.1

Design a logic circuit that lights an LED whenever the input is the binary expression for a prime number. Identify the primes between 0 and 15. This calls for a 4-bit input because with 4 bits you can represent $2^4 = 16$ states. Remember that a prime number is one that is only divisible by itself and 1. Note: 0 and 1 are not primes. Now do the rest yourself. Don't look at the solution in App. D until you are finished.

Design Project 11.2

Design a circuit that translates a 3-bit standard binary code into a 3-bit Gray code. Gray code is a binary sequence where only one bit in the 3-bit input sequence changes from one state to the next.

Determine the exact number of different logic states. Assign some names or letters to each input and output. The input signals will be a 3-bit standard binary code with bits A, B, and C. With three bits in, there are $2^3 = 8$ different states. The Gray code output signals we will call D, E, and F.

List all possible cases of the inputs and the corresponding outputs. For example, if there are three inputs, there will be $2^3 = 8$ possible input combinations. You define what you want the outputs to be for each input case. To get you started, the following truth table is shown.

Inputs	Outputs
Binary Code	Gray Code
A B C	D E F
0 0 0	0 0 0
0 0 1	0 0 1
0 1 0	0 1 1
0 1 1	0 1 0
1 0 0	1 1 0
1 0 1	1 1 1
1 1 0	1 0 1
1 1 1	1 0 0

Your job is to create the logic circuit for each of the outputs in the truth table (D, E, and F).

Be sure to read the solutions to the design projects in App. D. The topic is Karnaugh maps, but the solutions contain added information and procedures on design. Do try to develop a solution of your own before looking at the answer. You should be aware that the solution presented is only one of several that may be possible. Yours may be different but just as good.

Design Project 11.3

Another practice project is to design and implement a digital cable tester. This is a good example of how to design using discrete IC logic. Design a cable tester for a four-wire cable with connectors on each end. The test should identify any opens in the wires or bad connector pins and connections. This is just a continuity test that you might perform manually with an ohmmeter. It should also identify any shorts between wires or adjacent connector pins. This project automates the test to shorten the time required to check out a high volume of cables.

1. The inputs are logic signals applied to the connector that travel down the cable to the other connector. There is one signal per wire (4). Outputs are LEDs that signal which wire/pin is open or which two wires/pins are shorted. Figure 11.24 shows the diagram. The test pulses are developed in the transmitter (TX), and they are sent down the cable to the inputs to the receiver (RX). The LEDs indicate the wire status.

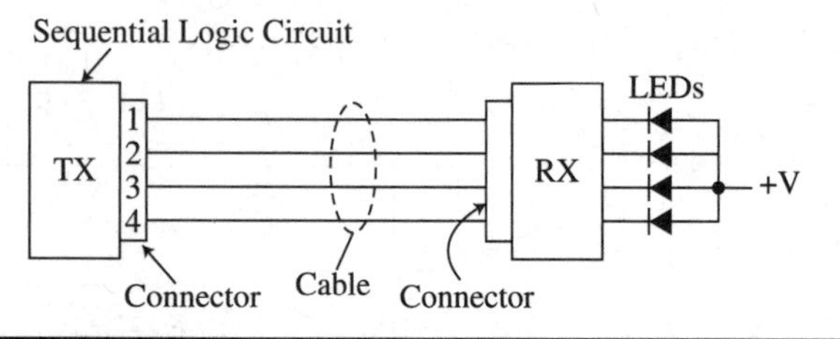

Figure 11.24 Simplified block diagram of the cable tester project.

2. Here are the basic steps to be implemented.
   a. Apply a binary 1 to the first wire at the cable input to be tested. Apply binary 0 to all others.
   b. Test the cable output for the reception of the binary 1.
   c. Turn on an LED, signaling that wire 1 is OK.
   d. Keep this LED on for 2 s (or more) to give you time to determine if that wire is good, then turn it off.
   e. Apply a binary 1 to wire 2 at the cable input and binary 0 to all others.
   f. Test the cable output for the reception of the binary 2.
   g. Turn on an LED, signaling that wire 2 is OK.
   h. Repeat steps b and c for wires 3 and 4.
   i. If a short exists between wires or connector pins, two (or more) adjacent LEDs will turn on at the same time instead of individually. Be careful to design to eliminate the problem of a short causing the outputs of the ICs supplying the pulses to be damaged.
   j. When all four LEDs have sequenced on, reset the program and prepare for the next cable test.
3. Build your circuit and test it. Solution is in App. B.

Design Project 11.4

Design a frequency divider that produces an output of 1.25 MHz from a master clock oscillator operating at 50 MHz. Use standard TTL or CMOS ICs. Draw the schematic and give part numbers. Solution in App. B.

Design Project 11.5

Design a binary counter with a display that shows the 4-bit count sequence on LEDs. Another display is a seven-segment LED that will show the numbers 0 through 9. Use available functional ICs where possible rather than individual gates.

Design Project 11.6

Build the circuit in Fig. 11.23. This will be used to test the ADC. The 10 k pot serves as the 0- to 5-V dc input to be digitized. The LEDs connected to the outputs will tell you the binary value related to the specific dc input. Set the input pot to 2 V. Measure the output of the op amp. What is the binary number displayed on the LEDs? Explain.

Design Project 11.7

Remove the 10 k pot from the ADC input and connect an ac sine wave of 1 V pp at 1 kHz, using the circuit shown with Fig. 11.23. Then observe the translated output from the DAC. Describe your results.

Next, adjust the analog input sine wave from 1 to 2 kHz then 5 and 10 kHz. At each clock frequency, observe the output of the op amp. Can you describe what you are seeing?

Design Project 11.8

Disconnect the ADC, and replace it with a 74LS393 dual 4-bit counter connected as an 8-bit binary counter. See Fig. 11.25. Build a 555

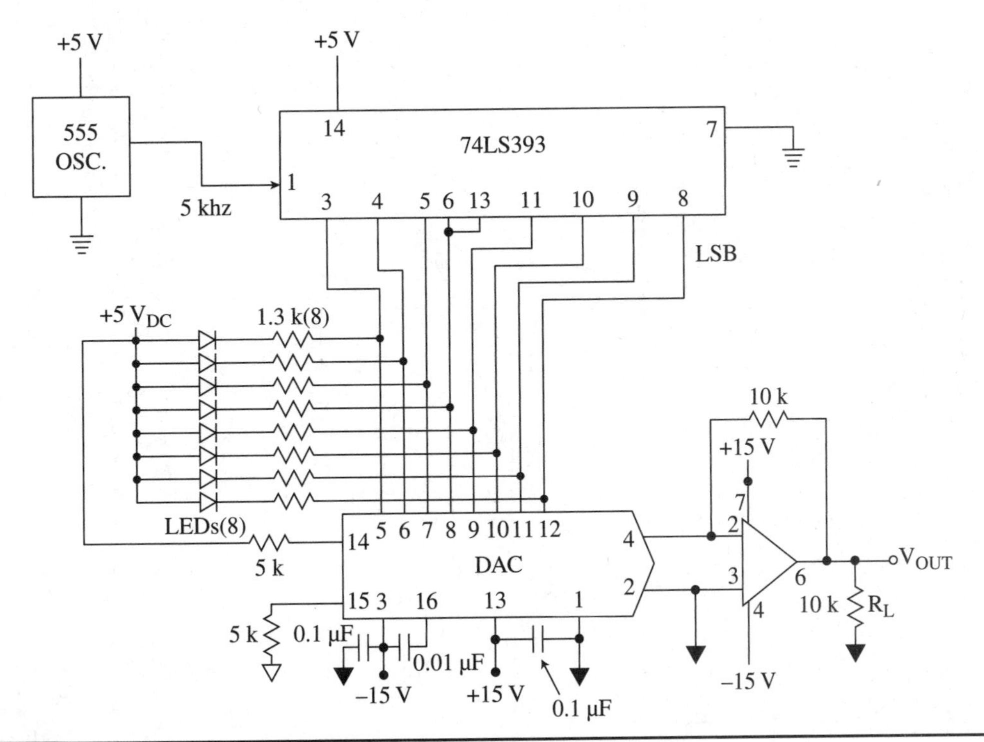

Figure 11.25 A DAC driven by a binary counter.

oscillator to drive the counter. Set its frequency to 5 kHz. See Chap. 8. The counter outputs will drive the LEDs and DAC. Turn on the circuit and monitor the DAC output at the op amp with your oscilloscope. Describe what you see.

Design Project 11.9

A classic example that others have used to show the digital design process is a digital die. The goal is for you to design a single die using LEDs and implement a way to simulate the "roll" of the die to get a random outcome. Figure 11.26 shows the die where each LED is assigned a letter designation. Use the format shown. The letters define a signal for each LED. Create a truth table for these letters. The design process will create the logic circuits that drive the LEDs. Check one solution in App. B.

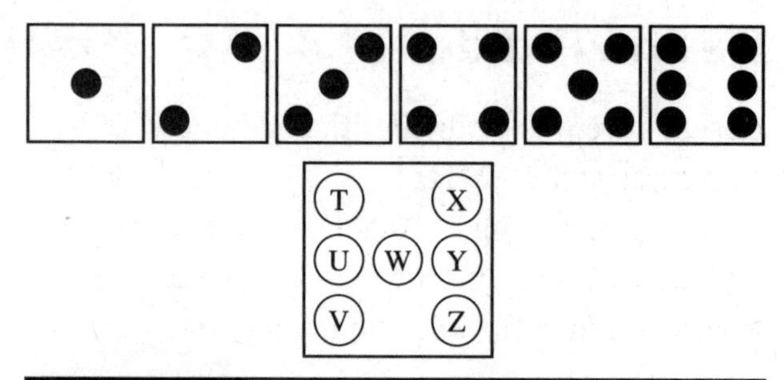

Figure 11.26 A digital die made with LEDs.

CHAPTER 12

Programmable Logic Devices (PLDs)

An alternative to discrete TTL or CMOS logic ICs is a programmable logic device (PLD). This is an integrated circuit containing some prewired logic circuits that can be further programmed to implement your boolean equations or other logic design. These logic circuits use the sum of products (SOPs) format in which the AND gates or the OR gates are programmable, or both. Different methods have been used to program these chips. While these chips greatly reduce the size of a design, some of them have disappeared from the market. The focus instead has been on a larger and more complex PLD called a field programmable gate array (FPGA). The FPGA can implement almost any digital circuit. This chapter summarizes these devices and offers guidelines for experimenters.

Programmable Logic Types

Three PLD configurations were developed and used successfully for a while until they were upstaged by the FPGA. These are generally referred to as simple programmable logic devices (SPLDs). There are four basic types:

1. **Programmable Read Only Memory (PROM)**—You have already seen how these chips implement logic functions. See Chap. 11. The inputs are used as the addresses where the desired outputs are stored in the relevant addressed memory locations.
2. **Programmable Logic Array (PLA)**—This device features a group of programmable AND gates and a group of programmable OR gates. This gives you great flexibility in implementing sum-of-product (SOP) circuits.
3. **Programmable Array Logic (PAL)**—This device features a fixed array of OR gates and a group of programmable AND gates.
4. **Generic Array Logic (GAL)**—This is similar to the PAL in that it features a fixed OR array and a programmable AND array. GALs also make it possible to program some of the outputs. GALs were invented by Lattice Semiconductor, but the company no longer makes them. Some of the GALs or similar replacements are still available from Atmel/Microchip Technology.

Programming PLDs

There are basically two ways that the programming is accomplished, with fuses and with memory switches. Figure 12.1 shows a fuse-programmable AND array as used on the older PALs. All of the logic inputs are initially connected to all of the AND inputs with fuses (Fig. 12.1). To program the gates, a high voltage is applied to pins on the IC that blow selected fuses, leaving the desired inputs connected to the selected gates. Figure 12.1 shows an example. Trace it out and verify the output C for yourself.

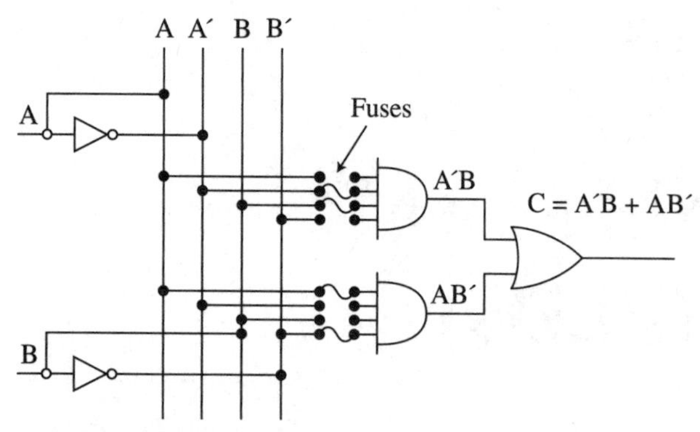

Figure 12.1 One segment of a PLD programmable with fuses—a common PAL or GAL configuration.

More modern GALs use programmable read-only memory technology that uses MOSFET storage cells like those in an electrically erasable programmable read-only memory (EE-PROM) instead of the fuses. The MOSFETs can store a bit as either an ON or OFF MOSFET. A special external programmer device turns on the MOSFET or turns it off. The beauty of this technology is that you can always erase your connections and reprogram them.

Figure 12.2 shows a simplified GAL arrangement. The AND array is programmable by turning the MOSFET switches off or on. The OR array is fixed. An example of a logic implementation is given. Note the simplified way to show the input connections to the AND gates. Rather than show every input, only a single line is shown, but the input connections are designated by the circles. Note the slash across the AND inputs, indicating that there really are three inputs.

Just as a check on your understanding, write the boolean equation for J yourself. GALs also include an output logic macrocell (OLMC) where the outputs are programmable, allowing you to select normal or inverted outputs, just the OR gate outputs, or outputs from flip flops that can be used to form storage registers or counters. The typical GAL has less than eight inputs and eight outputs.

Finally, the GAL is programmed by a special programing device that is configured by software that lets you enter your logic design. Check with Microchip Technology for GAL details and availability. These are Atmel products that are a part of Microchip.

Two of the most popular GALs are the GAL16V8 and the GAL22V10. The Atmel/Microchip part numbers are different, but the devices are equivalent functionally. An internet search will turn up other sources.

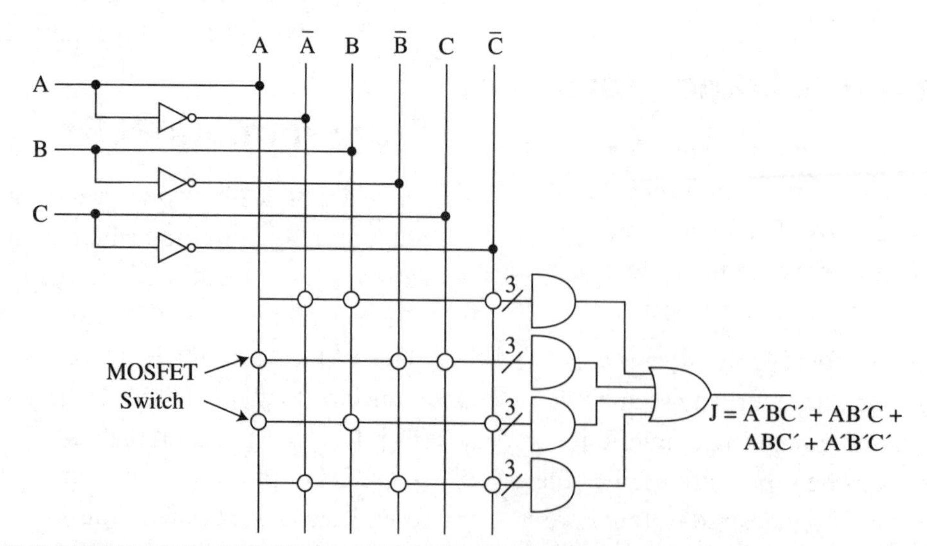

Figure 12.2 A simplified method of showing the programming of a PAL or GAL, where each circle represents a MOSFET memory switch.

The software to program SPLDs is a hardware description language (HDL). The two types widely used with PLDs are ABEL and CUPL. There are other specialized or proprietary types that you may encounter.

Changes in the PLD technology and availability occur often and rapidly. To keep up with what is current, do an internet search, and check with companies like Microchip Technology for the latest situation.

Complex Programmable Logic Devices

Complex programmable logic devices (CPLD) are an intermediate form of PLD. That is, their size is between that of a SPLD/GAL and an FPGA. Essentially, a CPLD is what amounts to a group of multiple GALs on the same chip. Each GAL section can be internally connected to any other GAL section using EE-PROM–like circuits. Output configurations are programmable. Again, the interconnections are by internal MOSFET switches. This is accomplished by a programmer device and software. You can see where this trend was headed—to the FPGA.

FPGA Dominance

Chapter 11 mentioned field programmable logic arrays (FPGAs) as another option for implementing digital logic. The FPGA is a programmable logic IC that contains multiple logic circuits and selectable interconnecting links. You do your design and enter the details with a software tool. This software is called a hardware description language (HDL). It is used to program or encode the circuit "wiring." The software lets you enter the circuit in boolean equation form. You use the HDL to program digital functions. The software then processes the coding to produce the final form. The software then programs the FPGA by interconnecting all the logic circuits in the FPGA to finalize your design. Instead of a bunch of logic circuits with multiple ICs, your finished product is literally one chip. The issue is getting to that end result.

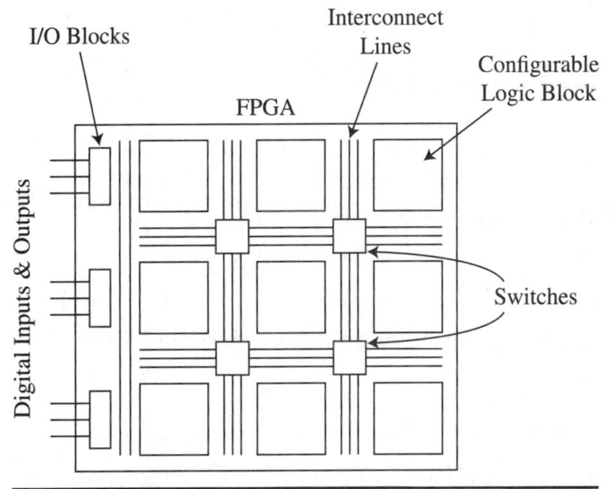

Figure 12.3 The concept of an FPGA showing configurable logic blocks, connection switches, interconnect lines, and selectable I/O blocks.

An ultra-simple conceptual diagram of an FPGA is given in Fig. 12.3.

The basic element of the FPGA is the configurable logic block (CLB) or cell. Modern FPGAs have hundreds of thousands of these cells. A cell can be programmed to emulate other logic circuits. It usually consists of a lookup table (LUT), some multiplexers, flip flops, and miscellaneous programmable logic circuits. A simplified logic block is shown in Fig. 12.4. These cells are interconnected by an internal matrix wiring that uses programmable switches to make the connections.

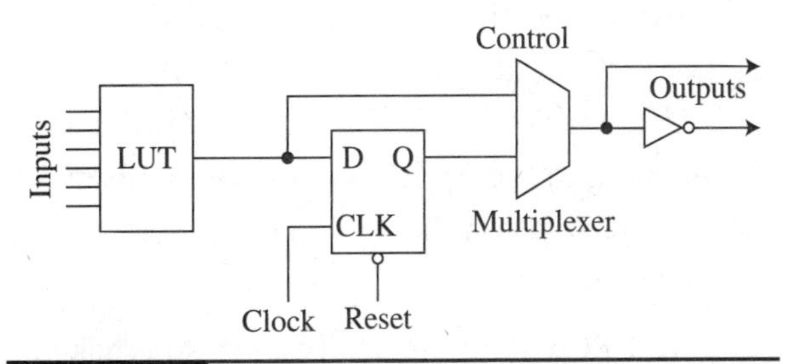

Figure 12.4 A common programmable logic block with LUT and output circuits.

The LUT is a small PROM into which you store the desired outputs based on the given inputs. The LUT does not do logic as such but stores the needed outputs for any given input condition, as described in Chap. 11. Programming consists of stating the inputs, desired logic operations, and desired outputs. All this is entered by way of the HDL. The output is a file or program that is sent to the FPGA for programming.

FPGAs also have internal standard I/O blocks or interfaces to use in connecting to other components. Some FPGAs may also have on-chip cores or MCUs that can be used in complex circuit implementations. Other features on some FPGAs are on-chip random access memory (RAM) blocks, phase-locked loops (PLL), or digital signal processing (DSP) blocks. Standard serial interfaces are also available (SPI, etc).

There are multiple FPGA vendors. The largest is Xilinx. Second is Altera, now a part of Intel. Others include Atmel, now a part of Microchip Technology, and Cypress Semiconductor. Most of their FPGAs contain thousands or even millions of circuits, and they are only available in surface-mount packages with hundreds of pins. Some also require multiple power supplies with voltage sequencing requirements. Supply voltages as low as 1 V are typical on the fastest of these components.

Application Decisions

Remember your three choices for realizing digital design: old-fashioned TTL/CMOS ICs, an MCU, and PLDs like FPGAs (or GALs if you can find them). If you have not chosen a hardware solution, here are a few things to consider.

First, FPGAs are big devices, not so much physically but in sheer volume of circuitry. Most have hundreds of thousands of logic blocks. These are surface-mount devices with over a hundred pins. FPGAs are primarily used in creating large, complex products.

These devices are also super-fast. If your best MCU or microprocessor is not fast enough to do the job, use an FPGA. Most run clocks at hundreds of megahertz. Another use case is big projects that require multiple parallel operations with multiple outputs with bit manipulation. FPGAs are increasingly being used for digital signal processing (DSP) and machine learning that implements long, complex, simultaneous algorithms.

It has been my experience that FPGAs are overkill for most experimenter projects. While they have gotten better and cheaper over the years, you can probably afford one, but you still need to learn one of the HDLs to do the programming. I recommend taking a formal online course before you attempt your design.

Just So You Know

Please note that the SPLD, CPLD, and FPGA or their HDLs do not do the logic design for you. You still need to go through the processes described in Chap. 11. The SPLD, CPLD, and FPGA are just hardware options for digital logic implementation. You do not have to do any minimization. There are hundreds or thousands of gates available for use, so any simplification is not a factor in saving money. Some reduction may take place in the software when you program FPGAs, but you do not see it.

FPGAs are not that practical for small projects. Stick to basic logic ICs or an MCU for most projects unless you just need the speed and large-scale capabilities. If you are a digital junkie and hell-bent to learn FPGAs, buy one of the many development boards and learn VHDL or Verilog, then you can experiment and learn.

An Introduction to VHDL and Verilog

There are two major HDLs: VHDL and Verilog. Both are widely used. The basic process of implementing your design begins with your final logic design diagrams and boolean equations. Next, acquire one of the programming languages. You can usually get one from the FPGA vendor. You will also need a good text editor. You may get one along with the language. These are installed on your PC. Here you will develop the code that describes your design.

You are going to need to spend some time in learning the software. Try to access any documentation or tutorials from your vendor. You can take an online course from Udemy. I also found an amazing number of online courses or tutorials. Partake. I also discovered several good basic books you can find online that you can buy via Amazon. References are given later.

There are essentially four parts to the FPGA design process.

- Programming or coding. This is where you enter your design into the software with the text editor. You keyboard it in just as you would a C program. Like any software, it has its quirky syntax and symbols. Some software lets you enter your design by logic diagram. The process is similar to using a circuit simulator.
- Simulation. This process takes your design and tests it to see if it will work. It will catch any errors and lets you fix them before going on.
- Synthesis. This is a software translation from your code to a netlist that codes the actual FPGA circuits.
- Compiling. This is where the program or design description actually gets converted to bits. Some call it place and route. To complete the process, the final file is downloaded to the chip.

Development Boards

You probably don't want to start with a chip and make up your own platform to carry out your programming and testing. Therefore, you will need a development board where the FPGA chip is mounted. These boards contain all sorts of switches, displays, and other support devices that help you evaluate the final product. Figure 12.5 shows an example.

There are multiple development products available from Digilent, Numato, and other companies. Some are older and feature FPGA chips that are dated and software that may no longer be available. See the recommendations later in this chapter.

As a safeguard, choose the development board first, and identify the specific FPGA used, but don't buy it yet. Go to the vendor (like Xilinx) Web site to see if the software is available. Then you can buy the board.

Most of the process of using the FPGA development environment is installing the

Figure 12.5 A popular FPGA development board called the Elbert V2 by Numato. It uses a Xilinx FPGA and connects to your PC with a USB cable.

software and learning to use it. If you are determined to learn FPGAs, plan to invest plenty of time in the project.

Coding the Digital Circuit with an HDL

Here is a simple example of coding your design. See Fig. 12.6. You do need to give the inputs and outputs names as shown. The coding procedure is for Verilog.

Programming is broken into short segments called modules. A module is just a file to be stored and used later. The first step is to name the module, then define or declare the inputs and outputs.

Module_Logic_Example

```
(
    // Inputs
    sig_x
    sig_y
    sig_z
    //Outputs
    log_j
    log_k
);
//Port definitions
input    Sig_x;
input    sig_y;
input    sig_z;

output   log_j;
output   log_k;
//  Design logic
assign log_j = sig_y | sig_z;
assign log_k = sig_x & sig_y & sig_z;
endmodule
```

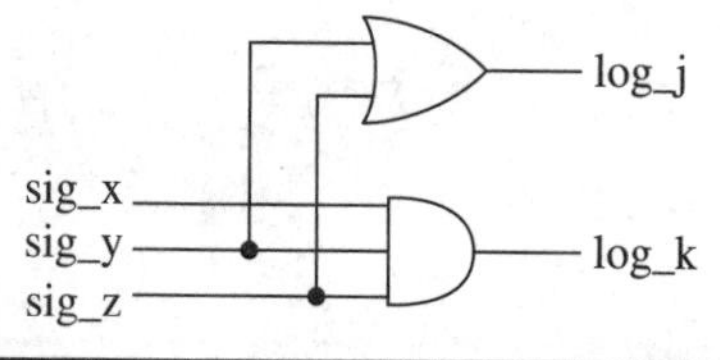

Figure 12.6 A simple logic circuit to illustrate Verilog programming with signal inputs x, y, and z, as well as logic outputs j and k.

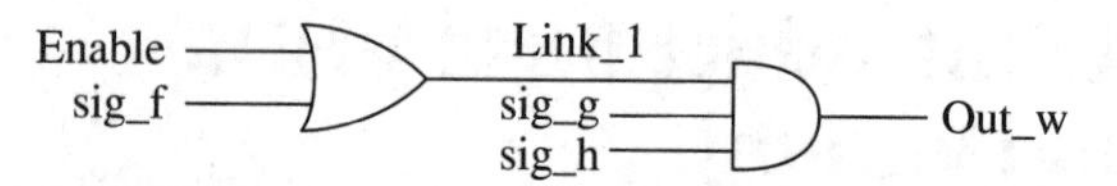

Figure 12.7 A logic circuit showing the inter-logic connection link_1.

The double slashes are used for comments and programmer notes. Also, all lines in the Port definitions and Design logic must have a semicolon after them. These are typical programming language syntax details you will need to learn. A vertical line (|) designates an OR operation and an & is used for an AND operation. The word "assign" states the actual boolean equations of the logic.

In most circuits, there are multiple layers of logic, so you need a way to code those intermediate links. The circuit in Fig. 12.7 and the following code show how this is handled. The intermediate connection signal is called link_1. It is called a "wire" by Verilog.

Module_Two_Layer_Logic

```
(
    // Inputs
    sig_f
    sig_g
    sig_h
    enable
    //Outputs
    out_w
);
//Port definitions
input    sig_f;
input    sig_g;
input    sig_h;
input    enable;

output   out_w;
wire     link_1;

//  Design logic
assign link_1 = enable | sig_f;
assign out_w = link_1 & sig_g & sig_h;
endmodule
```

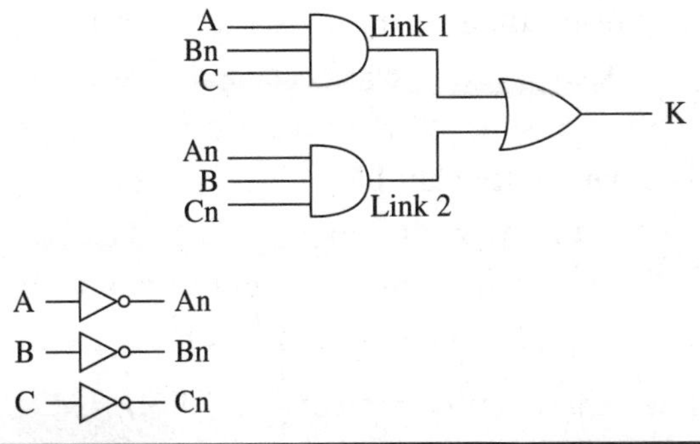

Figure 12.8 A logic circuit example for VHDL.

Those simple examples are intended to give you a feel for the FPGA coding process in Verilog. There is much more to it as you take your code through the simulate, synthesize, and compile steps before loading the FPGA chip. VHDL is different. Here is a simple example to show the differences.

Refer to Fig. 12.8. This circuit is called Logic_SOP. The first step is to define this circuit as an entity rather than a module. The VHDL coding for this is

```
entity Logic_SOP is
   port (A, B, C : in bit;
         K       : out bit);
end entity;
```

The logic we want to implement is

$$K = AB'C + A'BC'$$

Since some complements of the A, B, and C inputs are needed, these need to be defined. We can call these An, Bn, and Cn, where the n means NOT or complement.

As you can see from Fig. 12.8, we need to define the AND gate outputs. Let's call them Link1 and Link2.

Now the circuit can be described in the coding.

```
architecture Logic_SOP_arch of Logic_SOP
is
   signal An, Bn, Cn : bit;
   signal Link1, Link2 : bit;
begin
   An <=not A;     -- NOT's
   Bn <=not B;
   Cn <=not C;
   Link1 <= A and Bn and C;     -- AND's
   Link2 <= An and B and Cn;
   K <= Link1 or Link2;     -- OR's
end architecture;
```

Select between VHDL and Verilog based upon which software is used with the chip you choose.

FPGA Learning Resources

I strongly urge you to get some formal training in FPGAs and the HDLs before you tackle them. I like and recommend the online course site Udemy. Check it out. There are also lots of online tutorials you can tap into. Do a search on FPGA or HDL tutorials.

Udemy Online Courses

Udemy offers thousands of courses online at www.udemy.com. There are probably a dozen or more associated with FPGAs, VHDL, Verilog, and related topics. Prices are very reasonable—usually less than $20. However, most require that you buy the recommended development board. Prices run from about $50 to over $100.

Books

Simon Monk, *Programming FPGAs, Getting Started with Verilog*, McGraw-Hill, New York, 2017.

David Romano, *Make: FPGAs*, Maker Media, San Francisco, 2016.

Blaine Readler, *Verilog by Example*, Full Arc Press, 2011.

Blaine Readler, *VHDL by Example*, Full Arc Press, 2014.

Development Board Sources

I have only identified two vendors, Digilent.com and Numato Labs, that I have had a positive experience with, but there are others. Check with distributors like Adafruit, DigiKey, and SparkFun. Do a search on "FPGA Development Boards." In all cases, be sure you can get the software first and then buy the board. Even Xilinx and Altera have their own development boards to go with their software, so check there, too.

Summary

FPGAs are widely used in industry. They do teach FPGAs and coding in college EE programs. As indicated earlier, their primary applications are for very large blocks of logic and/or when super speed is needed. Digital signal processing (DSP) and artificial intelligence (AI) are modern examples. The choice of implementing your design with an FPGA is usually made when you discover that your MCU is just not fast enough to do the job.

For experimenter projects, an FPGA is probably the most costly and complex solution. GALs may be a solution, but they will require a programmer device. For some intermediate-size projects, you may wish to give GALs a try. Otherwise, stick with MCUs as a solution, or use the retro TTL/CMOS ICs that are cheap and abundant. Once you learn digital design, you will probably want to upgrade your knowledge and skills to include CPLDs and FPGAs.

CHAPTER 13

Designing with Microcontrollers

Your first choice when designing any digital product should be to use a microcontroller. In most cases, you can do it with one chip (the micro) and a few peripheral components. It will usually be the least expensive approach hardware-wise. The development cost lies primarily with the programming time and the attendant debugging and testing. This chapter promotes this approach and outlines a general procedure to follow. You can use this as a guideline then modify it to fit your own needs and knowledge as you gain experience and expertise in software.

In-depth coverage of micros in one chapter is not possible. This chapter is a summary of the subject and offers some guidelines on how to approach designing with micros. This chapter assumes that you are relatively new to the subject and need some guidance on how to begin and follow through with a microdesign. This approach is for those of you just starting to use micros or for those of you who have put it off for years and have at last come to face the reality of modern digital design.

Just So You Know

I use the term "micro" to describe most microcontrollers. Other equivalent terms are "embedded microcontroller, microcontroller unit (MCU), or μC." I use these interchangeably, so don't get confused. In all cases, the term describes a single chip that includes the processor with RAM, ROM, various interfaces, and I/O features. The primary feature of each is the basic processor word size, usually 8, 16, 32, or 64 bits. The micro is a flexible, configurable digital chip that is versatile enough to implement almost any digital product.

The term "microprocessor" is another thing altogether. It is called an MPU and, unlike the MCU, it is just a processor. It needs external memory, I/O circuits, and other peripheral devices. Examples of MPUs are the Pentium in a PC and the ARM processor in a Raspberry Pi computer. The MPUs are not covered here.

Embedded Controller Design Process

Here is a general procedure to follow in designing a product with a micro.

1. Define the product. As usual, start with a verbal description of what you are designing. Write it out and be as detailed as possible.
2. List the inputs and outputs. Typical inputs are logic levels from something else, such as switches, keyboards, and sensors. Common outputs are logic levels to other circuits, LEDs, LCD or 7-segment displays, or control of some mechanical thing like a motor, relay, solenoid, or servo. Draw a black box diagram with inputs and outputs as a starting point.
3. Choose a microcontroller. What is your target chip or board? For personal projects, you may be using an Arduino or Raspberry Pi or some other board product. You may be using some specific chip like one of the Microchip PICs, a Texas Instruments MPS430, the old standby 8051, or one

of the dozens of other possibilities. Get yourself familiar with the board or chip you choose. Acquire the data sheet and any app notes from the manufacturer. Check for the availability of books on the subject to use as reference. I identified a half dozen books, each on the Arduino, PICs, and Raspberry Pi. A more detailed look at selecting a micro is given later in this chapter.

4. Identify your software. You will need the development software for your board or chip. This is usually in the form of an integrated development environment (IDE) that includes an editor, monitor, simulator, debugger, and a compiler for a language like C or an assembler for assembly language for your MCU. Definitely confirm the programming language you plan to use like assembler, C, or other. It may come with the IDE or be available from the same source. Find out if there are any subroutine libraries you can tap into. The IDE and other software you can usually get from the chip maker or board supplier.

5. Summarize the operation of the product. Break the process down into discrete steps as much as possible. Be sure to put the operations in sequence. Drawing the state diagram as described in Chap. 11 on digital design is a good way to describe what you want to do. Identify any special steps like math.

6. Draw a flowchart of the process. Use the standard symbols like those in Chap. 1, Fig. 1.1.

7. Try to divide the program up into separate modules of related code. Partitioning the problem into separate operations makes the whole software development process simpler. The divide-and-conquer approach will identify modules for segments of the design. These will be shorter and easier to program and test. Then you can link them together later to form the whole program.

8. From all of the above, you may be able to deduce an algorithm to solve your problem. It should emerge as you do steps 5 and 6. Again, try to divide your program into smaller logical groups because they will be easier to write and test. For example, one segment may be interfacing to the inputs, another devoted to processing those inputs, and another defining the outputs.

9. Define any interfaces. If you need external input and output devices, define them now. Will you need extra transistors, diodes, LEDs, or ICs to connect your (I/O) devices? Do you need an analog-to-digital (ADC) converter? Many micros have ADC built in. What about PWM capability? Many newer micros have that capability. Are timers needed? Also determine whether you need USB or other interfaces like I²C, LIN, CAN, RS-232 (UART), or SPI. Design any special interfaces now using the procedures outlined earlier in the book.

10. Write your code. The software you are writing for a micro is generally referred to as firmware, mainly because it will be stored in some form of ROM, meaning it is more hardware than software. Anyway, your language is most likely to be C, but use what you know. Some microproducts use BASIC. It is easier to learn than C. Don't overlook the assembler that you program with by using the native instruction set of your micro. It gives you ultimate bit-by-bit manipulation capability. The process steps are often more easily deduced if you know just what the processor can do. If you are a hardware-oriented person, you may like assembler better than C or some other abstract language. Give it some consideration if you have not looked at it before.

11. Test out your code in your IDE if it has a simulator. This will provide an initial evaluation and an opportunity to fix problems before the big final test.
12. Build a prototype. Use an available board or evaluation module and breadboard any interfacing and peripheral devices.
13. Test and debug. Run your code on the prototype and debug and fine-tune it until it works as desires.
14. Design the packaging. Do a PCB layout.

Figure 13.1 One version of the BASIC Stamp2 (BS2).

Choosing an MCU and the Software

Even if you are experienced in electronics, you may find choosing a micro a challenge. There are hundreds of choices out there in the real world. How does one select? Here are a few ideas that should help. For example, let software be your guide. If you are still learning, base your decision upon the availability of good documentation, such as books, manufacturers' literature, and training sources.

- Beginning coder. If you are new to programming, try something simple to get you used to the idea of writing code. Perhaps the easiest language to learn is BASIC. It has been around for decades but is no longer widely used commercially. I learned it years ago, and I still use it for some personal projects. The best platform for this is Parallax's BASIC Stamp. See Fig. 13.1. This is an inexpensive (<$50) module product featuring one of the popular Microchip Technology PIC processors. It has a BASIC interpreter embedded in it. You write some code with the development software on your PC and download it to the programmable flash ROM in the PIC microchip, and then the micro runs it. The BASIC Stamp board just has basic I/O lines for interfacing to switches or LEDs. Get the best and latest version of the Stamp2.
- An interesting option is the PICaxe. It also uses a version of the PIC microfamily and has been preprogrammed with a BASIC interpreter. You can then program the chip and incorporate it in a design of your own.
- Experienced coder. If you are a pretty good programmer, go for the C or C++ language. These are the most popular languages used by embedded microdesigners. Virtually all micros support them. As for a microchoice, the PIC processors are very popular and come in an amazing array of models and options. The low-end micros are 8-bit micros and come in an 8-pin DIP. The upper-end micros use 32 bits, and some incorporate digital signal processing (DSP) features.
- Another good choice is the Texas Instruments MPS430 series. This is a 16-bit micro with dozens of memory and I/O options. I highly recommend their LaunchPad development board and platform. See Fig 13.2. This processor uses very low power, so it is a good choice for battery-powered products.

Figure 13.2 The popular Texas Instruments LaunchPad using the MPS430 micro.

- Choose by popularity. In the DIY experimenter community, the Arduino (Fig. 13.3) has become the favorite for low-end, entry-level boards and micros. This board is super cheap and uses the Atmel (now a part of Microchip) ATmega328 8-bit processor. It uses a variant of the C language for programming. Programs for Arduino are called "sketches." There are all sorts of software and accessories available for Arduino. For example, you can buy a variety of "shields" that plug into the main board. These are I/O interfaces for working with different external devices, such as displays, motors, or the like. There are many books for learning and support.

Figure 13.3 The micro of choice for many experimenters, the Arduino Uno.

- Experienced programmer. If you are more experienced, go for the other big popular micro called the Raspberry Pi. It comes in several models and versions up to Raspberry Pi 4 as of this writing. The original unit uses a Broadcom variation of the ARM 32-bit processor that gives it super powers that can handle almost any complex design. Later models use faster versions, and the latest Raspberry Pi 4 unit uses a quad core (processor) with 64 bits. This is overkill for most projects, but it is available at a reasonable price. All versions use the Linux operating system or some similar OS. There is lots of software available. C is the most common language. Again, there are plenty of accessories and books to support it.
- Your personal choice. If you know another micro, definitely go for it. I learned micros on early Intel 8080s and Zilog Z80s and then Motorola's 6800 and 68000 series. Motorola became Freescale and now NXP. Their micros morphed from the 6800 into the 68HC11/12 series and today a massive line of 8-bit and 16-bit processors. Some of the 8-bit series are HC05, HC08, HC11, RS08, and S08. I did many projects with the 68HC11 using assembler, so I am partial to it and still use it occasionally. If you have a favorite and it is still available and appropriate to the task, I recommend that you stick with it.
- Another longtime favorite is the 8051. It was one of Intel's earliest products in the late 1970s and early 1980s (MCS51). It is still kicking around and available from multiple sources. Most of its use is as an embedded MCU on a single chip along with some other circuitry. An example is a wireless system

on a chip (SoC). The 8051 is just a support circuit for a radio transceiver.

I have worked with the 8051 at the assembler level, and it is easy to learn and use. I recommend it, too, as a viable choice for some projects. There is lots of open-source software available for it, too, and that can save you time and aggravation.

- micro:bit. An unusual board product. See Fig. 13.4. This board was developed by the BBC to be distributed to millions of students in the U.K. for learning computers and software. It is very small but uses a 32-bit ARM M0 MCU. Programming is in JavaScript. It can also use MicroPython, a subset of the popular new language Python. I have one but have not explored its capability. Like most of these board products, its primary value seems to be teaching you some programming.

(a)

(b)

Figure 13.4 The tiny micro:bit uses a powerful 32-bit ARM processor. (*a*) The processor and some interface connections on one side of the board and (*b*) a 25 LED matrix display and two push buttons on the other side.

- Commercial or personal product. For personal projects, you have the widest choices for a micro. For a commercial product, you must be more careful in selection. If a large volume of chips is required, cost will be a big factor. Another factor is whether you will require a second source of chips. That may be critical. Since most of the cost of developing an embedded product is software coding and debugging, try to choose a chip with a good IDE and available software from libraries or open-source sources to help shorten development time.

Just So You Know

If the previous discussion was not clear enough, let me be even clearer here. Do not start with the actual microchip or IC. Instead, acquire a board product. This is just a small micro on a printed circuit board with related memory, I/O circuits, and other hardware like switches, LEDs, LCD readouts, and plenty of connectors. All you need to do is connect dc power to the board and attach the board to your personal computer with a USB cable. Any other external connections are your ideas.

Later, when you become a good microdesigner, you can start with the chip and put it on a board of your own design.

More About Selecting an MCU or Processor

Consider the following as a checklist before you make your final decision on an MCU.

1. Define your product. How many times have I already mentioned this? The repetition is deliberate because many designers really

do skip this step or do very little definition. There are some brilliant engineers with lots of experience who can get away with this. For us normal humans, skipping this step generally leads to problems later in the project. Take some time and do a thorough job here. It will help you make some of those tough design decisions later. You don't want to find out later that the MCU you chose does not implement a function that you need or is not capable of being used in a follow-up product. Oh yes, don't forget to write all this down in your notebook.

2. Do you need a microcontroller or a microprocessor? Remember, a microcontroller is a complete computer on a chip with processor, memory, and lots of I/O interfaces. A microprocessor is just the processor. It needs external RAM, ROM, and interface circuits. The Intel or AMD chip in your PC is a microprocessor. Microprocessors need external memory, I/O, and support ICs. Designing with a microprocessor is a different process not covered in this book.

3. Select a word length. Micros or MCUs are generally categorized by the basic word size they process. These are 8, 16, 32, and even 64 bits. The greater the number of word bits, the more powerful the processor. The 8- and 16-bit micros are less powerful but are ideal for small projects. They are cheap, but given the state of semiconductor technology, even the lowly 8-bit micro can do a great deal. The 32- and 64-bit devices are for major projects because they have higher speeds and more general computing power and have the ability to support more memory for large, data-intense applications. These big chips have gotten cheaper over the years thanks to semiconductor manufacturing technology, and you may want to consider them. In some cases, they may not cost that much more. The ARM-based chips are a good example.

4. Speed. This is usually stated as a clock speed. The higher that speed, the more powerful the micro. Clock speeds run from about 4 MHz to over 1 GHz. A better measure is instruction execution time, but that is not always readily available information unless you really dig into the specifications and instruction set and try to match those up with your application. This is a tough one to decide upon. Generally, the faster the better, but don't let that influence other factors like cost. Try to estimate your minimum need based upon your application.

5. RAM and ROM. Most MCUs have onboard RAM and ROM. The flash ROM or electrically erasable PROM is used to store your programs, boot routines, or other software that initiates the operations or performs always-needed actions. The RAM is used for the general computing operations. The amount of RAM is usually small, and the ROM is bigger. Various models of the same MCU can be had with different amounts of RAM and ROM. Try to guess how large your programs will be so you can get the amount of memory you need. This is usually an estimate, so be generous or conservative and default to more rather than less memory. Try to select a processor that is available in different versions with differing memory sizes. You don't want to have to change MCUs (and the software!) later if you find that an upgraded product is needed or future applications cannot be accommodated.

6. Interfaces. Does your product call for a given type of interface like USB or UART (RS232)? Other widely used serial interfaces are I^2C, CAN, SPI, micro-USBs, Ethernet, and multiple GPIOs (general purpose input outputs). Specify what you need.

7. Does your product require an analog-to-digital converter (ADC) or digital-to-analog converter (DAC)? Will you be using an external chip? Many MCUs have integrated ADCs. Those with 10 or 12 bits are common, and many come with an input multiplexer that allows the ADC be shared with multiple inputs. Embedded DACs are not very common, so most applications requiring them use an external device.
8. Pulse width modulation. PWM operations are widely used today. If you intend to use that technique in your design, you should choose a micro with PWM built in. PWM provides an inexpensive and flexible digital-to-analog conversion, as well as light dimming and motor speed control.
9. Timers. Will your application require internal timers? These are counters that let you easily implement timing operations. These are also used in implementing the PWM.
10. Math capability. Does your product need major math capability? All MCUs will add and subtract. Only a few can do multiplication and division. Check your application to confirm what you need. Will you need floating-point math? The larger, faster micros will be the most likely to have what you need.
11. DSP. Will you be doing any digital signal processing (DSP)? Maybe you will be implementing some digital filters, or you may need to run a fast Fourier transform (FFT). While special DSP chips are available, you may not need their full capability. However, there are a number of micros with special instructions, such as multiply and add, that make DSP faster and easier to do.
12. Artificial intelligence. Will you be implementing some AI functions? Most micros will accommodate some types of AI, but you will need a special chip if you plan to use deep machine learning. Look into graphic processing units (GPUs) from AMD, Intel, and Nvidia.
13. Packaging. Do you need special chip packages? Most micros are contained in a multipin surface-mount package. You can still get some MCUs in the older dual in-line packages (DIP). These are mostly available for the older, smaller, less-expensive micros.
14. Environmental specifications. Will your product have to operate in extreme temperature conditions or have to endure major shock and vibrations?
15. Power supply. What supply voltages will you need to operate the micro? A single low-voltage like 3.3 or 5 V is typical, but some micros are an exception. Estimate your current needs and those of any interfaces or peripheral devices. Is low power consumption a critical need? It probably will be if your final product is battery powered.
16. Dig deep for microvendors. Some microvendors are well known and popular. Examples are Microchip Technology, known for their low-end PIC series, or ARM for their widely used 32-bit devices. The temptation is to go with the familiar, but you may discover just the right configuration from some other vendor. Take a look at companies like Cypress, NXP, ST Micro, TI, and others—all with extensive MCU product lines.
17. Cost. This is always an issue, and it is primarily based on the volume you intend to use.
18. Information sources. MCU vendors have everything you need to build a product around their chip. They all feature massive data sheets and applications guides, as well as additional software documentation. It is overwhelming. Independent sources, such as

books, online material, tutorials, magazine articles, and other data, generally provide a better, simpler approach to designing with a specific chip.

19. Software. You must be serious about the software because that is what you will be using most of all in designing your product. Make sure you have a good IDE and the related packages like editor, debugger, and emulator. Ascertain that you will have the programming language of your choice, either compiler or interpreter. Also consider software sources other than your microvendor. Other questions to ask are the availability of program libraries or an operating system (Linux?) if you need one. How about the requirement for a real-time operating system (RTOS)? Is security software necessary? It has been said that the best approach in selecting an MCU is to find your software first, then buy the micro that runs it.
20. Evaluation boards. Are evaluation boards or other board products available so that you can do some early evaluation and development? This will save you time. Most vendors have these, and even some others make a board you can use. These evaluation boards are also referred to as single-board computers (SBC). Most of the chip vendors offer such boards, but third-party sources are also available.
21. Micro alternatives. If your product has special needs, you may have a hard time finding an appropriate MCU. For example, available micros may just not be fast enough, or you are asking for an MCU to accommodate too many parallel, concurrent operations. If this is the case, maybe you need an FPGA. They are faster and run parallel operations easily, but they are not easy to reprogram. See Chap. 12 if you have not already done so.

Software and Programming

This book does not teach you how to program. That is something you need to learn on your own. Maybe you already know how to program. This chapter mainly is about selecting a programming language and showing you examples of how to implement specific operations related to designing a product. Some short coding segments will hopefully illustrate selected operations that occur in many applications.

Programming Languages

It has been said, and there is general agreement within the electronics industry, that the programming language of choice is C. More embedded controllers are programmed in C than any other language. There are several reasons for this. First, it is taught in colleges and universities both in engineering and computer science, so many already know it. Second, it is possible to learn it on your own. Third, C code is generally portable or transferrable. That is, a C program written for one micro will typically run on another micro. Minor syntax adjustments may be necessary, but you generally do not have to start from scratch or reinvent the wheel. If you have not yet learned a language, definitely consider C.

That said, do keep in mind that you have other options. As mentioned earlier in this chapter, there is assembler language to consider. BASIC is also mentioned but may not be a long-term solution if you have to design multiple commercial products and support them. However, it is a good language to start with because it is easy to learn, and programming is almost intuitive. Another language that is gaining acceptance is Python.

Learning a Language

Probably the best way to learn a programming language is by way of a formal college course. It will cover the fundamentals, and you will get some real practice. If you do not have access to such a course, you should teach yourself. It is relatively easy if you are committed and willing to learn.

One starting point is books. There are dozens of books on languages and programming. Go to your local bookstore, select a few books that look good to you, and start reading. Search for books on the languages of interest to you on the internet, and see what turns up. Get ready to be surprised at how many there are.

An alternate route is an online course. Many universities offer courses completely online. You may want to check out the availability at a college or university near you. Another good source is Udemy. This online source offers multiple programming courses in several languages. C and Python are popular choices. Prices are reasonable. Other online courses are given by edX and Coursera. Check out both sites to see what is available. I have used Udemy courses several times in the past with success. The courses are college level and not easy, but you do learn from them.

The main learning requirement is to buy a language and IDE for your PC and start writing code for your board micro. Most of the online courses will require you to have the language and software so that you can practice. Programming is like writing. You learn by doing it. It is hard and tedious. Keep it up, and you will begin to write useable code in a short time. Patience and persistence pay off eventually.

One recommendation is to select a course that includes an embedded controller. I learned C in an online course that included a Texas Instruments' MPS430 micro. A parts kit is available so you can perform some experiments with hardware controlled by software.

Between an online course, some reference books, and the software, you will be able to learn programming.

Programming Language Examples

A popular way to illustrate how the different languages are used is to do basic operations. For most applications, you will need to generate some outputs to flash an LED off and on or operate other devices, such as a relay or motor. Another basic operation is inputting a signal from an external circuit or components. Here are some programming examples using BASIC Stamp2 (BS2) and Arduino Uno.

BASIC

Here is what BASIC code looks like to blink an LED off and on.

```
DO
    High 7
    Pause 500
    Low 7
    Pause 500
LOOP
```

An LED with resistor is connected to pin 7 of the BS2 board. See Fig. 13.5. High 7 sends a logic high (1) to I/O pin 7 that turns on the LED. A pause 500 allows pin 7 to stay high for 500 ms. Low 7 sends a logic low (0) to pin 7 turning off the LED. Another 500 ms pause occurs. The LOOP command causes the program to loop back to the start and repeat the same code. As a result, the LED continues to go off and on.

Simple, huh? Look into this option. The prices are low and the language almost intuitive to learn. It is a good starting point.

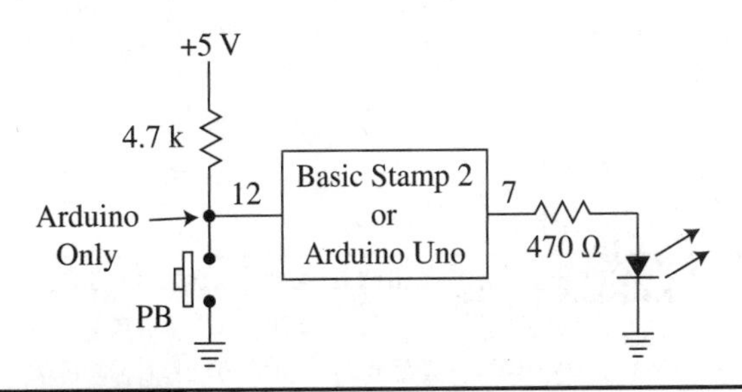

Figure 13.5 Connecting an LED and push button (PB) to a BS2 or Arduino.

Next, here is the code for reading the state of a push button (PB). Figure 13.6*a* shows the circuit for connecting a push button to the BS2. With the PB not pushed, the input at pin 3 is a binary 0 or low, thanks to the resistor on pin 3 to ground. Pressing PB connects the supply voltage (+5 V) to the pin through a 220 Ω resistor, causing the Stamp2 to see a binary 1 or high.

Figure 13.6*b* shows another connection that produces the opposite effect. With the PB not pressed, the input to pin 3 is a binary 1 or +5 V. Pressing PB brings pin 3 to ground or a binary 0.

The program code for reading the input state is

```
DO
    DEBUG ? IN3
    PAUSE 250
LOOP
```

The DO/LOOP just repeats the code. DEBUG refers to the Stamp2 software and what it displays on the video monitor. This command asks what is the status of pin 3 and pauses for 250 ms, then sends the state of pin 3 to the video monitor screen and MCU. Then the code repeats.

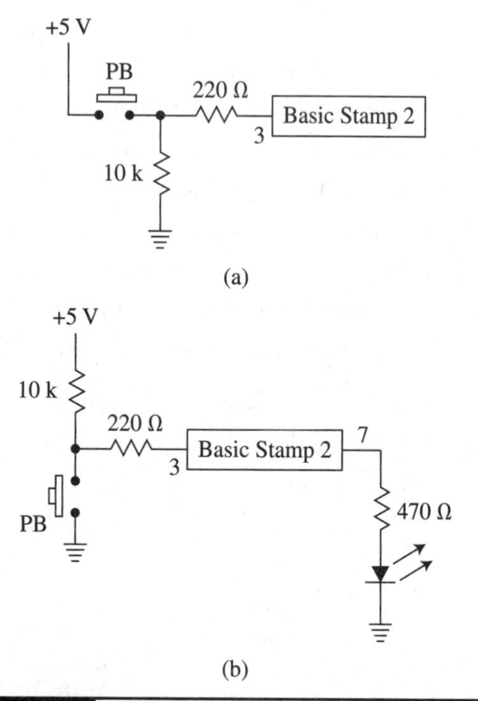

Figure 13.6 Connecting a push button and LED to a BS2. (*a*) Pressing the button produces a +5 V input. (*b*) Pressing the button produces a ground or 0 V input.

Here is an expanded example.

```
DO
    IF (IN3 = 1) THEN
        HIGH 7
        PAUSE 100
        LOW 7
        PAUSE 100
    ELSE
        PAUSE 100
    ENDIF
```

This illustrates the IF THEN ELSE command that makes decisions based upon an input. If the PB is pressed and pin 3 is binary 1, the LED on pin 7 will blink very fast for as long as the PB is pressed. If the PB is not pressed, the ELSE command is paused, then the IF THEN cycle is ended. The program loops back to wait for another PB push.

You can see how easy BASIC is. You just have to learn the commands and how they work.

Arduino Uno

Here is what the C-like code for the Arduino looks like to flash an LED. Arduino programs are called sketches. Again, an LED and resistor are connected to pin 7 on the Arduino.

```
//Flash LED
int ledPin = 7
void setup ()    {
pinMode (ledPin, OUTPUT); }
void loop()
    digitalwrite(ledPin, HIGH);
    delay (500)
    digitalwrite(ledPin, LOW)
    delay (500)
```

The line //Flash is a statement or explanation. You can notate the program with statements beginning with //.

```
Int led Pin = 7
```

The LED is connected to this pin with a resistor. It is similar to the connection for the BASIC Stamp2 in Fig. 13.5. This line tells the micro to send the output to pin 7.

The void setup() line is a necessary part of the syntax that tells the C language what to do. The pinMode line assigns the output to the term ledPin. The next line, void loop(), forms a programming loop that will repeat. When the last statement has been executed, the program loops back to the beginning to repeat the operation. The digitalwrite sends either a HIGH (1) or LOW (0) to ledPin 7 with a 500 ms pause or delay between the LED flashes.

Another example shows how an input is used to turn an LED off or on. The LED is still connected to pin 7. A push button (PB) is connected to pin 12. See Fig. 13.5.

```
int ledPin = 7;
int inputPin = 12;
int val = 0;
void setup() {
    pinMode(ledPin, OUTPUT);
    pinMode(inputPIn, INPUT):
    }
void loop() {
    val = digitalRead(inputPin);
    if (val = HIGH) }
    digitalwrite (ledPin, LOW);
    }   else {
    Digitalwrite (ledPin, HIGH);
    }
}
```

This sketch, as Arduino programs are called, turns the LED off or on, depending upon the state of the input button. The IF ELSE coding makes the loop repeat.

Some Takeaways

There are a couple of things you can learn from this. First, the Arduino program versions are different in command set and syntax than BASIC and a bit more complex, but they do the same thing.

Second, you don't always have to write your program from scratch. Find a program in books, online, or wherever and copy them, modify them, and use them. Search for subroutine libraries and other sources. In many cases, you can find some code that will do something similar to what you need. All you have to do is merge it into your software and possibly modify it a bit.

Learning to Love BASIC

If you are a serious programmer or professional coder, you probably hate BASIC. You no doubt think it is old and unsophisticated, not cool or current, and limited in its capability. You would think this way if you have never really used BASIC. I am sure you would rather be programming in C, C++, Python, even Java, or whatever the latest hot language trend is. If you are writing code for big commercial projects, use whatever your organization uses. And if you are so experienced, why are you reading this chapter anyway?

For you experimenters and makers, particularly those of you just learning to design, BASIC is a great starting place. It is cheap, easy to learn, and can handle all but the most sophisticated projects. I learned it years ago, and I still use it. I can and do program in Arduino and C, but it is not as much fun as working with BASIC. My recommendation is, don't knock it until you have tried it.

I have used two micros that integrate BASIC right into the MCU. The Parallax BASIC Stamp2 (BS2) and the PICaxe are single-board micros you can build into your projects. Both use a version of the Microchip Technology PIC MCUs. They have a built-in BASIC interpreter that reads your program code and executes it one line at a time. They are not particularly fast but fast enough for almost anything you want to do. The PICaxe is also available as an IC that you can build into any of your electronic projects.

If you have not made up your mind about programming and micros, consider BASIC and these micros. They are affordable and will get you off to a fast start. Later, when you get some experience, you can always learn a new language like C. Try the Arduino language next. Or if you are serious, try assembly language. If you just want to get things done, finish a project fast, and speed up the process, take a look at BASIC first.

The Case for Assembly Language

Most embedded microprojects are programmed in C. Or maybe C++. That is the main reason for learning that language. If you already know C then you probably don't want to hear about assembler. But if you are at an early stage of learning microcontroller development, you may want to consider assembly language. I learned it early because that was the only language initially available for most micros. BASIC came next but faded, but it is still around. Pascal was popular for a while. Python is getting more popular now. Assembler has survived.

The big benefit of assembler is that you work with the micro's actual processor instruction set with instructions like store, move, add, decrement, jump on zero, output, read input, and the like. You can control things at the bit and register level. If you have designed digital circuits at the gate, flip flop, counter, register, and logic levels, you will find assembler intuitive. I certainly did and prefer it to this day. I can program in C but seem to resist it because of its abstractness and all its fussy syntax. Maybe that is just me. You may not feel the same. All I am saying here is, if you have not considered assembler, you may want to look into it as an option. It is the ultimate digital designer's tool.

Here is a hypothetical example based upon a generic microcontroller (MCU). Figure 13.7

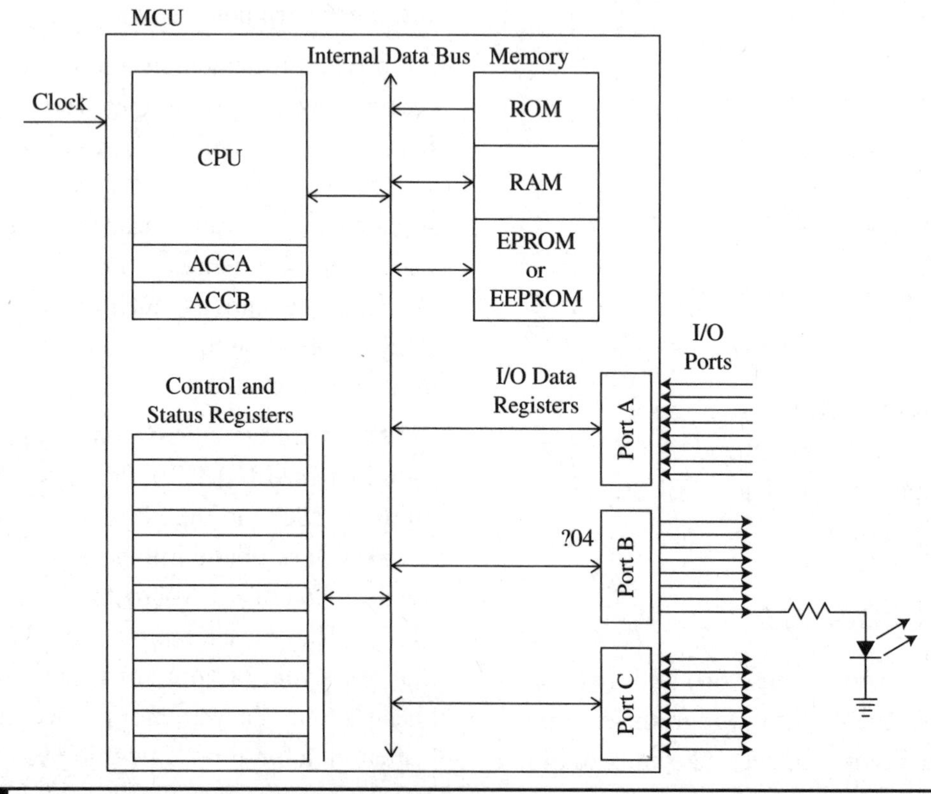

Figure 13.7 A generic block diagram of an embedded microcontroller. Note I/O port connects to the external LED, external switches, etc.

shows the MCU and all of the memory, control and status registers, and I/O ports. The program will be stored in RAM. The MCU fetches and executes each instruction one at a time in sequence.

The main processing register in the MCU is 8-bit accumulator A (ACCA). Another register is accumulator B (ACCB). The MCU has several 8-bit input/output ports. These ports are also memory locations as defined by a hexadecimal (hex) value such as $04 (where the $ means a hex number) that is also the address of the output port. Let's call it port B. When one of the output lines goes high, it turns the LED on. The program generates a pause or delay then turns the LED off by outputting a low value. The operation repeats. The program loops back until ACCB is decremented to zero. Then the program jumps the STAA instruction that now contains all zeros thanks to the COMA instruction. The program continues to repeat flashing the LED. The value in accumulator B determines the speed of flashing. With all 1s to start, the value is 255. The speed of the countdown depends on the clock frequency and the instruction execution time.

Here is a simplified assembler program that will flash an LED as an example:

PORTB EQU $04	The assembler assigns the address $04 to the name PORTB. $ indicates a hex number.
ORG $B000	The assembler designates that the program will begin in memory location $B000.
LDAA #FF	This is a load register instruction that puts all binary 1s into ACCA.
LDAB #FF	This instruction puts hex FF or all binary 1s into ACCB.

Note: The # sign means immediate. This is a type of instruction where the data to be manipulated is included with the instruction itself.

REPEAT STAA PORTB	This stores the content of ACCA in address $04 that sends it to output port B. This turns on the LED.
COMA	This complements ACCA, changing all binary 1s to all binary 0s.
JSR PAUSE	This is a jump to a subroutine instruction called PAUSE (for a delay).
PAUSE DECB	Decrement ACCB by one. Initially ACCB is set to hex $FF (decimal 255 or binary 11111111), so it decrements to $FFFE or 254. The goal is to decrement it to zero.
BNE PAUSE	Branch if not zero. Go to repeat PAUSE subroutine if ACCB is not zero.
BRA REPEAT	Loop back to the instruction at REPEAT.

The key to understanding this program is knowing what each instruction does. That is usually learned over time as you write code in assembler. Table 13.1 describes what each instruction does.

Developing an assembly program requires a board-level micro and what is usually referred to as a monitor program. It is a piece of software, like a limited operating system, that lets you enter and test programs. It runs on your PC. It also contains the assembly language assembler. You write the program on your PC using this software. This program is called the source program. The assembler processes the source code and makes an object coder version. This is the code that gets downloaded to the micro, and you run the program on the board computer.

Table 13.1 An Assembler Program Listing with Comments and Explanations

Assembler directions	Instruction	Comment, explanation
PORTB EQU $04		The assembler assigns the address $04 to the name PORTB.
ORG $B000		The assembler designates that the program begins in memory location $B000. $ Means hex.
	LDAA #00	This is a load register instruction that puts all binary 0s into ACCA.
	LDAB #FF	This instruction that puts hex FF or all binary 1s into ACCB.
REPEAT	STAA PORTB	This stores the content of ACCA in address $04 that sends it to output port B. This turns on the LED.
	COMA	This complements ACCA to change all binary 0s to all binary 1s or all 1s to 0s.
	JSR PAUSE	This is a jump to the subroutine called PAUSE that introduces a delay that keeps the LED on for a fixed time.
PAUSE	DECB	This decrements ACCB by one from FF (decimal 255) to 254. The goal is to decrement to zero.
	BNE PAUSE	Branch if ACCB is not zero. Go back to PAUSE.
	BRA REPEAT	Branch back to previous code.

Microdesign Considerations

Anything that you can do with legacy digital logic ICs you can do with a micro. Consider all the basic digital functions. The following list summarizes them and indicates how they are implemented on a micro.

- AND, OR, NAND, NOR, XOR, invert, miscellaneous logic functions
- Add/Subtract
- Multiply/Divide (Only some micros)
- Count, up, down, BCD, etc.
- Shift, right, left, random number of bits
- State machines
- Multiplex
- Demultiplex
- Compare
- Encode
- Decode
- Code conversion
- Input
- Output
- Pulse width modulation (PWM)
- Time
- Analog-to-digital conversion (ADC)

The thing to be aware of is that these functions can only be performed one at a time in sequence. They do not occur simultaneously as they can or do with discrete logic circuits. In some cases, the operations defined previously may actually correspond to one of the micro's native instructions. For example, addition is carried out by an ADD instruction, count is implemented with an increment (INCR), and shift is executed by a SHIFT instruction. In some programming languages, the previously listed operations may correspond to an available function. Otherwise, you will write code to create a subroutine that performs the desired function and have to string all of these instructions, subroutines, or software language functions in sequence to implement your end objective. Remember, micros execute instructions sequentially, so timing is important.

Most MCUs operate at megahertz speeds, so while operations are being performed fast and sequentially, they can give the appearance of doing multiple things simultaneously.

What you must be aware of is the time it takes to do each operation. Micros are pretty fast these days. Clocks run from a few megahertz to over 1 GHz, and instructions execute in only a few microseconds or less. Just be aware that for some applications that may not be fast enough. The solution is a faster processor or to dedicate some of the high-speed functions to faster external digital circuitry. Perhaps your application requires multiple parallel operations. If that is the case, you probably should be using an FPGA instead.

Another factor to be aware of is that functions or statements in higher-level languages like C are translated into multiple sequential native micromachine language instructions by the programming language compiler. This occasionally leads to longer or slower programs than desired. In some cases, rewriting parts of your program in assembly language produces a shorter, faster result, since fewer instructions are required.

Are You a Hardware Person or a Software Person?

Designing with a micro tends to be about 20 percent hardware and 80 percent programming and software. Those percentages change a bit depending upon the project, but they are in the ballpark. I was disappointed in this during my first microprojects. I am a hardware person and like to fool around with the chips and other components. Truth be told, I hate programming. I have learned to do it and have done it for years. I am generally competent at it. I go by the old Seal Team 6 tenet that says, "You don't have to like it, you just have to do it." Many other electronic engineers feel the same way. Here are some of the quotes I have heard over the years (paraphrased, of course):

"I'd rather be soldering."

"Solder is my programming language."

As you gain practice designing, you will quickly discover if you are a hardware or a software engineer. I will leave you with one key fact: The future is software. Learn to code and develop algorithms, and keep up with the various processors, languages, and other software products.

Microinterfacing

A major part of building products with a micro is interfacing, that is, connecting it to the outside world of switches, sensors, displays, and actuators like relays and motors. It is not difficult, but you should be aware of how this is done. Here are the basics.

Micros usually have some general-purpose input output (GPIO) lines or pins that permit the attachment of other digital circuitry or devices. Some pins are dedicated to either input or output, and others can be configured as either with an instruction or two. These generate logic output signals or accept logic input signals. Output or input levels are commonly 0 and +5 V or 0 and +3.3 V. Each output pin can usually deliver several milliamperes as a sink or source. Most output pins can drive an LED directly without an interface circuit other than a current-limiting resistor. Refer back to Fig. 13.5.

Input pins can also usually accept inputs from TTL or CMOS logic circuits. Since all micros are CMOS, those inputs typically do not require any signal conditioning. In some cases, TTL inputs may have to be modified before they are acceptable to a CMOS device. Figure 13.8

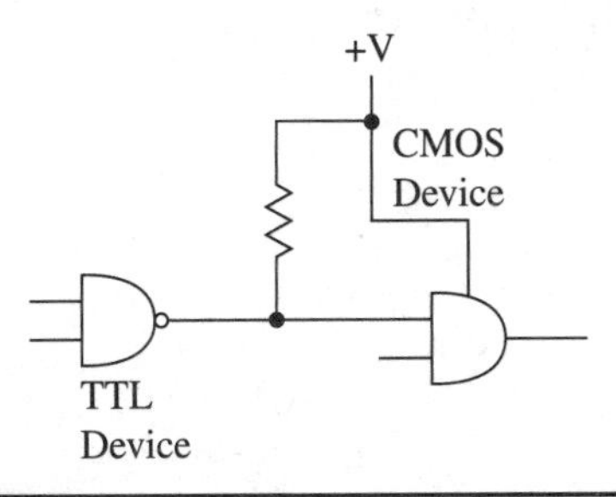

Figure 13.8 One way of interfacing a TTL output to a CMOS input.

shows an example. In some cases, the problem can be solved with a pull-up resistor from an open collector TTL output to CMOS input. Something in the 1- to 10-k range usually works. This does not always work.

The best practice is to avoid the problem. Just do not mix TTL and CMOS ICs, period.

When using the outputs of the MCU or any logic circuit for that matter, the available current may not be sufficient to operate a bunch of LEDs. In that case, use some single-transistor driver circuits like those described in Chap. 5. In this way, you can operate motors, relays, solenoids, and multiple LED strings. Seven-segment LED displays are easily accommodated. LCD displays require the most complex interface.

If you plan to use the PWM feature, you may need to buffer the output with a transistor switch to provide the needed current.

Many MCUs also include an ADC. The micro may or may not have an associated multiplexer that supports several input lines that allow the ADC to be shared.

The widely used 7-segment display makes a good example of interfacing. These displays are just multiple segments made up of individual rectangular LEDs. Figure 13.9 shows the usual segments plus a decimal point.

These LEDs work just like any other. However, they come in two formats, with either a common cathode or a common anode connection. See Fig. 13.10*a*. To turn on the LEDs in this common anode display, the micro IO pins have to go low. Note: The required resistors are external to the display. If a common cathode display is used (Fig. 13.10*b*), the IO outputs from the micro need to go high to turn the LEDs on. Again, external resistors are needed to control the brightness of the display.

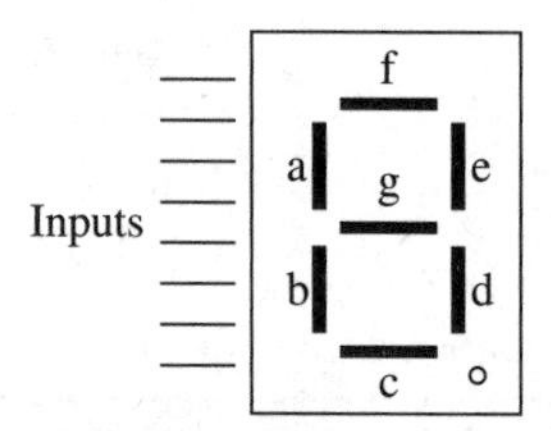

Figure 13.9 The LED layout in a 7-segment display.

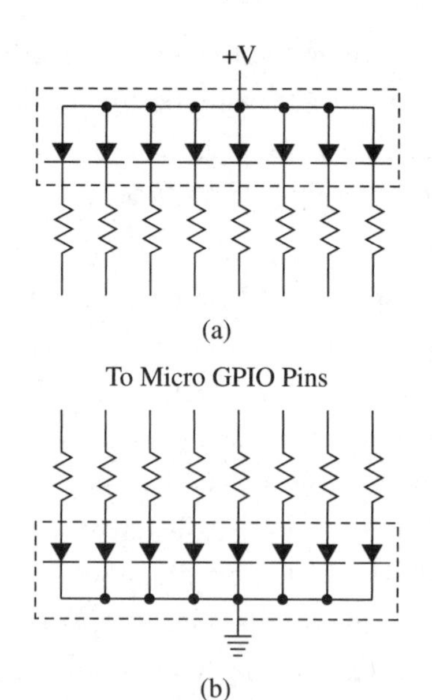

Figure 13.10 Both common anode (*a*) and common cathode (*b*) connections are available for 7-segment displays. External current-limiting resistors are required to protect the LEDs and control brightness.

Many microapplications require outputting program results to a display. This can be done several ways. Figure 13.11 shows BCD to 7-segment decoder driver ICs connected to the general-purpose I/O pins on the micro. These ICs take a 4-bit BCD code and translate it into a code that can drive the individual LED segments in the display. Typical devices that do this are the TTL 74LS47 and the CMOS 4000 series CD4511.

The purpose of the EN (enable) pins on the driver ICs is to turn an IC on when a signal is received from the micro. Since both decoder-drivers get the same BCD code from the MCU at the same time, they will display the same value. The EN signals turn the related display

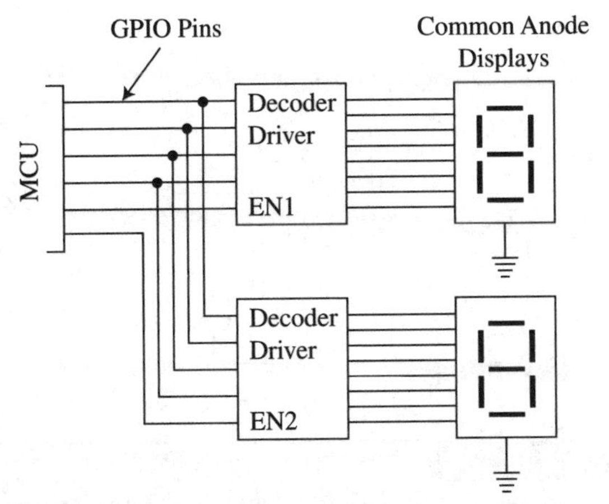

Figure 13.11 Creating a two-digit LED display using BCD to 7-segment decoder/drivers operated by the MCU output. The two digits are turned on and off quickly so that the digits do not seem to be flickering or fast blinking.

on when the BCD code is to be displayed. This is called a multiplex display. The two digits are turned off and on alternately so quickly that the human eye cannot distinguish the switching. Instead, the eye perceives a steady nonflickering two-digit display.

Another way to do it is to connect the I/O bus pins to the 7-segment displays directly so that one I/O port bus can control the display. Inverter drivers are usually needed between the I/O outputs and the LED inputs. With arrangement, each I/O port pin controls one of the LED segments.

Next, make a truth table showing the bit pattern to be sent to the display for each BCD input. Store that group of binary patterns in sequential memory locations, creating a lookup table.

When the MCU program computes a two-digit number to be displayed, it will output the bit pattern for the desired value in one digit then another BCD value for the other display. The program to do this looks at the binary value it computed, and a software routine converts it by looking up the value in the truth table and sends the appropriate bit pattern to the enabled display. The code is repeated again with the next binary number. The two output patterns are outputted alternately at a high rate, so a human's persistence of vision makes the flickering go away.

In Case You Forgot

BCD code divides up a decimal number into individual 4-bit code words instead of one continuous binary word. Micros process binary data so that it has to be translated into BCD format by a subroutine for an external display. For example, the decimal number 73 in binary is 01001001. In hexadecimal, it is $49, where $ means a hex value. In BCD, the number 73 would be coded as 0111 0011. Each digit gets its own code. Notice the separation between the 4-bit BCD digits. Another example, the decimal number 391 would be expressed in BCD as 0011 1001 0001.

Here are the codes in case you forgot. The hex code represents the decimal values 0 through 9 and from 10 to 15 with the letters A through F. To express a binary number in hex, you divide the bits into 4-bit groups starting on the right side of the word. The binary code for 391 is 110000111 or hex $187.

BCD code only represents the 10 decimal digits. An X in the following table means an invalid code.

Decimal	Hex	BCD
0	0000	0000
1	0001	0001
2	0010	0010
3	0011	0011
4	0100	0100
5	0101	0101
6	0110	0110
7	0111	0111
8	1000	1000
9	1001	1001
A	1010	X
B	1011	X
C	1100	X
D	1101	X
E	1110	X
F	1111	X

A Plan for Learning Micros

As you have seen, this is primarily a hardware book. As I have mentioned several times, however, the future is micros and software. You must learn to program. It can be done through self-study, but it is not easy. Get some books, a microdevelopment board, and some software. Put in some time learning how to use the software. Practice by writing your own code for simple functions. For example, learn to blink an LED, read the status of a switch, and implement PWM. Get some books and copy the code given as examples, and modify it to do something similar but different. For larger projects, partition the program into shorter modules. The modules will be easier to develop, and you can link them later to do the big project. Have patience. Don't let failure stop you. Try again. A recommended approach is to associate with someone who already knows how to code. Get advice and answers from that person.

I know I am repeating here, but if you are just beginning, I suggest BASIC. It is easier to learn than C or any other. Get some hardware that supports BASIC, such as the BASIC Stamp2 or the PICaxe, or get both. They are inexpensive, and there is lots of material to learn from. Once you are competent with BASIC, move on to an Arduino that uses a version of C. Then move on to assembler.

Just So You Know

Because micros are so cheap and flexible, they regularly replace older analog or other digital circuits. A good example to illustrate this is shown in Fig. 13.12. This is the popular 555 timer IC connected as a clock oscillator to generate a rectangular wave at a desired frequency set by R_A, R_B, and C.

This circuit can be replaced by a micro, in this case, a PIC 12F508 as shown in Fig. 13.13, or some other similar micro like the 12F1572. Both are 8-bit micros in an 8-pin DIP, just like the 555. It may be a bit more expensive than the 555 circuit, but it uses an external crystal for much greater frequency precision and stability. The downside of the micro is that you have to program it to generate the same output that you would get with the 555 circuit, and the frequency is difficult to change in software. That may be easy for an experienced programmer, but not for a hardware person. You could probably build the 555 circuit on a breadboard and get it working faster than the programmer could write the code, test it, and debug it.

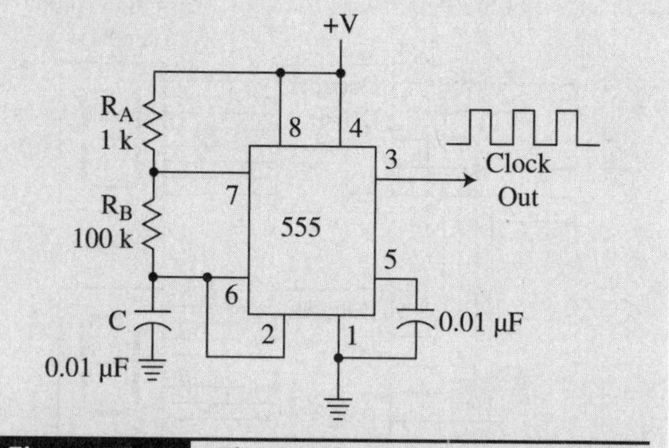

Figure 13.12 The popular 555 timer IC connected to supply a rectangular wave of a desired frequency.

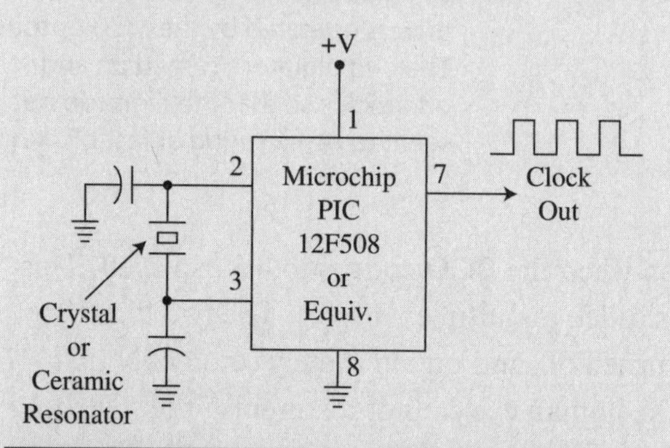

Figure 13.13 A microcontroller replacement for the 555 using one of the small PIC MCUs.

Committing to a Microfuture

1. Make a formal commitment to spend some money and time on learning micros and especially programming.
2. Choose a processor to learn. Follow the guidelines outlined earlier in this chapter.

3. Buy a development board for that processor.
4. Choose a language to learn. Acquire books, tutorials, take formal courses, participate in online courses, and learn on your own.
5. Get the software, including any operating system, IDE, language, simulator, compiler, debugger, and libraries.
6. Practice programming. Short, simple code to begin, then gradually more difficult and longer programs.
7. Design a product of your own with a micro you choose and a program that you write.

Design Project 13.1

As a practice project, try designing and implementing the digital cable tester described in Chap. 11. This is a good example of how to translate a discrete IC design into an MCU-based version. Use the process described at the beginning of this chapter. Use any language you want: BASIC, C, or assembler.

Design Project 13.2

Develop a truth table showing the decimal value to be shown on a 7-segment LED display and the bit pattern for each. Use the segment designations in Fig. 13.9. Then develop a step-by-step explanation that outlines how the numbers would be displayed.

Design Project 13.3

One operation that comes up repeatedly in programming is creating a delay and keeping track of how many times a particular function takes place. The algorithms typically involve counting. Since MCU operational registers can be counters, shift registers, or just storage locations, you can use them to implement some of those operations. For this project, using the previously explained assembly language program, develop a short subroutine that outputs the numbers 0 through 9 to output port $04 then stops. Use the assembly language instructions given earlier, but add one more: INCA increments the A register by one.

CHAPTER 14

Component Selection

Once you design something, you need to build it. That means finding the right components. There are thousands to choose from, making the selection difficult. This chapter will help you select and buy components for your designs. The previous chapters generally assume that you know how to choose parts, yet most newcomers to design don't really know which parts to select. It is not something usually taught in school. The chapter covers most common parts. The recommendations made here are based on actual use and experience.

The key components of interest are resistors, capacitors, inductors and transformers, and semiconductors (diodes, transistors, ICs).

Resistors

The most commonly used electronic part is the resistor. They are available with resistance values from a fraction of an ohm to about 22 MΩ.

Resistor Packaging

Resistors come in two basic forms, a cylindrical type with axial wire leads and a tiny surface-mount device (SMD). See Fig. 14.1. Both are widely used. The leads of the axial resistors are passed through holes in the printed circuit board (PCB) and soldered. The SMD resistors are soldered directly to the top of the PCB.

The axial resistors sizes vary in length from about 4 to 9 mm for the most common ¼- and ½-W units. The 2-W units are about 1 in long. The SMD resistors come in several standard sizes. See Table 14.1.

The main specifications are resistor type, meaning the materials used to make the resistor, the resistance value in ohms, the tolerance, and the wattage rating.

Resistor Types

Resistors are made from different materials. The following are the most common types:

- Carbon composition. A mixture of carbon powder and an insulator is packaged into a cylindrical shape. Very noisy with poor temperature characteristics. Cheap, but not widely used.
- Carbon film. A fine layer of carbon deposited on an insulating cylinder and laser-cut into a spiral to set the value. Widely used because of its low cost.
- Metal film. A metal film deposited on an insulating cylinder and cut into a spiral to set the value. These resistors are more stable with temperature and contribute less noise to a circuit than the carbon film resistors. In applications with small signals that can be masked by or interfered with by noise, this is the best choice.
- Wirewound. Resistance wire like nichrome wound on an insulating form. Used for high power. Units are available in sizes from 5 to 1000 W.

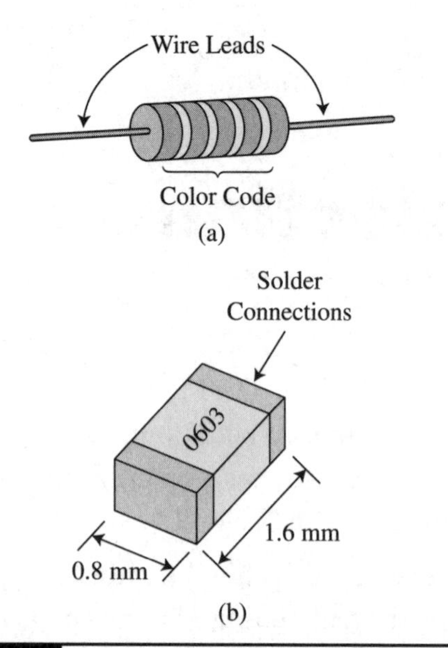

Figure 14.1 Common resistor form factors. (a) Axial, size varies with wattage, color code indicates value. (b) Surface-mount device (SMD), dimensions shown are correct.

Table 14.1 Standard SMD Resistor Sizes

SMD Standard Sizes	Dimensions: L × W × H (mm)
0201	0.6 × 0.3 × 0.25
0402	1.0 × 0.5 × 0.25
0603	1.6 × 0.8 × 0.45
0805	2.0 × 1.25 × 0.5
1206	3.2 × 1.6 × 0.6
1210	3.2 × 2.6 × 0.6
2010	5.1 × 3.2 × 0.6
2512	6.5 × 3.2 × 0.6

Standard Values

Resistors are made in standard electrical values. The values were decided upon by the Electronic Industries Association (EIA) standard groups. The groups are based upon their tolerance: E-96—1 percent, E-48—2 percent, E-24—5 percent, and E-12—10 percent.

There is also an E-192 list for 0.5, 0.25, and 0.1 percent tolerances. These are not widely used but are available if needed. The most popular types are the E-24 and E-96, 5 and 1 percent values, respectively. The following list shows the available sizes. The E-48 and E-12 are not widely used.

E-24 Resistor Values—5 percent

10, 11, 12, 15, 16, 18, 20, 22, 24, 27, 30, 33, 36, 39, 43, 47, 51, 56, 62, 68, 75, 82, 91

To get the actual value, you multiply the numbers above by 0.1, 1, 10, 100, 1000, or 10,000, or 100,000.

Remember: k = 1000

For example, select 27. You can get resistors with values of 27, 270, 27 k, and 270 kΩ. Select 27 × 1000 = 27 k. The color code is red-violet-orange-gold. (See Fig. 14.2.)

E-96 Resistor Values—1 percent

1.00, 1.02, 1.05, 1.07, 1.10, 1.13, 1.15, 1.18, 1.21, 1.24, 1.27, 1.30, 1.33, 1.37, 1.40, 1.43, 1.47

1.50, 1.54, 1.58, 1.62, 1.65, 1.69, 1.74, 1.78, 1.82, 1.87, 1.91, 1.96, 2.00, 2.05, 2.10, 2.15, 2.21

2.26, 2.32, 2.43, 2.49, 2.55, 2.61, 2.67, 2.74, 2.80, 2.94, 3.01, 3.09, 3.16, 3.09, 3.16, 3.24, 3.32

3.40, 3.48, 3.57, 3.65, 3.74, 3.83, 3.92, 4.02, 4.12, 4.22, 4.32, 4.42, 4.53, 4.64, 4.75, 4.87, 4.99

5.11, 5.23, 5.36, 5.49, 5.62, 5.76, 5.90, 6.04, 6.19, 6.34, 6.49, 6.65, 6.81, 6.98, 7.15, 7.32, 7.50

7.68, 7.87, 8.06, 8.25, 8.45, 8.66, 8.87, 0.09, 9.31, 9.53, 9.76, also 2.2 M

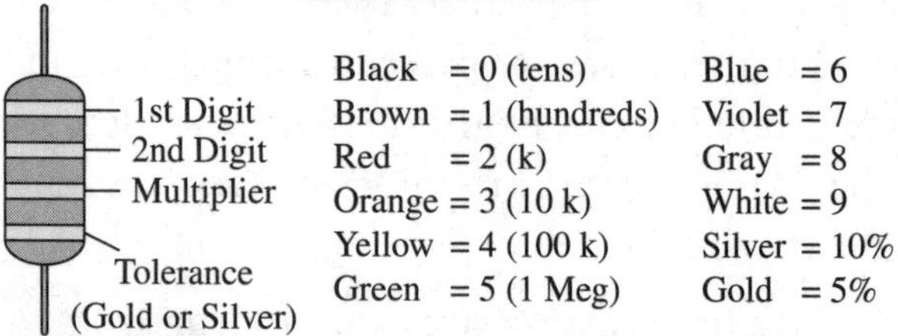

Figure 14.2 The well-known resistor color code.

To get the actual value, you multiply the numbers in the E-96 listing above by 0.1, 1, 10, 100, 1000, 10,000, or 100,000.

Tolerance

This is the range over which a resistor value may vary. At 10 kΩ, a 5 percent resistor could have an actual value of ±5 percent. Five percent of $10{,}000 \times 0.05 = 500\ \Omega$. The actual value will be between $10{,}000 - 500 = 9500$ and $10{,}000 + 500 = 10{,}500\ \Omega$.

Coding

This refers to how the resistor value is presented. For common cylindrical axial resistors with wire leads, a standard color-coding scheme is used.

To determine the value, record the numbers of the first two bands. Then observe the third band that is a multiplier. Add the number of 0s designated by the third band to the first two digits. The fourth band is the tolerance. As an example, determine the value of a resistor marked orange-white-orange-gold. The first two digits are 39. Orange third band says add three 0s. The value then is 39,000 Ω or 39 KΩ, 5 percent. If the multiplier band is black, the value is just the first two digits.

For SMD resistors, the markings are different. First is the size. All are tiny rectangular devices with metal ends that are soldered to the surface of the printed circuit board. The smallest package is designated 0201 and has the dimensions 0.6 mm × 0.3 mm. The largest is the 2512 with a size of 6.35 mm × 3.2 mm.

As for ohmic value, the SMD resistors follow the EIA standard sizes. The E-96 and E-24 are for 1 and 5 percent values. The value of the resistor is marked on the package. A three- or four-digit code designates the ohmic value.

The three-digit code is two digits followed by a third that is a multiplier. Some examples follow:

$$470 = 47 \times 10^0\ (1) = 47\ \Omega$$

$$103 = 10 \times 10^3 = 10{,}000\ \Omega$$

$$2R2 = 2.2\ \Omega$$

The four-digit code is similar. The first three digits are the base value, and the fourth digit is the multiplier.

$$3300 = 330 \times 10^0 = 330\ \Omega$$

$$2202 = 220 \times 10^2 = 22{,}000\ \Omega\text{s}$$

Most three-digit code resistors have a tolerance of 5 percent, while four-digit code resistors are 1 percent.

Wattage

The most common wattage rating is ¼ W for axial resistors. The standard values are also available in 1/8-, 1-, and 2-W ratings. SMD resistors have power ratings from 1/20 W for the smallest and ½ and 1 W for the largest sizes. Common wirewound resistors are available in 5- and 10-W versions and larger cylindrical units that can handle up to 1 kW.

Selecting a Resistor

For basic experimenting and project building, stick with the 5 percent axial lead resistors of ¼ W. Select the 1 percent values for critical circuits. If you are going to experiment with SMD resistors, select the large sizes because they are easier to solder.

Special Resistors

Other resistors you may encounter are thermistors and photoresistors. These resistors are really sensors that change their ohmic value in response to heat or light, respectively.

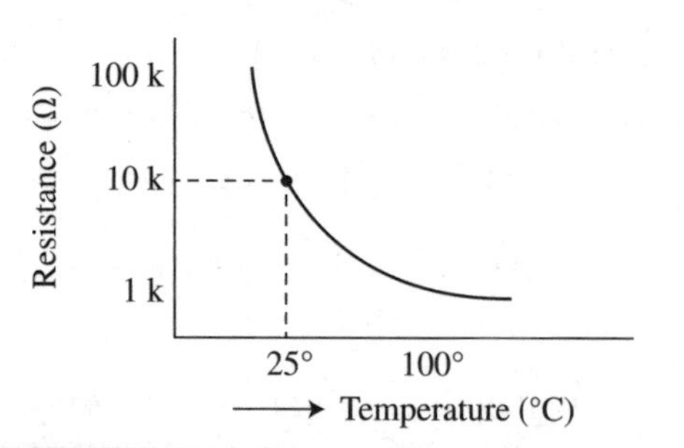

Figure 14.3 The resistance value of a negative temperature coefficient (NTC) and how it changes with temperature.

Thermistors

Thermistors change their ohmic value over a wide range with temperature variations. These are available in negative temperature coefficient (NTC) and positive temperature coefficient (PTC) variations. An NTC resistor's value decreases with an increase in temperature. See Fig. 14.3. A PTC resistor's value increases with a temperature. There are only a limited number of values from 1 k to 50 kΩ.

Photoresistors

Photoresistors or photocells are resistors that change value over a wide range when exposed to a variation in light level. They are usually made with cadmium sulfide (CdS). There are no standard values. Minimum resistances are about 1 kΩ to a maximum of over a megaohm. The resistance variation is something like Fig. 14.4.

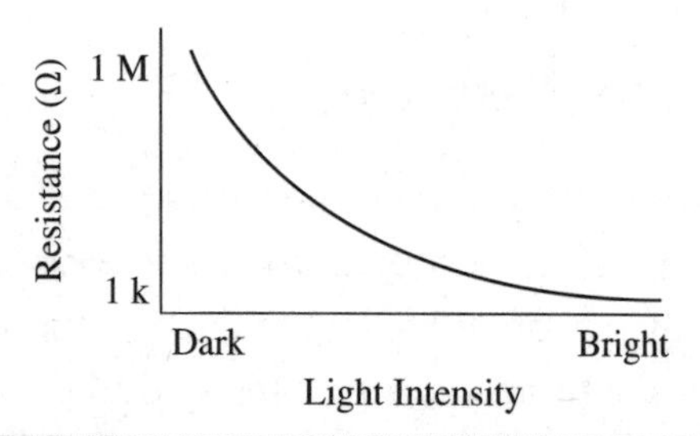

Figure 14.4 How the resistance value of a CdS photocell changes with light intensity.

Potentiometers

These are variable resistors or voltage dividers. Most are housed in metal containers and have a rotary shaft. See Fig. 14.5*a*. Smaller units called trimmers are housed in a plastic package and use a rotary screwdriver adjustment. Even linear pots with a linear or straight-line slider is provided to adjust the resistance. The resistive elements may be carbon or cermet. Carbon is unacceptable for most applications. Cermet pots are made with a metal film deposited on a ceramic base. Figure 14.5*b* shows the internal resistive element and the wiper arm that rotates while making contact with the element. Figure 14.5*c* shows the schematic symbol.

These pots are available in just a few values from 500 Ω to 1 MΩ. Typical values are 1 k, 2.5 k, 5 k, 10 k, 50 k, 100 k and 1 MΩ. Additional sizes may be available such as 2 k, 25 k, 250 k, and 500 k. Common tolerances are 20 and 5 percent. Power ratings run from a few milliwatts to standard ¼- and ½-W devices. All have three connection terminals. On circular metal pots, the terminals are designed for soldering wires. Connections on the smaller trimmers are made for through-hole PCB mounting. The pin spacing on these trimmers may allow them to plug into a breadboard socket.

Some specialty pots are wirewound and multiturn. Wirewound pots have wire resistances made for high-power applications. Another is

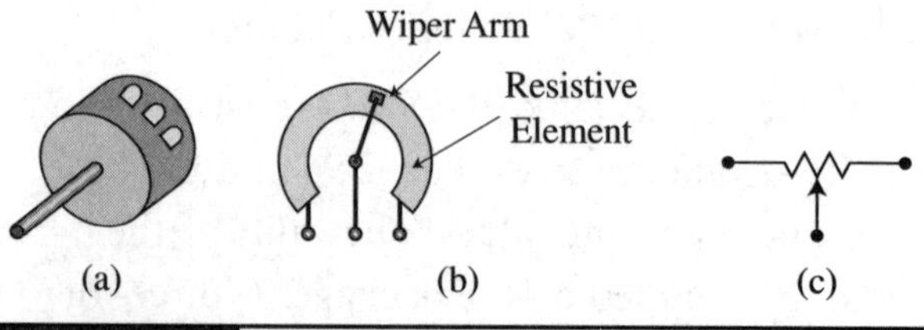

Figure 14.5 (a) A popular pot type, (b) resistive element and wiper, and (c) schematic symbol.

the multiturn pot. Most pots only have about a 120° rotation limit. Multiturn pots may have as many as 10 turns for very precise adjustments.

Another pot specification is its taper. Taper refers to how the resistance value varies as the shaft rotates and the slider passes over the resistive element. The most common taper is linear since the resistance value changes in a straight line as the arm is rotated on the resistive element. The other common taper is the audio taper. The resistance element is made so that its variation over the rotation range of the pot varies logarithmically. These pots are used as volume controls because the loudness of audio varies in a logarithmic way to match the response of the human ear.

Capacitors

Like resistors, there are dozens if not hundreds of types and sizes of capacitors. Remember, a capacitor is essentially two metallic plates separated by an insulating material called the dielectric. It is the dielectric material that determines the capacitor type. Capacitor values range from a few picofarads (pF) to about 30,000 microfarads (μF). Figure 14.6*a* shows how common values are marked on the capacitor.

Packaging

Sizes and configurations are wide ranging. See Fig. 14.6. Most are small rectangular devices with wire leads. Others are flat circular disks. Some are larger cylinders with axial as well as radial leads. Capacitors are also available in standard SMD packages.

Capacitor Types

The most common types use a metal film deposited upon the dielectric. The metallic film forms the plates of the capacitor. The dielectrics vary widely, including polyester, polycarbonate, polypropylene, and polystyrene. Mica capacitors are available for critical temperature stability applications. To achieve a large amount of capacitance in a small package, the internal construction is usually multiple plates separated by the dielectric and stacked.

Figure 14.6 shows some popular capacitor types.

Poly-type capacitors use a mylar dielectric. See Fig. 14.6*b*. Values range from 0.001 to 10 μF. Voltage ratings are up to the 100- to 600-volt range.

A widely used favorite is the ceramic disk capacitor. See Fig. 14.6*c*. It is made with metal deposited upon the ceramic dielectric. Wire leads are common. Values range from 1 pF to 1 μF. Voltage ratings range from 50 to 1000 V.

Another category is the electrolytic capacitor. See Fig. 14.6*d*. This capacitor is constructed by depositing a really thin layer of dielectric by chemical means on thin aluminum plates. This permits very high capacitor values, roughly

Value	Code
10 pF	= 100
100 pF	= 101
1000 pF	= 102
0.001 uF	= 102*
0.01 uF	= 103
0.1 uF	= 104

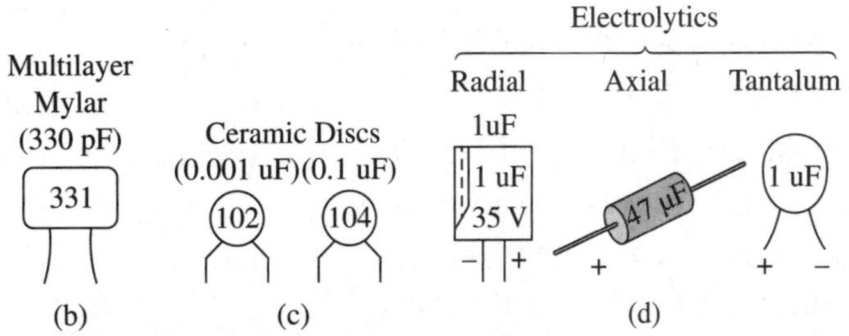

Figure 14.6 (a) Capacitor marking codes determine its value. The packages are (b) rectangular, (c) disc, and (d) electrolytics with radial and axial leads, and a tantalum.

from 1 to 30,000 μF. Voltage ratings are 16, 25, 35, 50, and up to several hundred volts. The dielectric is typically an aluminum compound or tantalum. These capacitors are also polarized. This means they can only be connected in the circuit one way without damage. These capacitors are used mainly in power supplies and audio equipment. One of the leads is usually marked with a minus sign to ensure that the capacitor is connected correctly with the desired polarity.

A special category of capacitor is the supercapacitor or ultracapacitor. It is available with values from 1 to hundreds of farads—yes, farads. The voltage rating is very low, only about 2.7 or 5.4 V. They can hold a charge for a very long time and can actually replace some types of batteries in low power consumption devices.

Another capacitor specification is ESR or equivalent series resistance. While capacitors aren't supposed to have resistance, in reality they do. It is a small amount of resistance and made up of the resistance of the plate material and the connecting leads. ESR is a critical factor in capacitors used in power supplies and other high-power circuits. With ESR and current flowing during charge or discharge, power is dissipated so the capacitor may heat up, changing its value.

Capacitance Value Coding

Refer again to Fig. 14.6. Electrolytic capacitors have their capacitance and voltage rating printed right on them. Most also show a minus sign indicating which lead is negative. Some tantalum capacitors mark the positive lead with +.

Ceramic and most film capacitors are marked with a three-digit code. Figure 14.6*a* shows the codes for the most popular sizes from 10 pF to 0.1 μF. Other three-digit codes use the same format. For example, a 332 marking means 33 plus two zeros or 3300 pF.

Just So You Know

If you do a great deal of design and experimenting, you should think about investing in resistor and capacitor kits. This is a collection of all the major standard sizes of components, usually packaged in a plastic housing with separators or drawers. Nothing is more frustrating than when prototyping a circuit you do not have the needed resistor sizes or capacitor values. With these kits, you have most common parts right at hand. You will always have to special order some special part, but that's the way it goes. These kits are not too expensive and are a good investment. Kits of diodes, transistors, and integrated circuits are also available. Check with the vendors for what is available.

Inductors

Inductors are also called coils or chokes. They are essentially wirewound devices with many form factors. Small commercial inductors look like axial resistors. Values of inductance range from a few nanohenries (nH) to 1 microhenry (μH) and then to 100,000 μH (100 mH).

An important specification is the Q. Q is the quality factor that is calculated with the expression:

$$Q = X_L/R = 2\pi fL/R$$

This is the inductive reactance divided by the resistance. The Q indicates the energy storage capability compared to the energy dissipation. Q is a major factor in filters and tuned circuits.

Most small value inductors look like a ¼-W resistor. Check with the manufacturer for coding of values. I could not confirm any standard. A popular inductor form is the toroid. See Fig. 14.7. Values up to several hundred millihenries are typical. Many toroid inductors are made for the application by buying an

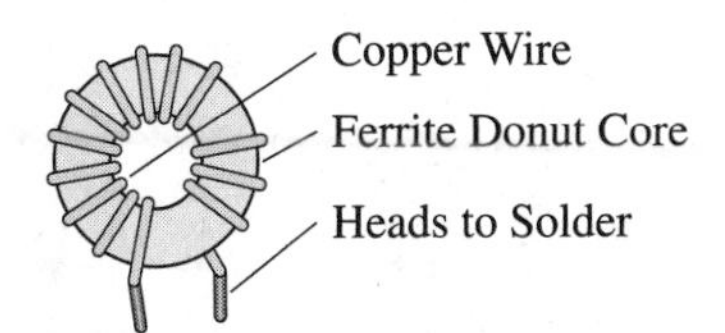

Figure 14.7 A popular inductor package: the toroid. The magnetic core greatly multiplies the inductance value even with a small number of turns of wire.

appropriate size donut-shaped magnetic core and wrapping the correct number of turns of wire to get the inductance you need.

Semiconductor Selection

When designing transistor or IC-based circuits, the new engineer always seems to ask, "What component do I use?" There are so many to choose from. It is a real problem. Here are some guidelines to help you zero in on a part for your design.

1. If your company uses lots of transistors and ICs, they are probably in stock and available for any prototyping. Furthermore, the company buys them in bulk, so they are cheap. These devices are also probably found to be reliable. Choose one of these parts unless you really do need to use something special.
2. Search for a device online. A good starting point is to go to known transistor or IC vendors' Web sites and use their selection tools. This software helps you pick out a device that will meet your needs. Alternately, you can ask one of the companies' application engineers to help you choose a device.
3. Follow the industry trends. Use parts that seem to show up in the designs of others. Read magazines, search for circuits online, and find out what is being used. Many parts are used again and again just because they work.
4. Ask experienced engineers what they use or recommend.
5. Use the lists of parts given here as a guide. Most of these parts have been around for years and have proven inexpensive and reliable. They are still available and cheap.
6. Don't forget to get the data sheet for the part you choose to be sure that the needs of your circuit and application match the specifications of the part (voltage, current, speed, frequency, packaging, etc.).

Diodes

As with other devices, there are dozens to choose from. Table 14.2 shows some popular devices. Remember, diodes are polarized. They have an anode and a cathode. On the axial diodes, the cathode is marked by a band at one end. The cathode on an LED is the shorter lead. Some

Table 14.2 Popular Diodes for Design

Part Number	Type	Application
1N4001–1N4007	Silicon rectifiers	Power supplies, power loads.
1N914, 1N4148	Switching diodes	Digital designs
1N5817,1N5818, 1N5822	Schottky diodes	High-speed switching circuits, SMPS
1N4728–1N4745	Zener diodes	Reference, power supplies
LEDs	Light-emitting diodes	Illumination
1N34, 1N60	Germanium diodes	Low signal detection

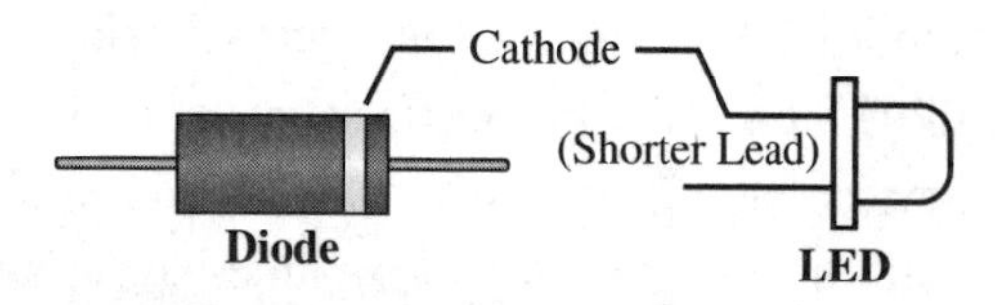

Figure 14.8 Standard diode packages, axial and LED. Note the critical cathode marking that dictates the polarity of the voltage that must be observed.

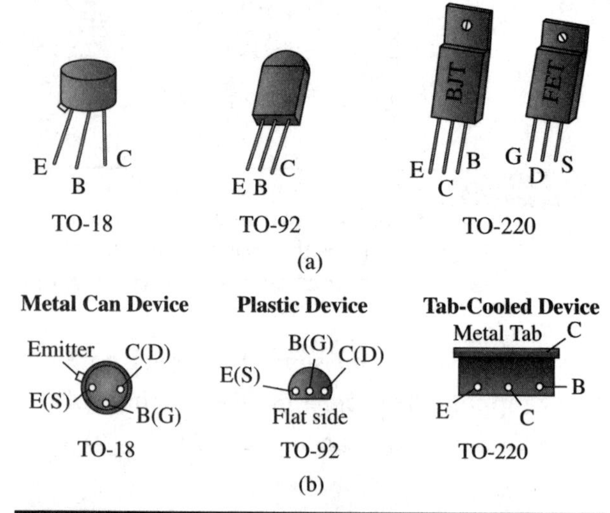

Figure 14.9 The most common transistor packages and lead identification. (a) The packages for small signal transistors are TO-92 and the TO-18. The TO-220 package is typically for power transistors and some integrated circuits. (b) This is a bottom view showing the transistor lead connections. E-B-C refers to the emitter, base, and collector for bipolar junction transistors (BJT) and S-G-D in parentheses refers to the source, gate, and drain of a junction field-effect transistor (JFET). Pin connections on the TO-220 vary, so check the data sheet.

LEDs also have a flattened place on the side body designating the cathode. See Fig. 14.8.

There are literally thousands of diode types. It is tough to select one. Table 14.2 lists some of the most popular and widely used diodes. It is always good to stick with high-volume parts. Costs are low, and these devices have been around for years and have proven themselves.

Transistors

Again, there are thousands to choose from. The ones in Table 14.3 are widely used. Figure 14.9 shows the standard packages and lead identifications.

Table 14.3 Popular Transistors for Design

Part Number	Type	Application
2N2222	NPN, switching, power	Switching applications, power amplification
2N2907	PNP, complement to 2N2222	Switching and power
2N3053	NPN	Power, RF. TO-39 package
2N3055	NPN	High power, TO-3 package
2N3819	JFET	General purpose
2N3904	NPN, general purpose	Switching or amplification
2N3906	PNP, complement to 2N3904	Switching or amplification
2N4401	NPN, general purpose	Switching or amplification
2N5457	JFET	General
MPF102	JFET	RF, high frequency
IR510, IR530	N-type MOSFET	Switching, audio, low RF
2N7000	MOSFET, TO-92 package	Switching

Table 14.4 Popular Digital ICs for Design

Part Number	Type	Application
7400, 74LS series	TTL	General and high speed
74HC series	CMOS	General, CMOS versions of 7400 devices
CD4000 series	CMOS	General

Integrated Circuits

Tables 14.4 and 14.5 summarize the most popular digital and linear ICs.

When doing digital design, you need access to the data on many different logic ICs. The semiconductor companies used to provide massive data books with details. These are no longer generally available. You can actually find some old used ones online for sale. Three other sources I found for TTL follow:

http://www.ti.com/lit/sl/scyd013b/scyd013b.pdf *Digital Logic Pocket Data Book*

http://ecee.colorado.edu/~mcclurel/ON_Semiconductor_LSTTL_Data_DL121-D.pdf

https://en.wikipedia.org/wiki/List_of_7400-series_integrated_circuits

For the 4000 series CMOS chips, check out the following:

https://www.cmos4000.com/

https://en.wikipedia.org/wiki/List_of_4000-series_integrated_circuits

Table 14.5 Popular Linear ICs for Design

Part Number	Type	Application
LM324	Quad op amps	Designed for single power supply, general purpose
741	Op amp	Low frequency, general, probably the most widely used op amp in history
TI07x (x = number of op amps in the package, 1, 2, 4)	Op amp	JFET input, general purpose
TI08x (x = number of op amps in the package, 1, 2, 4)	Op amp	JFET input, general purpose
LF411	Op amp	Oldie but goodie. General purpose, JFET input
LM1458	Dual op amp	General purpose
LM386	Power op amp	Audio amp for speaker or headphones
78xx	05, 09, 12 V linear regulator	Power supply regulators
LM317 LM337	Linear regulator Negative version of LM317	1.2–37 V, 1.5 A, adjustable voltage.
LM2576	Step-down switching regulator	Easiest to use switch mode regulator
LM311, LM393	Comparator	Linear input, digital output
LM339	Quad comparator	Linear input, digital output
LM555	Timer IC	Bipolar and CMOS versions
LM34Z/35Z	Temperature sensor	Celsius or Fahrenheit

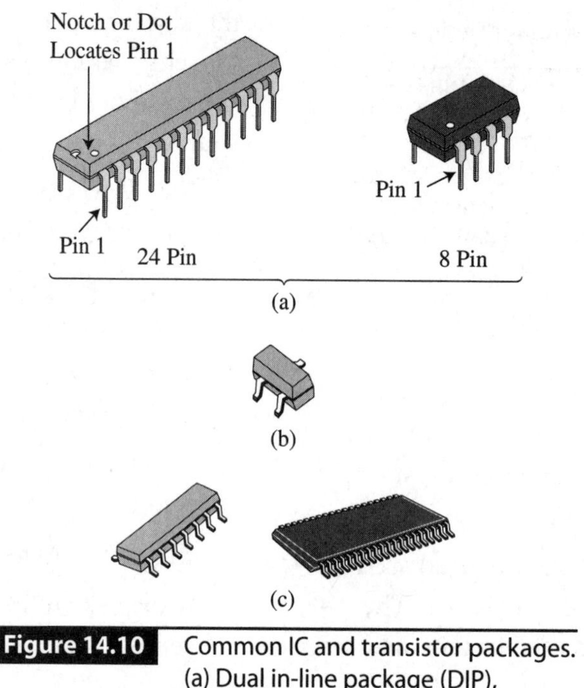

Figure 14.10 Common IC and transistor packages. (a) Dual in-line package (DIP), (b) transistor SO-23 surface mount, (c) small outline surface-mount ICs.

Packaging of ICs varies widely. The oldest and one that is still widely used is the dual in-line package (DIP) available in 8-, 14-, 16-, 20-, 24-, 28-, and 40-pin sizes. See Fig. 14.10*a*. It is best for breadboarding. This package is often identified by the letter N at the end of the part number. Multiple surface-mount packages are also available, including the popular transistor package SOT-23. See Fig 14.10*b*. Common small outline packages for ICs are shown in Fig. 14.10*c*.

CHAPTER 15

Troubleshooting and Debugging

What could possibly go wrong? Once you begin designing, you will soon discover that you actually spend at least as much time troubleshooting as you do designing. Much of it will come during the prototyping stage where you first build the real circuit and test it. Chances are the prototype won't work. If your design seems good, you validated it in a simulation, and you were careful building the prototype. It may work. If not, don't get discouraged. It happens to all engineers and designers. That's why you need to learn troubleshooting. Troubleshooting is the process of finding out why some hardware does not perform as you designed it.

The same is true of writing a program. Your first pass at the code will probably not work right. So you need to learn to debug your code. Debugging is the software equivalent of hardware troubleshooting. Most integrated design environments (IDEs) contain some debugging software to assist you.

Troubleshooting and debugging are best learned by doing. This chapter will give you some guidelines and procedures to get you started.

Mitigating the Need for Hardware Troubleshooting

There are things you can do to avoid troubleshooting or to make it less necessary. The main thing is knowing how your circuit or device works. If you do not know how it works, how do you know it is performing as you wish? You need to know what results to expect when you go to test the circuit or product. Use data sheets, books, or other references to determine what results to expect.

A major part of this process is knowing how the components work. Since you are probably designing with one or more ICs, make sure you understand how those devices work. If you have not already done so, get the data sheets for each device and review its use. Pay special attention to the signals and voltage into and out of the IC to be sure you are using them correctly. ICs have all sorts of special needs, such as grounding certain pins, keeping some pins open, adding pull-up resistors, connecting external parts, and other guidelines.

Another way to minimize the need to troubleshoot your circuit or device is to simulate it first. Use one of the electronic simulator programs to build your circuit in software, and try it out. You may encounter troubles doing that. Here you can solve them in software before you commit to a full hardware prototype. If you can make it work correctly in the simulator, it will most likely work in real-life hardware.

The same advice applies to software. Run your program on a simulator or emulator before you test it out on a real hardware micro. Find an IDE with an emulator for your micro. Your PC will act like your microcontroller and tell you if the program did as you wished.

Test Equipment

The importance of having the right test instruments was explained in Chap. 4. I cannot overemphasize it here again. As it turns out, you really cannot do troubleshooting without a few good instruments. These are a digital multimeter (DMM), an oscilloscope, and a function generator. The DMM will mainly be used for measuring dc voltages. Occasionally, you will use it to check ac voltage or measure a resistor value. If you have one of the latest full-function DMMs, you may also have capacitor measurement capability and transistor/diode test functions. I balked at the extra cost of a capacitor test feature in a DMM, but once I got it, I must confess that I really use it. It proved worth the extra cost.

Your real troubleshooting tool is an oscilloscope. The scope is your eyes into the circuit. It will trace signals, measure their characteristics, and perform other functions related to your specific circuit. It is hard to design without a scope but even harder to troubleshoot without one. If you have not already done so, bite the bullet and invest in a good scope. Finally, you should have a good function generator. This will give you test signals to stimulate your circuit. While individual test instruments are the most desirable, virtual instruments are just as effective.

Prototype Troubleshooting

As I have mentioned several times, you must actually build a working prototype of your circuit or product. A good starting point is to use a breadboarding socket and then graduate to perf board or a PC board. If your initial prototype does not work, it could just be a faulty design. You got it wrong. It happens. I know I am repeating here, but it is important. Try simulating your circuit before you go on, and correct the design there before building the prototype.

There are some mechanical and electrical reasons why it may not work.

Mechanical Problems

The following is a checklist of conditions to look for in troubleshooting a hardware prototype:

- Incorrect wiring. Component lead, pin, or connecting wire in the wrong hole in the breadboarding socket, or wire or pin not fully inserted. When using breadboarding sockets, this is the main problem. Check your wiring again.
- Broken wire, lead, or pin. One occasional problem is an IC pin that has been bent and does not fit into the socket, or the pins in a connector are broken, missing, or dirty.
- Component mounted incorrectly. The most common problem is a polarized component connected backward. Such components are diodes, LEDs, electrolytic capacitors, and most ICs. Transistor leads are another problem. Check the data sheet to be sure of the connections.
- Component shorts. Bare wires or leads touching. Some complex circuits have messy wiring on a breadboarding socket, and components can get in the way of one another. Inspect the wiring to see if you can see bare wires or leads touching one another where they shouldn't.
- Problem with connectors and related wiring. Poor contacts, broken, dirty, etc.
- Bad solder joint on a perf board or other wiring surface. Resolder the suspected joint, or just reheat it if you suspect that it is a cold solder joint.

Electrical Problems

The following are the electrical problems to look for:

- No or wrong dc power supply voltages. First, set the desired dc power supply voltage before you connect it to your circuit. Connect it to your prototype. This voltage will come from a bench power supply, a virtual instrument, or batteries. If the supply has an adjustable output voltage, be sure that it is not set too high because it could destroy an IC or other device.
- No ac power line voltage. If you have no dc, obviously the circuit will not work. Check the power supply ac connection. Is it plugged in? Did you blow a fuse or trip a breaker? It sounds stupid to even mention this, but many have found that something simple and obvious is often the problem.
- No input signals present. Examples are clock signal for digital and microdevices or test signals from a function generator. Use the scope to verify that all inputs are there.
- Look for the outputs. You should know what to look for from the design. Use the scope and DMM to validate their presence and correct amplitude and shape.
- Incorrect digital circuit logic levels. The data sheets on the ICs you are using specify the supply voltage. This could be a problem if the power supply voltage is set incorrectly. Mixing TTL and CMOS ICs sometimes results in incompatible logic levels.
- Defective IC, transistor, diode, capacitor, resistor, etc. If the circuit does not work, it could be a bad component. Trying to identify that bad part is the difficulty. If you suspect a bad component, replace it with a new one or an equivalent that is known to be good. Component substitution is the best way to determine what is working or not. Always buy multiples of each IC, transistor, and diode just in case one is bad or if you destroy one.
- Wiring error. Chances are this is the main problem on a prototype.
- Excessive wiring length producing excessive stray capacitance or inductance causing high-frequency (> about 5 MHz) circuits not to function or function incorrectly. When working with high-frequency circuits, long leads add stray capacitance and inductance that can change the circuit behavior. Even the small capacitance between the connectors in a breadboarding socket can negatively affect some high-frequency signals. Shortening all leads usually helps.
- Incorrect value of resistor, capacitor, zener diode, etc. (incorrect reading of a resistor color code). Occasional errors are reading a capacitor marking incorrectly and misinterpreting a resistor color code. Resistor color codes are not perfect, and some band colors look different under different lighting conditions. Reds can look like orange or vice versa as an example. Measure the resistor with your ohmmeter if in doubt.
- Polarized components backward. This is a serious no-no. Be extra careful in connecting electrolytic capacitors, diodes, and LEDs. Serious damage is the usual effect. A tantalum electrolytic will go off like a gunshot if you put it in backward.
- Miswiring of IC pins. Wiring to the wrong pin of an IC is a common error. Check your wiring. Be sure you have identified pin 1 on an IC. It is to the left of the notch or dot at one end of the chip package. Check the data sheet if in doubt.

The solution to many, if not most, of these problems is to check and double-check your wiring. Incorrect wiring of components or interconnecting wires is probably the problem in most circuits built on a breadboarding socket.

Troubleshooting Procedures

Here is a list of standard procedures to follow when troubleshooting a circuit or device. If you are satisfied that your circuit is wired correctly, begin the following tests.

1. Verify the problem. Are you sure it's a problem? How do you know? Is the circuit or product not working at all, or is the operation incorrect? Do you really know what to expect for verifying proper operation?
2. Note the symptoms. Some examples: completely dead circuit, distorted output, input but no output, smoke, overly hot components.
3. Did the circuit ever work? If so, what have you changed? That change may have created a problem. Think things through.
4. Check for power. Measure the dc supply voltage to verify that it is correct. If it is not there, troubleshoot back to the power supply.
5. Conduct a visual inspection. Check your wiring. Are any components running hot? Are there any loose components or connectors?
6. Verify that all your inputs and outputs are present. Inputs include signals from a generator, switches, sensors, etc. Outputs operate displays, LEDs, relays, motors, etc. Use the DMM or scope as appropriate.
7. If your circuit has multiple stages, trace a signal through them stage by stage. See Fig. 15.1. Check for an input and then the output of each stage. The bad stage is the one with no or incorrect output. Separate the various stages. Isolate them so you can test them individually.
8. Try to isolate the problem to a specific thing. Replace suspicious components. Substitute good new components for those that may be bad. Retest.
9. Make repairs. Retest.

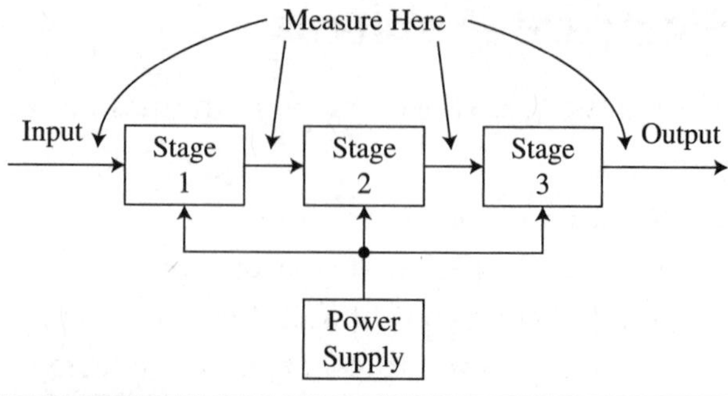

Figure 15.1 A block diagram illustrating the signal tracing process of isolating a bad stage or circuit.

As you are troubleshooting, it is a good idea to keep any relevant data sheets on hand. Be sure you know how all the parts work so you can determine if they don't work at all or work incorrectly. Sometimes there is no problem. It may be that you are the problem—pilot error. Not recognizing a problem or proper behavior leads to the conclusion that the circuit is bad. It may not be so. Learn what to expect.

Component Failure Likelihood

Some parts are more likely to fail than others. Based on years of experience, it has been determined that most problems can be traced to components that are just not as reliable as others. The following list indicates the order of likelihood of component failure:

1. Fuses and breakers
2. Switches and relays
3. Wire, cables, and connectors
4. Diodes and transistors
5. Capacitors
6. Integrated circuits
7. Resistors
8. Inductors and transformers
9. Printed circuit boards

Note that the mechanical devices (switches, connectors) are more prone to failure than the electronic parts. Most ICs and transistors are amazingly reliable unless abused or used in a bad design. A good first move in troubleshooting is to do a visual inspection of the circuit and check all the wiring, cables, or connectors or other mechanical parts first.

A Troubleshooting Example

One type of failure that occurs more than others is a problem in a power supply. Power supplies are subject to heavy currents, high voltage, and other stresses like elevated component heating. It is handy to know how to troubleshoot a power supply. Here is one approach that uses signal tracing to locate a fault.

Refer to Fig. 15.2. This shows a common dc power supply. It does not seem to be working. What is the procedure to find the problem? Using the steps outlined earlier, here is how it is done.

1. Verify the problem. The power supply is supposed to put out 3.3 V dc. It is not present. Measurement with a DMM confirms that the dc is not present.
2. Note the symptoms. No output voltage.
3. Check for power. This starts with verifying that the ac from the power line is present. Is the supply plugged in? Did a fuse blow? Is the off-on switch broken? Is there an open in one of the ac noise filter lines? Use your DMM to measure voltages or continuity as needed. Given what you find, you may have solved the problem. If not, move on.
4. Conduct a visual inspection. If the power supply worked before, it is not a wiring error. Are any components running hot? Does anything look burned?
5. If your circuit has multiple stages, trace a signal through them stage by stage. Refer back to Fig. 15.1. Now refer to Fig. 15.2. The output comes from the IC voltage regulator. If it is not there, check for an input to the IC. If you have an input to the regulator but no output, the problem is most likely a bad regulator IC. Replace the IC and retest.
6. If there is no input voltage to the IC, the problem may be a shorted filter capacitor. This would short the rectifier output to ground. That would probably blow the fuse and/or destroy one of the diodes. Remove the capacitor and measure it with an ohmmeter to see if it is shorted. A very low resistance near zero signals a short. Replace the capacitor. Retest and continue. If there is no voltage on the capacitor, maybe the rectifier is bad. Do any of the rectifier diodes look burned or broken? Another possibility is an open filter capacitor. In that case, you will measure pulsating dc at the input to the regulator. Take a look with an oscilloscope to be sure. Normally, you should see a near straight line (the dc) with only millivolts or so of ripple riding on the dc.
7. Next, look for ac voltage across the transformer secondary winding. If it is not there, the transformer may be bad, having an open winding. Next, look for ac across

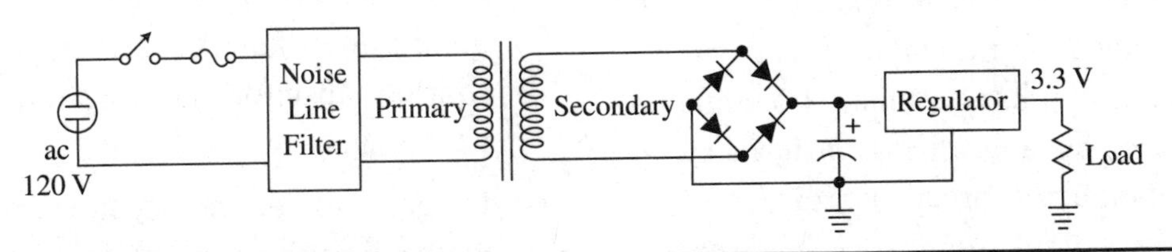

Figure 15.2 A standard line operated ac-dc power supply with an IC regulator.

the transformer primary winding. If it is not there, check the fuse, off-on switch, and input filter. Be careful when touching parts connected to the ac line. It is so easy to get a bad shock. Unplug the power supply and check all the parts on the primary winding side of the transformer with continuity measurements.

8. Make repairs. Retest.

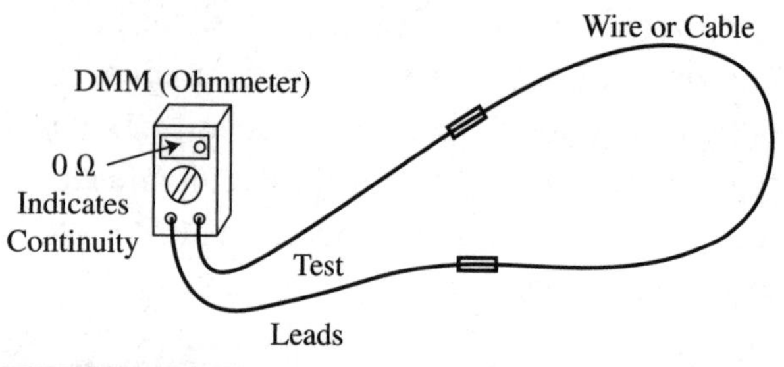

Figure 15.3 How to test for continuity with a DMM.

Just So You Know

Here is the effect when components fail.

- Resistors. Broken, open circuit, change in value, overheated and burned up. Resistors do not short.
- Capacitors. Broken, open circuit, change in value, short, incorrect polarity.
- Inductors. Open, wrong value, no shorts.
- Transformers. Broken leads, open winding, internal short between windings.
- Diodes. Open or shorted, broken leads, incorrect polarity.
- Transistors. Broken, overheating, opens or shorts between elements.
- Integrated circuits. Multiple ways of failure. Since ICs are made up of the kind of components described above, it could be any of them. Physical problems with pins broken or missing or not soldered, IC installed backward.
- Switches. Broken mechanically, dirty or burned contacts.
- Relays. Broken mechanically. Dirty or burned contacts, open coil.

Continuity Testing

The term *continuity* means the presence of an electrical connection, usually very low resistance between two points. For example, if a piece of wire is not broken, it has continuity from one end to the other. See Fig. 15.3. Transformer windings and inductors are made from long lengths of wire wound around a magnetic core. You usually cannot see them. If a winding has continuity, there is a solid path from one lead to another.

You measure continuity with an ohmmeter, that part of the DMM that measures resistance. Use the lowest scale (Rx1), because continuity generally means very low resistance. Some tests like those on a transformer, inductor, or relay coil will show some continuity and a low resistance value. If you are checking a switch, you will read 0 Ω when it is on and infinite ohms when it is off. Checking the wire in a cable with connectors on both ends, you will read near 0 Ω if the wire is good and infinity if the wire or connector is broken.

A continuity test also works on some types of components. Here is how to test each.

- Resistors. If you get a normal resistance reading corresponding to the color code or body marking, the resistor is good. If you get an infinite reading, the resistor is open.
- Capacitors. Use the resistance measurement mode of your DMM. A good capacitor should show an infinite resistance. Large capacitors like electrolytics will show an initial reading as it charges up to the internal DMM battery voltage, but it should read infinite after that. A common problem is a shorted capacitor that will show a reading of near 0 Ω.
- Diodes. A diode should show continuity in one direction and an open in the other direction. The leads of your ohmmeter are

polarized positive (+) red lead and negative (−) black lead. To test a diode, connect the black lead to the diode cathode (the end of the diode with the band) and the red to the other end. The diode is usually good if you get a low ohms reading. An infinite reading means the diode is open. Now reverse the test leads and measure again. A good diode should show an infinite reading. A low ohms reading means that it is shorted.

- Transistors. Bipolar junction transistors (BJTs) are actually the equivalent of two diodes connected back to back. One diode is between the emitter and base and the other between the base and collector. You can probably test the transistor with a DMM, but I have found the tests so variable as to be unreliable. However, many DMMs have a built-in transistor tester. Read the manual on your DMM for instructions. Better still, if you suspect that a transistor is bad, just replace it. Transistors are cheap, so keep a few extras on hand of the types you are using.

General Troubleshooting Suggestions

The following is a mixed list of items to consider when troubleshooting:

- Do you know how the circuit works? Figure out how each part of the circuit works and determine what the circuit can do. What is the expected output or performance?
- Do you have schematics and block diagrams of your circuit or device? How can you troubleshoot without this information? If you are troubleshooting your own circuit or device, you should have recorded block diagrams with inputs and outputs defined and complete schematics. If not, you are working blind; get that lack of detail taken care of so you will know how the signals flow in the circuit.
- What about other relevant documentation? Do you have data sheets for the components, relevant app notes, manuals, and other publications related to your design? This material may provide clues to the problem.
- Make a wild guess at what is wrong. Given your familiarity with the equipment, circuit, and design, use your gut to estimate the problem. Take a shot at this, because your own feelings enter into the search and repair.
- Did the equipment or circuit ever work? If it did, something went wrong. Ask yourself what is different. Look for clues as to what changed. Follow that up.
- Look for simple and obvious things first. Is the unit getting ac or dc power? Are wires loose? Did your IC melt?
- Your oscilloscope will be your best tool for troubleshooting because you can look in a circuit and see what is going on. Your DMM is also a key tool for tracking down dc supply voltages and related power issues. Examine all inputs and outputs with the scope. In a digital circuit, start with the clock signal that runs everything else.
- Try signal tracing by applying inputs to your circuit and tracing it through the various stages as defined by your block diagrams and schematics. By analyzing your circuit, you should know what to expect from each input and output. Examine them with your oscilloscope.
- One helpful process when all else seems to fail is to draw out the expected signals and waveforms. This is helpful especially in a digital circuit that includes counters, shift registers, and other logic. You must know how all your ICs and other devices work so that you get this right.
- A well-known telltale in troubleshooting is to look at the output of each transistor or IC. Measure its dc voltage. If that output voltage

is the same as the supply voltage, then that transistor or IC is not conducting. Maybe it is bad, not biased correctly, or not getting the correct input. If the output is zero, that is also usually a bad sign.

- One common problem in digital circuits is noise on the main dc power bus. Use your scope to look at it. It should be a smooth, straight line. If you see noise, random pulses, or ripple, then the regulator may not be working, or there is insufficient bypassing or filtering. A quick test is to temporarily connect a 10 μF electrolytic across the dc power bus. Also, it is generally recognized that you need to add a 0.1 μF disc capacitor across the power pins of each digital IC, including the MCU. Keep the leads short. High-speed digital circuits generate lots of noise and transients that get transmitted to other circuits by way of the power bus. Filter all this stuff out with heavy bypassing with capacitors on the bus. That will eliminate any false triggering of other circuits.
- Is there any way to take your circuit or product apart and divide it up into its main sections and circuits? If so, do that if all else fails. This "divide and conquer" approach lets you isolate individual circuits and components and test them faster and easier. Once you identify the problem, you can put it all back together again.
- Never assume anything. When searching for the solution, you undoubtedly assume certain things to be true. They may not be.
- Try everything. Forget the scientific approach and dive into the problem. Replace suspected parts, rewire the circuit, disconnect everything and start over. Work fast. Try everything to see what works. Strangely, this disorganized process works more times than not. Don't try to figure it out. Guess and work it. This is often the fastest way.

Software Debugging

It is rare for a program to work correctly the first time you try it. Programming mistakes are easy to make. The problem may be with the syntax and procedures of the language you are using. Your IDE (integrated development environment) and language software should identify these for you. If your procedures check out, your program logic or algorithm may be wrong.

A good next step is to review your program. One approach is to compare your steps or sequences of commands with your initial flowchart. Next, if your program was designed to be modular (and it should have been), isolate each module and execute each to verify that they run OK. Fix any errors. Once your individual modules are all operating, put them back together and try again. Your interconnection of the modules may be at fault.

If your IDE/language software allows it, try single stepping through each line of the program and noting the result. Your software may also have some special debugging processes that will help. Finally, if you know someone who programs, ask them to take a look or join you in debugging.

Troubleshooting Practice

Here are a couple of circuits to troubleshoot. One uses op amps and the other digital circuits.

A Linear Circuit

Figure 15.4 is a circuit using multiple op amps. It has been said that it is not working right. What is wrong? The output voltage is incorrect. It should be +3 V. This project uses all the basic forms of op amp circuits, so you should be able to determine the problem based on the inputs provided and the circuits used. The power supplies for the op amps are +9 V and –9 V. These supply voltages are present on each op amp.

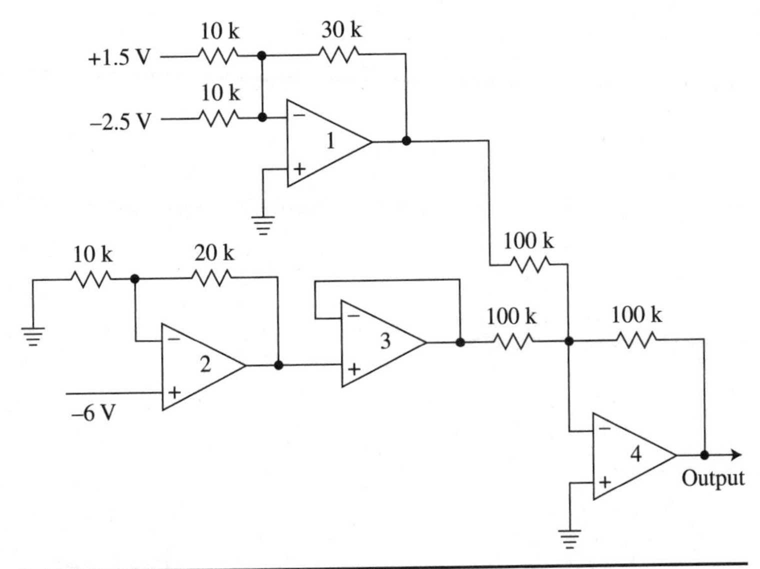

Figure 15.4 A linear op amp circuit for troubleshooting.

A Digital Circuit

Figure 15.5 is a digital circuit made up of 74LS series of ICs. It is a binary up counter made of 74LS112 JK flip flops (FF). They are negative edge triggered or toggled. The counter operates a 74LS138 3-line to 8-line decoder. The decoder outputs drive 74LS07 open collector inverters that in turn operate LEDs. Only one inverter driver circuit is shown to simplify the schematic. As the counter counts, the LEDs should come on one at a time in sequence. However, the circuit does not work correctly. The LEDs appear to light up in some random order. Why?

See the solutions in App. B.

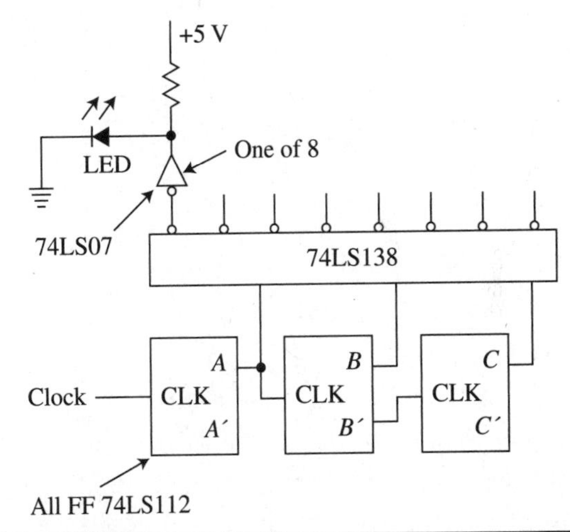

Figure 15.5 A digital circuit for troubleshooting.

APPENDIX A

Recommended Reference Books

1. Ashby, Darren, Bakeer, Bonnie, Hickman, Ian, Kester, Walt, Pease, Robert, Williams, Tim, Zeidman, Bob, *Circuit Design: Know It All Series*, Newnes/Elsevier, Oxford,UK, 2008. This book is another great reference with broad coverage. Multiple authors give you a broad view of electronic design.
2. American Radio Relay League (ARRL), Newington, CT, *Handbook for Radio Communications*, 2019 or latest edition. The handbook has been published for decades. It is a great electronic fundamentals source, and it is up to date because they revise it every year. Of course, its focus is ham radio and hands-on projects. It is a good wireless source.
3. Frenzel, Louis, *Electronics Explained,* 2d edition, Newnes/Elsevier, Oxford, UK, 2016. Yes, I am recommending my own book. It is up to date and a short read for anyone learning electronics for the first time.
4. Frenzel, Louis, *Contemporary Electronics: Fundamentals, Devices, Circuits and Systems*, McGraw-Hill, New York, NY, 2017. A college text covering electronic fundamentals. It was probably the first text to focus on the systems approach to electronic fundamentals and is full-color and hardcover.
5. Frenzel, Louis, *Principles of Electronic Communication Systems,* 4th edition, McGraw-Hill, New York, NYm, 2016. If you are looking for comprehensive coverage of wireless communications, this is a good introduction. It is a college text, full-color, and hardcover. I am currently working on the fifth edition update.
6. Horowitz, P., and Hill, W., *The Art of Electronics,* 3d edition, Cambridge University Press, New York, NY, 2015. This is probably one of the best electronic reference books available. It is very detailed and at the engineering level but readable. Everyone who designs electronic circuits and equipment should have a copy.
7. Maxfield, C., *Bebop to the Boolean Boogie,* 3d edition, Newnes/Elsevier, 2009. A great introduction to digital electronics, Maxfield provides a thorough and interesting coverage of digital fundamentals. He has written several other books just as good. His book on FPGAs is worth a look if that is a subject of interest.
8. Platt, Charles, *Make: Electronics and Make: More Electronics*, O'Reilly Media, 2014. Both of these books are for the beginning experimenter and DIY makers. This is a full-color paperbacks with lots of details. I don't like the way he draws schematics, but otherwise, these are excellent introductory books for experimenters.
9. Stadtmiller, D. J., *Applied Electronic Design*, Prentice Hall, 2003. An oldie but still valid book on basic design, this is a college text that will take you to the next level of design. I found a copy online because it may be out of print.

10. Scherz, Paul, and Monk, Simon, *Practical Electronics for Inventors,* 4th edition, McGraw-Hill,New York, NY, 2016. This is an excellent book and a massive collection of electronic fundamentals. It is also a good reference that you should have if you are an active experimenter. It is good complement to this book.

11. Vahid, Frank, *Digital Design with RTL Design, VHDL, and Verilog*, 2nd edition, John Wiley & Sons, Inc. Hoboken, NJ, 2011. A favorite book of mine that does a good job of summing up modern digital design. More advanced coverage in easy to assimilate form.

12. Wilson, P., *The Circuit Designer's Companion,* 4th edition, Newnes/Elsevier, Oxford, UK, 2017. It is a good reference for the practicing designer. It covers topics on design not covered elsewhere and is a go-to book when no other source has the answers.

If you are going to be a designer, build a library of these books that will save you time and frustration. Some are even good reads.

APPENDIX B

Solutions to Design Projects

The solutions presented here are just one way to do it. Your design may be different but still work. There are usually multiple ways to design from clever and elegant to traditional and pedestrian. The solutions presented here use the design procedures outlined in the various chapters.

Chapter 5

5.1

$$VS = 9 \text{ V}, V_{\text{LED}} = 2 \text{ V}$$

$$VR = VS - V_{\text{LED}} = 9 - 2 = 7 \text{ V}$$

$$R = VR/I = 7/15 \text{ mA} = 466.6\ \Omega.$$

A standard value of 470 Ω would work.

5.2

$$I_L = V_O/R_L = 1.2/10000 = 0.00012 \text{ A}$$

$$I_D = 10I_L = 10(0.00012) = 0.0012 \text{ A}$$

Next, calculate the value of R_2.

$$R_2 = V_O/I_D = V_O/10I_L = 1.2/0.0012 = 1000 = 1 \text{ k}$$

$$I_T = I_D + I_L = 0.0012 + 0.00012 = 0.00132 \text{ A}$$

Now calculate the value of R_1. This is given as

$$R_1 = (V_S - V_O)/I_D + I_L = (V_S - V_O)/I_T$$

$$= (5 - 1.2)/0.00132 = 3.8/0.00132 = 2878.8$$

Now specify standard resistor values. The closest standard 5 percent values are 2700 and 1000 for R_1 and R_2, respectively.

You may want to calculate the true output voltage with the specified resistor values. Use the actual values specified for R_1 and R_2 in the equation.

$$V_O = (V_S R_2/R_1 + R_2) = 5(1000/1000 + 2700)$$

$$= 5000/3700 = 1.35 \text{ V}$$

If you want, calculate the percentage of error.

%Error = 100(Actual voltage – Specified voltage)/ Specified voltage = 100(1.35 – 1.2)/1.2 = 12.5%

That is a high error value, so you may want to use a 1 percent resistor for R_1.

5.3

Using an LDR in the circuit of Fig. 5.5*d*. Shining a bright light on the LDR, its resistance will drop, so the output voltage should decrease.

In the circuit of Fig. 5.6*b*, if you use a 10-kΩ thermistor and you want the output to be half the supply voltage. That means that the other resistor also has to be 10 kΩ.

As you heat the thermistor, its resistance decreases, putting more voltage across the resistor. Therefore, the output voltage will go down.

5.4

With a 100 k load, the total divider resistance R_D can be one-tenth of the load or 10 k. It can also be less. The actual value is not critical as long as the divider resistance is no larger than 10 k.

The desired output voltages are 2 and 7 V. The voltage across R_3 is 2 V. The voltage across R_2 plus R_3 should be 7 V. That means that the voltage across R_2 is

$$V_2 + V_3 = 7\text{ V}$$

$$V_2 = 7 - V_3 = 7 - 2 = 5\text{ V}$$

Since R_2 is the 5 k pot, we can find the divider current.

$$I_D = V_2/R_2 = 5/5000 = 0.001\text{ A}$$

We can compute R_3.

$$R_3 = V_3/I_D = 2/0.001 = 2\text{ k}\Omega$$

Since the sum of the voltage drops should be equal to the source voltage, then

$$V_S = V_1 + V_2 + V_3$$

Then

$$V_1 = V_S - V_2 - V_3 = 9 - 5 - 2 = 2\text{ V}$$

$$R_1 = V_1/I_D = 2/0.001 = 2\text{ k}$$

The total divider resistance is

$$R_D = R_1 + R_2 + R_3 = 2\text{ k} + 5\text{ k} + 2\text{ k} = 9\text{ k}$$

This is less than one-tenth of the load of 100 k as required.

5.5

With a 6-V supply and 2 V across the LED, the resistor voltage should be

$$V_S = V_{\text{LED}} + V_R$$

$$V_R = V_S - V_{\text{LED}} = 6 - 2 = 4\text{ V}$$

Assuming a desired current of 20 mA, I_C will be the same. The resistor value will be

$$R = V_R/I_C = 4/0.02 = 200$$

Use a standard 220 value.

The 2N3904 is selected. The minimum h_{FE} is 70.

Using half the h_{FE} of 35, the base current is

$$I_B = I_C/h_{\text{FE}} = 20\text{ mA}/35 = 0.57\text{ mA} = 0.00057\text{ A}$$

Assuming that the V_{BE} is 0.7 V, the base resistor then is

$$R_B = (6 - 0.7)/0.00057 = 9298\ \Omega$$

The closest standard values are 9100 or 10 k.

For the MOSFET, with a supply of 15 V and a relay voltage of 12 V, the resistor voltage will be

$$V = V_S - V_R = 15 - 12 = 3\text{ V}$$

With 3 V across the drain resistor and a relay and drain current of 75 mA, the resistor value will be

$R = V/I_D = 3/.075 = 40\ \Omega$. You can use a 39- or 43-Ω standard value. Select 39 Ω to ensure that the relay current is just a bit higher then the 75 mA. The 2N7000 has a maximum drain current limit of 200 mA so we are safe.

The gate threshold voltage for the 2N7000 is 2.1 V. The driving turn-on signal on the gate should be at least that and more than say a 2.5 V minimum to ensure turn on—3 to 10 V would work.

Chapter 6

6.1

If you did simulate this power supply, you should have found it works just fine. No errors. That should have given you a real workout on using simulation software.

6.2

See Fig. B6.2. The data sheets for the LM317/LM337 regulators probably gave you most of what you needed. Following the procedure in Chap. 6, your own supply should look like Fig. B6.2. A standard available transformer with 12.6 VAC output should work. Half-wave rectifiers are used to simplify the design. These require a larger filter capacitor to get ripple down to a reasonable limit. The 3300 μF electrolytics should do the job. The 5 k pots are used to adjust the output voltage of each supply. It should be variable over about a 2- to 15-V range. The 10 μF output

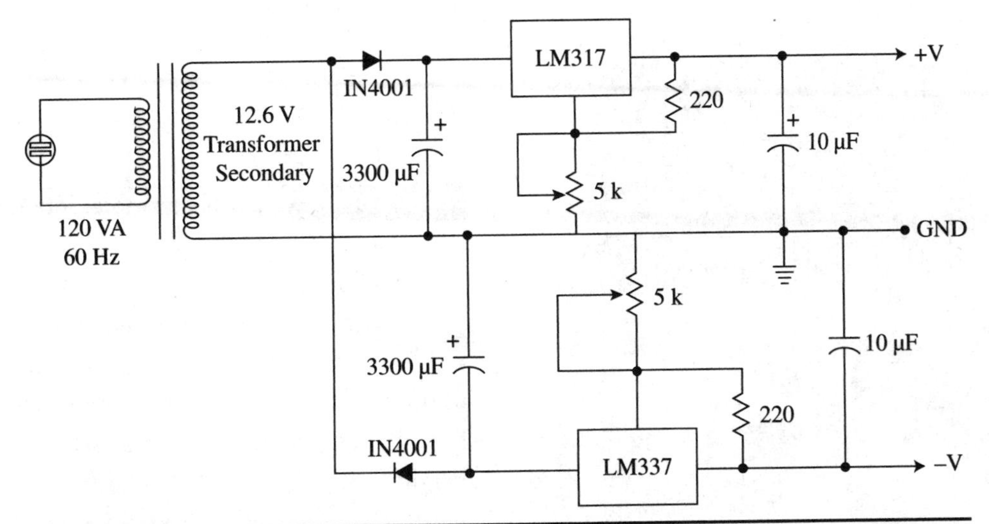

Figure B6.2 Dual plus and minus variable power supplies with an output voltage range from approximately 2 to 15 volts each.

capacitor provides additional filtering. Be sure the capacitor voltage ratings are high enough. Thirty-five volts ought to do it. You could build the circuits on perf board. This is a great supply to use when working with op amps that work best with dual ±V supplies.

6.3

Solar power supplies are generally used for portable, off-grid, or emergency conditions. Examples are hams at a field day event where all radio equipment is powered by batteries, gasoline generators, or solar. Another case is an emergency supply that will work during power outages, bad weather, or some other power-robbing event. For an emergency, you may want some light from a lamp other than a flashlight. You may also want a radio and a way to charge your smartphone.

Outlined here is the design process you should follow. It is generally not possible or at least very difficult or expensive to power ac devices from solar alone. The usual procedure is to use a battery as the primary voltage source and

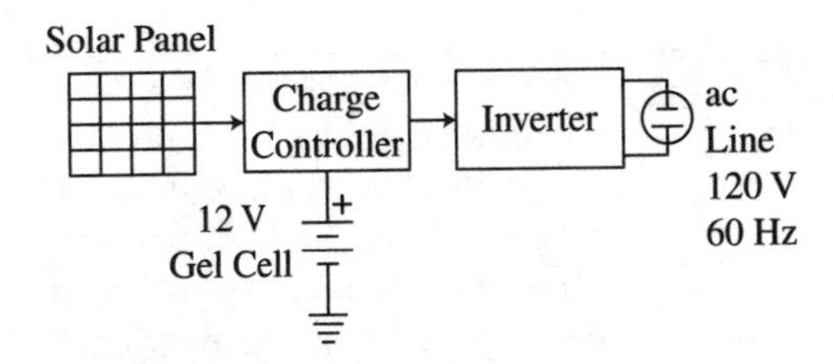

Figure B6.3.1 Simplified block diagram of a solar panel charging a battery that operates an inverter that converts 12 V dc to 120 V ac.

let the solar panel keep the battery charged. Here is an example. Fig. B6.3.1 shows a generic block diagram of such a system. The heart of this supply is a 12-V gel cell lead acid battery. Acquire the biggest Ah battery you can afford or the size that fits your packaging. Avoid car batteries because of their size and weight. Select one of the smaller gel cell batteries (<20 Ah).

Also select a 12-V solar panel. It does not have to be too big because you are only going to use it to keep the 12-V battery charged. Get one that will deliver about 10 W of power. Without any load on these solar panels, the output voltage in the bright sunlight is more than 12 V and more like 15 to 20 V. This is OK for battery charging.

Figure B6.3.2 The main components of a solar power system. The 12 V battery is on the left, the charge controller is on the right. The solar panel is obvious in the rear. The device in the front center is the inverter.

You will also want to get a charge controller. This is a circuit that charges the battery but limits the current so no overcharging and battery damage occur.

Finally, buy an inverter. An inverter is a dc to ac power supply. Most take in 12 V dc and put out 120 V ac 60 Hz. Look for one that has enough power to operate a small lamp or other devices. It is the power rating of the inverter that is the key specification. Something around 100 W or more is good. Be sure to add up the power consumptions of the devices you want to power and get an inverter to match that total. Figure B6.3.2 shows one solution.

6.4

Fig. B6.4.1 shows the simple black box diagram of the night-light. The input voltage source is

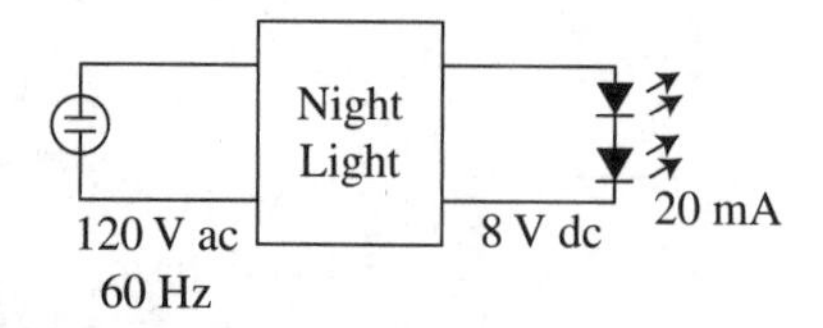

Figure B6.4.1 Simplified representation of a night-light.

a "scary" 120 V 60 Hz ac that can kill you if you are not careful. Your output is two white LEDs in series that use dc. Each will have an approximately 4 V across it when current is flowing. The current depends upon the desired brightness. Twenty milliamperes was chosen here.

No transformer is allowed in this design because they are large, heavy, and expensive. You need a way to drop the high input voltage way down, as well as rectify the ac into dc. Let's shoot for 12 dc to operate the LEDs.

To rectify the ac into dc, all you need is a single diode half-wave rectifier. A 1N4007 diode should work. A large electrolytic capacitor C_1 will filter the rectified ac into the dc voltage for the LEDs. When you rectify ac, the capacitor charges up to the peak of the ac input. You want to have a 12-V peak.

You could use a big resistor to drop some of the voltage as shown in Fig. B6.4.2.

Remember from ac theory that the peak value of an ac wave is 1.414 times the rms value. The peak of the 120 V rms input is 120(1.414) = 170 V. To get 12 V across C_1, the resistor has to drop 170 − 12 = 158 V. That resistor will have 20 mA (0.02 A) flowing through it, so its value according to Ohm's law will be

$$R = V/I = 158/0.02 = 7900\ \Omega$$

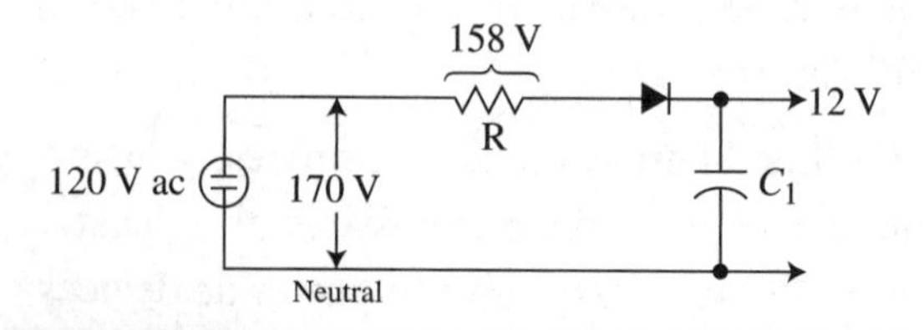

Figure B6.4.2 Lowering the high line voltage to a smaller value with a resistor.

The power rating will be

$$P = VI = 158(0.02) = 3.16\ \text{W}$$

There are no resistors of that value or that power rating. One possibility is to use 2-W resistors in some series-parallel combination to get close to those values. Understand that 3.16 W is a great deal of power (and heat) to dissipate. That is an undesirable characteristic of a night-light.

An alternative is to use a capacitor to drop that voltage. We would need a capacitor with a reactance (X_C) of 7900 Ω. Remember, capacitive reactance is

$$X_C = 1/2\pi fC$$

To find C, the formula is rearranged to be

$$C = 1/2\pi fX_C$$

The frequency f is 60 Hz.

$$C = 1/6.28(60)7900 = 0.336\ \mu\text{F}$$

A standard value capacitor is 0.33 μF.

The good news is that the capacitor will not dissipate power or heat. The voltage rating of the capacitor should be at least the peak of the ac or 170 V. Common voltage ratings are 500 or 100 V. A 500-V rating will do.

The filter capacitor C_1 value also needs to be determined. Using the formula from Chap. 6, the value is

$$C = I/fV_r$$

Some ripple can be accommodated. Let's use 1 V.

$$C = I/fV_r = 0.02/(60)(1) = 0.000333\ \text{F or } 333.33\ \mu\text{F}$$

A 330 μF value is standard, but it would not cost any more to go to the next largest of 470 μF. The voltage rating must be higher than 12. Typical values are 16 and 50 V.

With 12 V dc out of the capacitor, we need to drop it down to the 8 V needed by two LEDs, each with 4 V. So we need to drop 12 − 8 = 4 V at 20 mA.

$$R = 4/0.02 = 200\ \Omega.$$

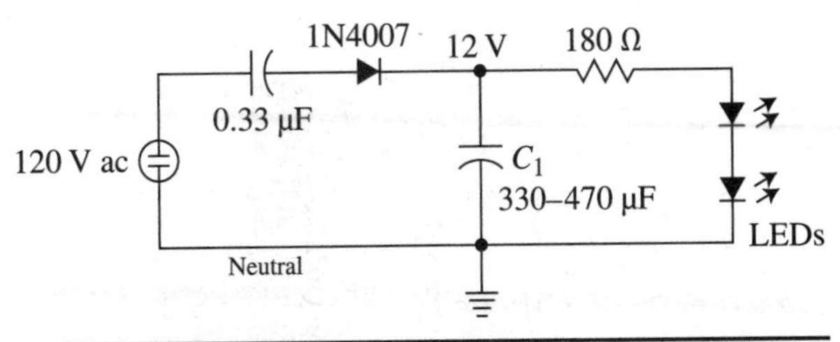

Figure B6.4.3 Using a capacitor to lower the line voltage to a smaller value.

A 180-Ω standard value could be used. A ¼-W resistor will work.

The circuit now looks like Fig. 6.4.3. Note that the ground side of the circuit should be connected to the neutral side of the ac power line for safety. This is the side with the wider prong on the ac plug.

With this arrangement, there is no regulation, so the brightness of the LEDs will vary with the line voltage. In a night-light, that probably does not matter. If it does, a refinement would be to add a zener stabilizer or a three-terminal regulator. In either case, you would need to make the voltage across the filter capacitor larger than 12 V to give the regulator some head room. Eighteen or 15 V would be enough, and you could use a 12-V zener or regulator IC.

Chapter 7

7.1

Probably the most useful solution is to design with the 386 IC power amplifier. Build the amplifier on a breadboarding socket or perf board. You may be able to use it in other projects. Use the circuit from the text that is repeated here as Fig. B7.1. Use a 12-V supply to get plenty of output. A small speaker is all you need. A 2-in speaker is OK, but do not go much larger than that because they require more power than this amplifier is capable of producing.

As for gain, use the basic setting of 20 as provided internally. Forget the extra 10 μF on

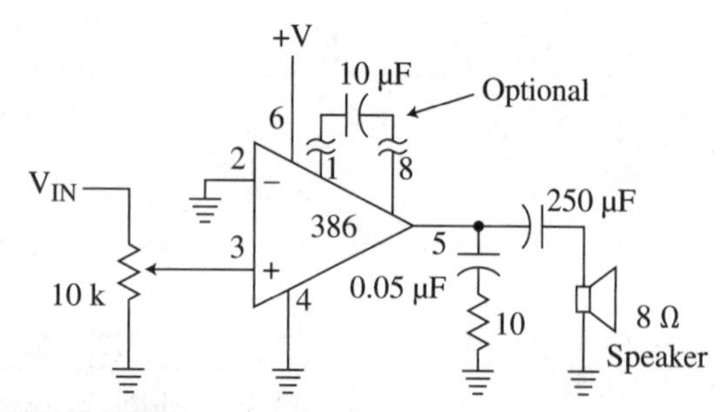

Figure B7.1 LM386 IC power amp can control an 8 ohm speaker up to 2 inches in diameter.

pins 1 and 8 unless you are really need the gain. In either case, the 10 k pot at the input is your volume control.

As for the tone control RC network on the output, it is not critical. You can omit it if you wish. If you want to customize the frequency response, follow the guidelines in the data sheet or any accompanying app note.

Build the circuit and test it. A good input signal is the rectangular output from a 555 timer IC oscillator. Be careful because it will be really loud.

If you are going to do a great deal of audio work, buy a complete amplifier or a kit. It will give you more power, and in some models, you can accommodate stereo (two channels) signals.

7.2

A basic inverter will do for this one. See Fig. B7.2. You can choose almost any op amp you want,

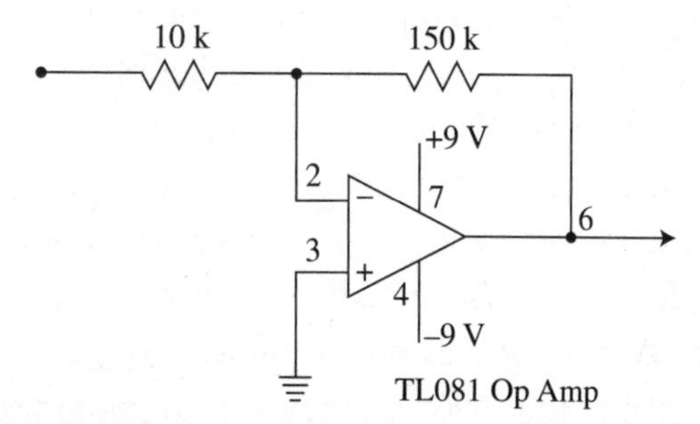

Figure B7.2 The TL081 op amp can be used for any of the circuits found in this book. The gain is 15, and the input impedance is 10 k.

but be sure your op amp can deliver a response of up to 20 kHz with the gain of 15. A TI TL081 was used as the example here.

Get the data sheet. Most data sheets will tell you the range of output voltage with different power supply values. Or use the slew rate calculation. The input resistor (R_i) should be 10 k because that is the input impedance of an inverter. To get a gain of 15, you need a feedback resistor of

$$A = R_f / R_i$$

$$R_f = A(R_i) = 15(10\text{ k}) = 150\text{ k, a standard value.}$$

To achieve the ±6-V output swing, you will definitely need two power supplies. A simple solution is to use two 9-V batteries.

The maximum output voltage of 6 V can be determined with the slew rate calculation. The slew rate of the TL081 is 13 V/μS.

$$V_o = SR/2\pi f = 13\text{ V}/\mu\text{s}/6.28\ (20{,}000)$$
$$= 13{,}000{,}000/6.28(20{,}000) = 103.5\text{ V}$$

The ±6 V output is easily achieved at 20 kHz.

7.3

See Fig. B7.3.1. Two op amp followers buffer the inputs. The 1-MΩ resistors set the input impedance to that value. The input impedance of the followers is very high and has negligible effect on the input impedance. The follower outputs send signals to the 1-k volume control pots. A two-input summing amplifier mixes the two signals. The 100-k feedback resistor along with the 10-k input resistors give a gain of 10 to each input.

When you apply input signals from a function generator or some oscillator, use a value of about 0.5 V pp. With a gain of 10 on each input, this could develop an output voltage of ±5 V. That is within the capability of the ±9-V supplies. However, you can reduce the input voltage levels with the 1-k volume controls.

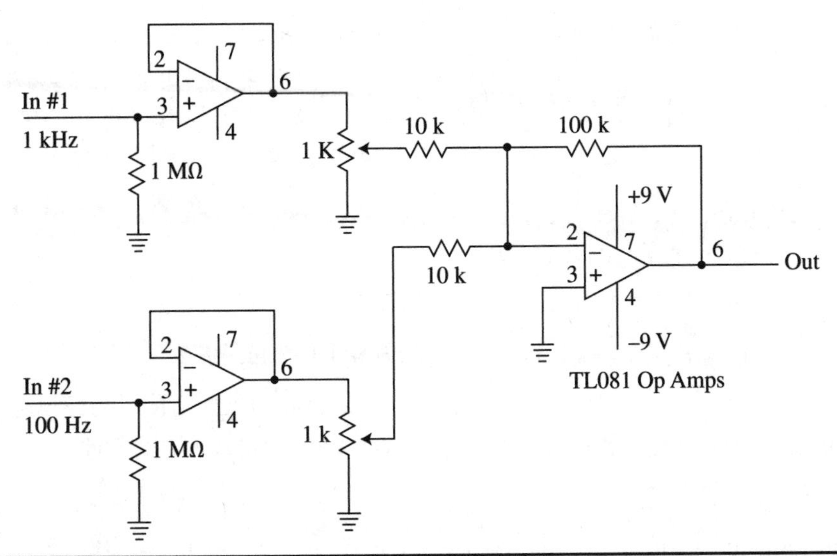

Figure B7.3.1 An audio mixer made with a summing amplifier, follower buffers, and separate volume controls.

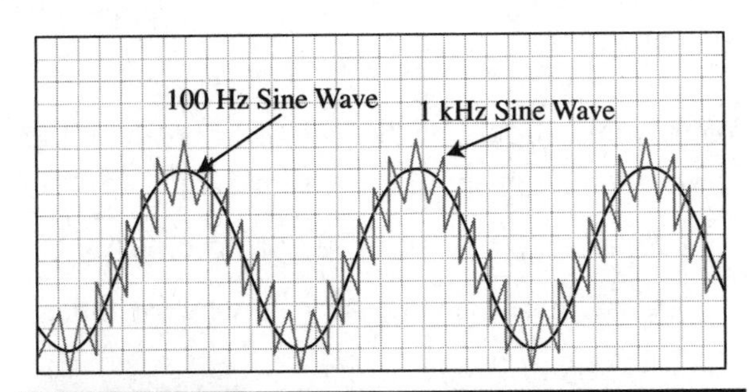

Figure B7.3.2 The mixer output should be a 100 Hz sine with a 1 kHz sine wave superimposed on it.

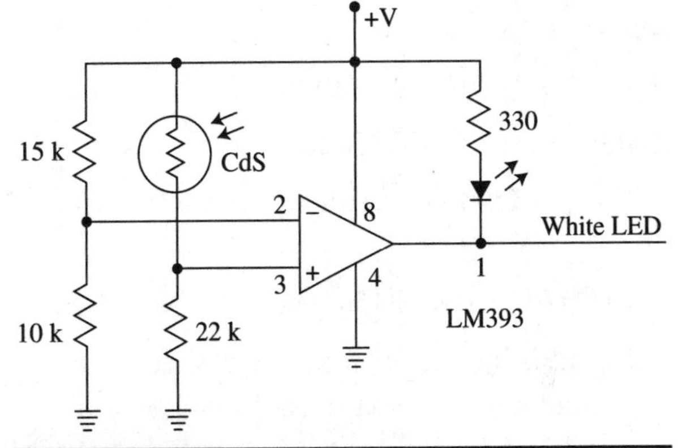

Figure B7.4 A comparator IC used as night light. The LED is off during light conditions. As it turns dark, the LED turns on.

What you should see on a scope if you monitor the output is a 1-kHz sine wave riding on top of a 100-Hz sine wave. See Fig. B7.3.2. The two signals have been linearly mixed. Experiment with the volume control settings to see how the output waveform changes.

7.4

Refer to Fig. B7.4. A fixed voltage reference is applied to pin 2 on the LM393 comparator by the 15- and 10-k resistors in a voltage divider. The CdS light sensor is also connected in a voltage divider with a 22-k resistor. With light on the sensor, the voltage at pin 3 on the comparator is less than the reference voltage. The comparator does not turn on the LED. As it gets dark, the resistance of the sensor increases, thereby increasing the voltage on pin 3 of the comparator. The comparator switches on the LED. The key to getting this circuit to work reliably is to have an adjustment for light sensitivity. One way would be to replace the 10-k resistor with a pot that could be adjusted for the desired light level. A 25-k pot replacing both the 15- and 10-k resistors would be better.

Chapter 8

8.1

See Fig. B8.1.

The frequency of oscillation of a Wein bridge oscillator and its gain are determined by the *RC* values.

$$C = C_1 = C_2$$

$$R = R_1 = R_2$$

The amplifier gain is set by R_3 and R_4.

$$A = 1 + R_3/R_4 = 3$$

To obtain a frequency of 1 kHz, use the following formula:

$$f = 1/2\pi RC$$

The design procedure calls for a $C = 0.1\ \mu F$ capacitor. Calculate *R*.

$$R = 1/2\pi fC = 1/6.28(1000)(0.1 \times 10^{-6}) = 1592.3\ \Omega$$

A standard value is 1500 Ω.

Using a 1500-Ω resistor will give a frequency of

$$f = 1/2\pi RC = 1/6.28(1500)(0.1 \times 10^{-6}) = 1062\ \text{Hz}$$

If that is too far off from the desired value, you could use 1 percent resistors. A 1 percent value of 1.58 k would get you closer to 1 kHz.

$$f = 1/2\pi RC = 1/6.28(1580)(0.1 \times 10^{-6}) = 1008\ \text{Hz}$$

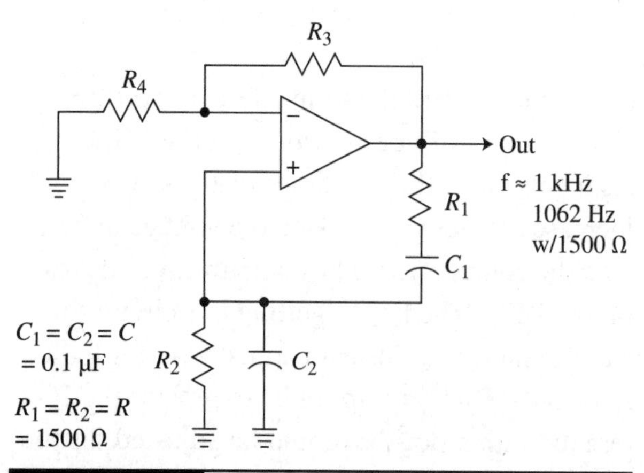

Figure B8.1 A Wein bridge sine wave oscillator with a frequency of about 1062 Hz.

Almost any combination or resistor values for the gain should work. It all depends upon what output stabilization method you used (light bulbs, zeners, etc.). Using 10 and 20 k, the gain is

$$A = 1 + R_3/R_4 = 1 + (20\ \text{k}/10\ \text{k}) = 3$$

8.2

Refer to Fig. B8.2.

Using a 1200 Ω feedback resistor, the capacitor value *C* for 5 MHz should be

$$C = 1.2/fR = 1.2/(5{,}000{,}000)(1200)$$

$$= 2 \times 10^{-10} = 200\ \text{pF}$$

Standard values of 180 or 220 pF will get you in the ballpark near 5 MHz.

8.3

Your circuit should be like Fig. 8.12 but with the resistor values given in the assignment.

R_A and R_B = 10 and 100 k, respectively.

You calculate *C* using the following formula:

$$f = 1.443/(R_A + 2R_B)C \quad \text{therefore,}$$

$$C = 1.443/(R_A + 2R_B)f = 1.443/10\ \text{k} + 2\ (100\ \text{k})(0.5)$$

$$= 1.443/210\ \text{k}(0.5) = 0.0000137\ \text{F} = 13.7\ \mu\text{F}$$

A standard value of 15 µF would get you the closest if available. The other available values are 10 or 22 µF. These are polarized electrolytics. For flashing an LED, usually frequency or rate is not critical.

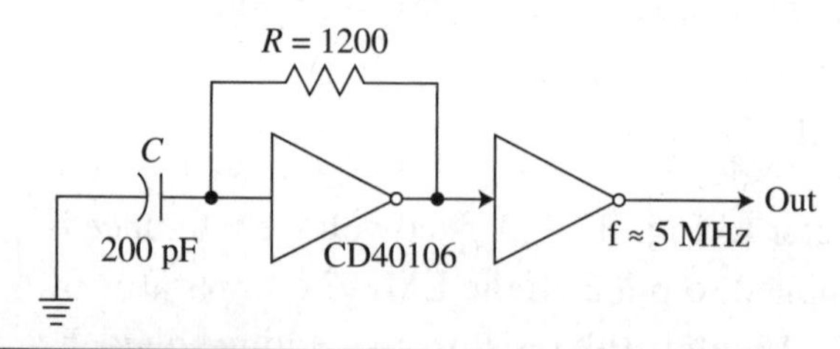

Figure B8.2 A CD40106 Schmitt trigger multivibrator. A 200 pF capacitor and 1200-Ω resistor should produce an output of about 5 MHz.

8.4

Varying the 100 k pot makes the output duty cycle vary from a few percent to near 100 percent. Low duty cycle will produce a dim LED, while a high duty cycle will make the LED brighter. You can also achieve any degree of brightness with the appropriate setting of the 100 k pot.

8.5

The 555 oscillator should produce a very loud 700 Hz tone that you can adjust with the 10 k volume control.

8.6

The metal can TTL clock device should plug right into a breadboarding socket. It uses a 5-V dc supply. Looking at the output on an oscilloscope, you should see a rectangular wave at the frequency you selected. You can check the frequency by measuring the period (t) and computing f, where $f = 1/t$.

The period (t) is the time for one cycle (one on and off interval). You compute the frequency (f) with the following expression:

$$f = 1/t$$

Also,

$$t = 1/f$$

For example, if you chose a 10-MHz crystal, the period should be

$$t = 1/(10 \times 10^{-6}) = 0.1 \text{ ms or } 100 \text{ ns}$$

The crystal is very precise and stable, and it can even be used to check the time calibration of your scope.

Chapter 9

9.1

The R value should be at least 10 times the driving impedance of 300 Ω. Use 3300 Ω. See Fig B9.1.

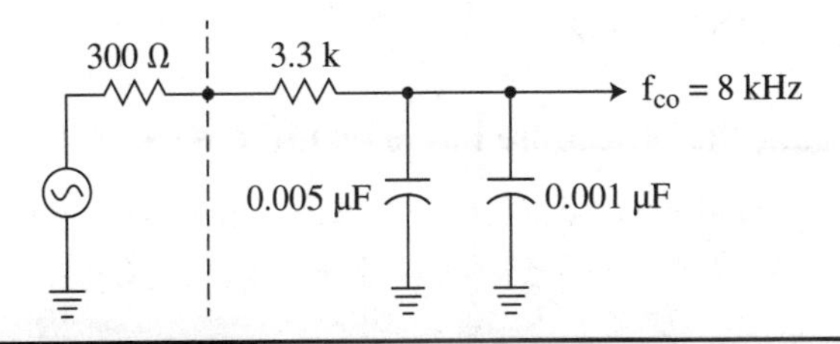

Figure B9.1 An RC low pass filter with a cut-off frequency of 8 kHz.

Calculate C for a cutoff of 8 kHz.

$$f_{CO} = 1/2\pi RC$$

$$C = 1/2\pi f_{CO}R = 1/6.28(8000)3300 = 0.006\ \mu F$$

A standard 0.005 µF capacitor in parallel with a standard 0.001 µF capacitor will give you the correct value.

9.2

Using the online calculator, the component values are given. The generator impedance is R1 and 150 Ω. The load is R2 and 150 Ω. These are the values for a low-pass 10-MHz Butterworth with 7 poles. See Fig. B9.2.

R1	150 Ω
L1	1.06236 µH
C1	132.311 pF
L2	4.30196 µH
C2	212.207 pF
L3	4.30196 µH
C3	132.311 pF
L4	1.06236 µH
R2	150 Ω

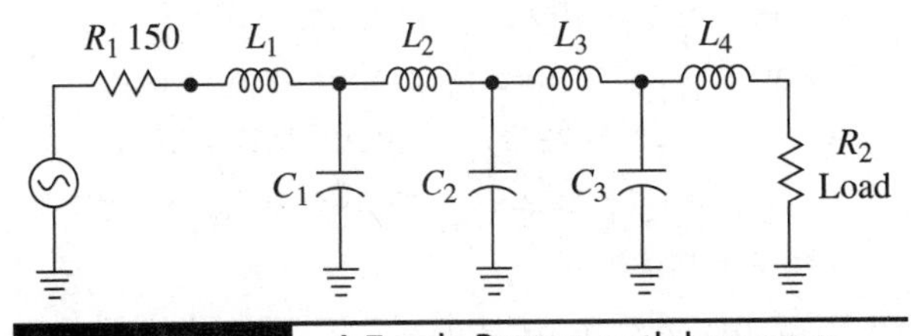

Figure B9.2 A 7-pole Butterworth low-pass filter with a cutoff of 10 MHz.

9.3

Using the procedure in the text, the result is

- The desired center signal frequency (f_C) is the third harmonic of 1 kHz or 3 kHz (3000 Hz).
- Select $C_1 = C_2$ = Choose 0.1 μF.
- Calculate $R_1 = R_4 = 1/2\pi C_1 f_C$
  $= 1/6.28(0.1 \times 10^{-6} \times 3000) = 531\ \Omega$.
- Calculate $R_3 = 19R_1 = 19(531) = 10{,}085\ \Omega$.
- Calculate $R_2 = R_1/19 = 531/19 = 28\ \Omega$.
- Use the closest standard resistor values. If necessary, use 1 percent resistors for precision in setting the center frequency.

9.4

The goal is to notch out 60 Hz.

$$f_C = 1/2\pi RC$$

- Select a value for R. It should be at least 10 times higher than the Z_O of the driving stage and lower than a factor of 10 or more of the load R_L. The Z_O of the of the op amp driving the filter is only a few ohms at best, so we can choose almost anything. Select 10 kΩ for R.
- Calculate C. Rearranging the formula to solve for C

  $$C = 1/2\pi f_{CO} R = 1/6.28(60)10{,}000 = 0.265\ \mu F$$

  A standard 0.22 μF capacitor would work but would shift the notch too much. You could put a 0.047 μF in parallel with the 0.22 to get 0.267 μF.

  $$f_C = 1/2\pi RC = 1/6.28(10{,}000)(0.047 \times 10^{-6})$$

  $$= 339\ \text{Hz}$$

  This is usually not suitable so a capacitor value closer to the calculated value must be used. You can experiment with putting capacitors in series or parallel to get closer to the desired value.
- Calculate $2C$ and $R/2$. Using the design results above:

$$2C = 2(0.267) = 0.534\ \mu F$$

$$R/2 = 10\ k/2 = 5\ k\Omega$$

This filter design is very sensitive to component values. They must be as close to the calculated value to achieve good attenuation at the center frequency. Typically, you must resort to 1 percent resistors and capacitors to get close enough to the desired values.

Another approach is to select the capacitor first. Choose a 2 or 5 percent capacitor then using that value calculate the resistor value. Then use the closest 1 percent value. Since there are more closely spaced resistor values, you are more likely to get close to the ideal.

A 5 k pot was used in the test circuit. A decibel calculation showed a 32-dB attenuation of the 60 Hz.

Chapter 10

10.1

There is not much to design here. You should have built the H-bridge circuit and demonstrated it in reversing motor rotation.

10.2

The 555 frequency of operation is about 130 Hz with the capacitor value shown in Fig. 10.12. The period is 7.7 ms. With a 50 percent duty cycle, the pulse width would be half that or 3.85 ms.

It should be clear that PWM is a great way to vary motor speed. Replacing the motor with an LED allowed you to see that brightness vary over a wide range as the duty cycle is changed.

10.3

The key to making the servo work is to get the frequency as close to 50 Hz as you can. Use your oscilloscope to determine the frequency by measuring the period. The period (t) is the time for one cycle (one on and off interval). You compute the frequency (f) with the following expression:

$$f = 1/t$$

Also,

$$t = 1/f$$

For 50 Hz, the period should be

$$t = 1/50 = 0.02 \text{ s or } 20 \text{ ms}$$

If you can get close to the frequency, you should be able to demo servo operation.

Chapter 11

Solutions for Design Projects 11.1 and 11.2 are covered in App. D.

11.3

A cable tester is shown in two parts, the transmitter (TX) in Fig. B11.3.1 and the receiver (RX) in Fig. B11.3.2.

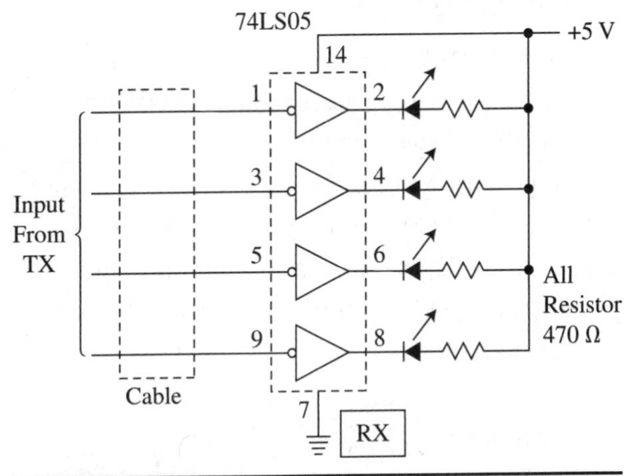

Figure B11.3.2 The complete cable tester receiver (RX) circuit.

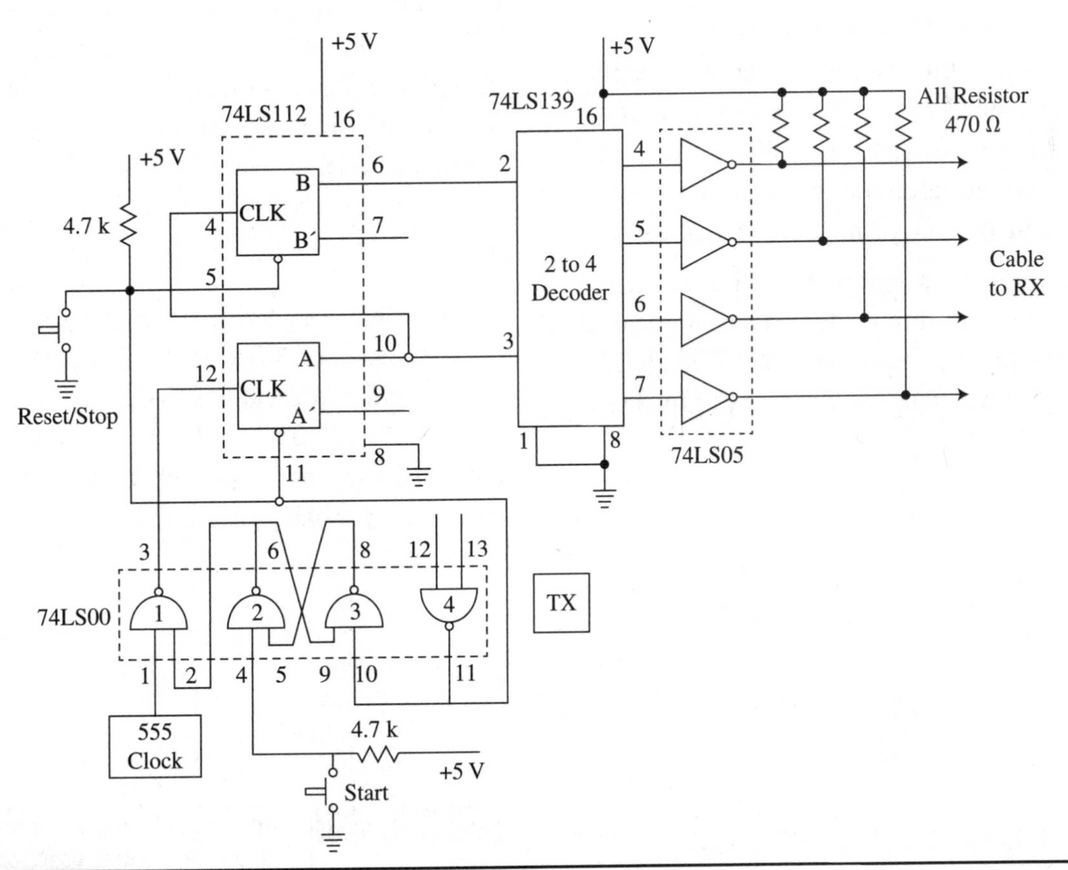

Figure B11.3.1 A complete logic diagram of the cable tester transmitter (TX).

The RX uses LEDs to indicate the status of the cable wires. These LEDs need a driver IC. This is a 74LS05 open collector hex inverter. Only four of the six inverters are used and shown. When the input to an inverter goes high, the LED in the RX turns on.

The 555 clock circuit is shown in Fig. B11.3.3. A large capacitor is used to make the timer slow so that it meets the 5-s requirement for turning each LED on. $f = 0.2$ Hz. You can make the clock faster by decreasing the value of the capacitor. A 22-μF capacitor will approximately double the speed. A 10-μF capacitor will about double it again.

Refer back to Fig. B11.3.1. The clock drives two JK flip flops in the 74LS112 IC that makes a simple counter to generate four states (00, 01, 10, 11). These four states from FF A and B are decoded into four states by the 74LS139 2-line to 4-line IC. This decoder outputs signals are sent to some 74LS05 open collector inverters then put on the cable. If two or more of the wires in the cable are shorted two of the inverter outputs would be connected together. No damage will occur with open collector ICs. When one of the decoder outputs goes low, the inverter output will go high and will turn on the related LED.

Here's how the circuit works. When the reset button is pushed, it puts the two FFs in the 00 state. This will also reset the latch FF made up of gates 2 and 3. This causes the input to gate 1 to block the signal from the 555 clock. At this time, the decoder output on pin 4 goes low and turns on the upper LED in the RX if the cable wire is good.

The test sequence begins when you press the start button. This sets the latch, and the output of gate 2 goes high, enabling gate 1. The clock pulses from the 555 are applied to the FF counter. The counter slowly steps through each state one at a time about every 5 s. The LEDs in the RX should light in sequence if the cable wires are good. A short will be indicated if two LEDs turn on at the same time.

When the counter sequences through its states, it will eventually go from state 11 back to 00. The circuit will continue to run until you press the reset/stop button. This disables gate 1, so the circuit stops. Just push the start button again to retest.

Your design will probably differ, but hopefully this solution will give you a better feel for how a digital circuit works. There are probably a half dozen other designs that will do the same thing. You could also do the same test with a single microcontroller as well.

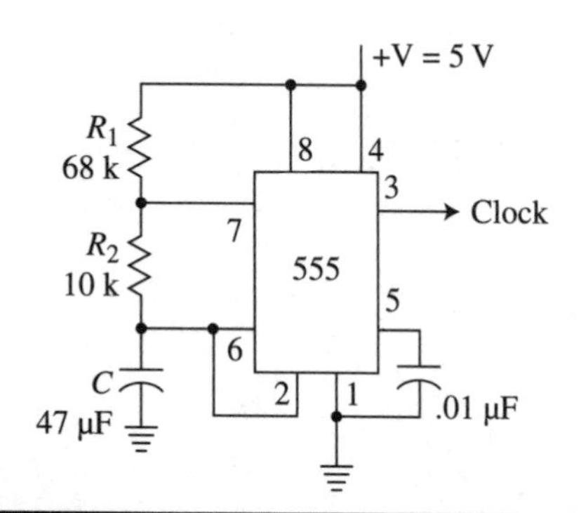

Figure B11.3.3 A 555 timer IC used as a clock circuit.

11.4

The problem is to derive a 1.25 MHz signal from a 50 MHz clock. A frequency divider will do the job. To find out, the divide factor just divides 50 by 1.25. 50/1.25 = 40. Using flip flops and counters, you can derive such a divider. The best solution is probably shown in Fig. B11.4.1.

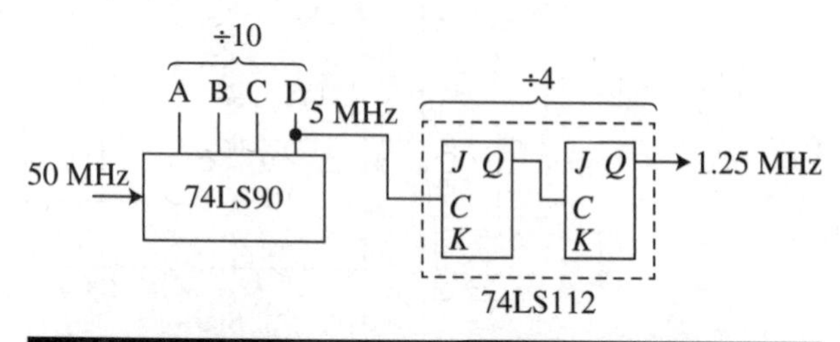

Figure B11.4.1 One possible implementation of the transmitter used for cable testing.

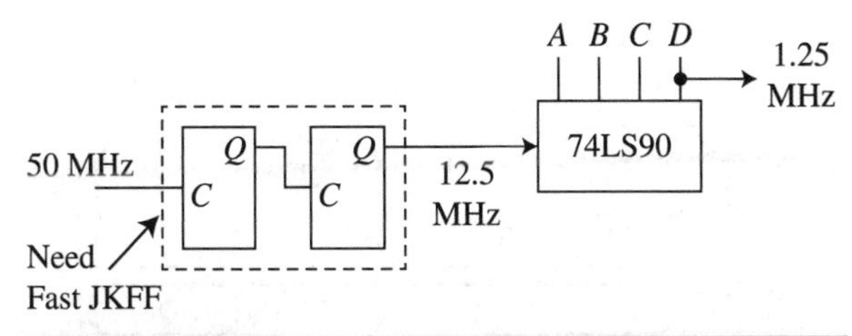

Figure B11.4.2 One possible implementation of the receiver used for cable testing.

A 74LS90 or 74LS192 BCD counter divides by 10. This is followed by two 74LS112 JK FFs that together divide by 4.

The problem here is that these ICs will not work at 50 MHz. You could switch to a faster 7400 series like the 74AS or 74ALS. Some faster 74HC variants may work at that frequency. One possibility is to put the two FF at the input to divide the 50 MHz down to 50/4 = 12.5 MHz. See Fig. B11.4.2. The 74LS90 or 74LS192 will work at that speed. What you need to find is a fast FF that will work at 50 MHz. Having the data sheets of parts or a good catalog is helpful in selecting ICs for your designs.

11.5

See Fig B11.5. A 74LS90 BCD counter counts from 0000 to 1001 (0 through 9). It drives a 74LS47

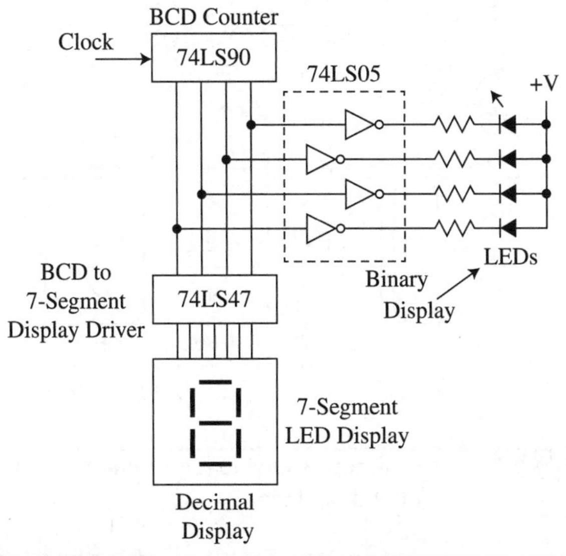

Figure B11.5 A divide-by-40 circuit with one BCD counter and two FF.

BCD to a 7-segment decoder-driver that connects directly to the 7-segment LED display. A 74LS05 open collector inverter serves as an LED driver to show the binary output of the BCD counter.

11.6

With 2 V applied to the ADC, you should read 2 V out on the DAC op amp. The binary value on the LEDs should be 01100110 or hex 66 or 102 decimal. With each increment equal to 19.53 mV, the output translates to $102 \times .01953 = 1.99$ V.

11.7

With the input sine wave at 1 and 2 kHz, you should have seen the same signal at the DAC op amp output. At 5 kHz, you may have seen a distorted replica at the DAC output. At 10 kHz and above, you get a condition called aliasing. The output becomes the difference between the sampling frequency and the input frequency. If the input is 13 kHz with a sampling rate of 10 kHz, you would see a 13 kHz – 10 kHz = 3 kHz sine wave at the DAC output.

11.8

As the binary counter counts up from 00000000 to 11111111, the output of the DAC rises linearly in 19.53 increments, producing an approximation of a straight line. At the maximum count of 11111111, the next clock pulse recycles the counter to 00000000 and the DAC output should go to zero. Then the cycle continues. What you should be seeing is a stepped approximation of a sawtooth wave.

11.9

The following truth table shows he die has six sides or states. Only six states are used 000 through 101 or decimal 0 through 5. Three FF gives us $2^3 = 8$ states. We will not use states 110 and 111. However, we can use these states to help minimize the circuits. The unused states 110 and 111 cannot occur with the circuits we are using. We call these "don't care" states.

Die Number	A, B, C Inputs	T U V W X Y Z
1	000	0 0 0 1 0 0 0
2	001	0 0 1 0 1 0 0
3	010	0 0 1 1 1 0 0
4	011	1 0 1 0 1 0 1
5	100	1 0 1 1 1 0 1
6	101	1 1 1 0 1 1 1
	110 "Don't care"	
	111 "Don't care"	

You should notice from the truth table that T column is the same as the Z column so T = Z; the U column is the same as the Y column, so U = Y; and likewise, V = X. Therefore, the same circuits can operate two LEDs.

The boolean equations for each group are

$$T = Z = A'BC + AB'C' + AB'C$$

$$U = Y = AB'C$$

$$V = X = A'B'C + A'BC' + A'BC + AB'C' + AB'C$$

$$W = A'B'C' + A'BC' + AB'C'$$

Many gates are necessary to implement these equations.

Transferring these states to Karnaugh maps, we get Fig. B11.9.1*a-d*. The boxes with 1s in them are from the equations. (Note: The bars over the letters represent the NOT state rather than an apostrophe.) The boxes with Xs in them are the "don't care" states. Circling 2 or 4 groups formed with the 1s or Xs, we get the following minimized equations:

$$T = Z = A + B$$

$$U = Y = A + B + C$$

$$V = X = A + C$$

$$W = C'$$

The logic circuits are shown in Fig. B11.9.2.

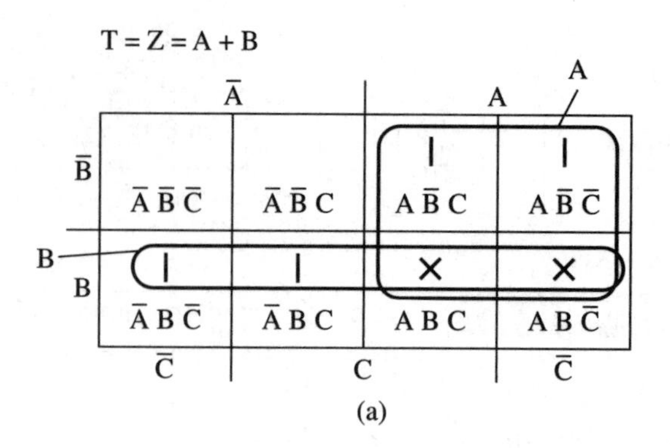

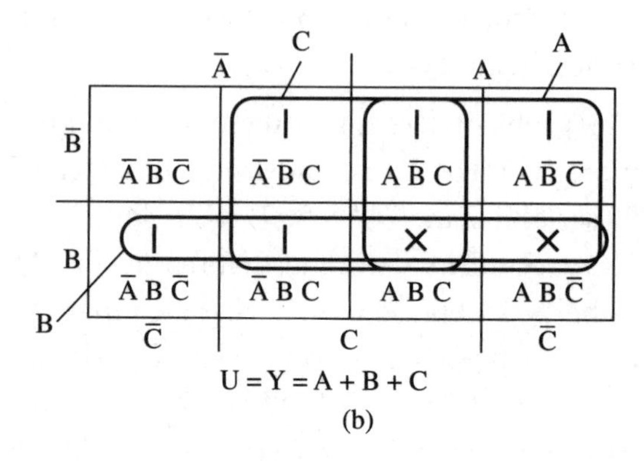

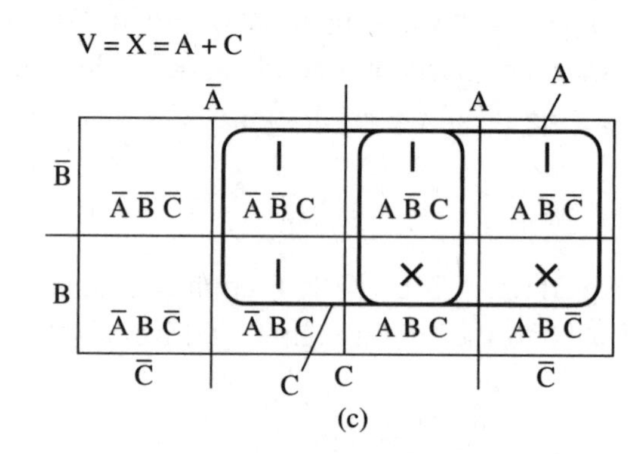

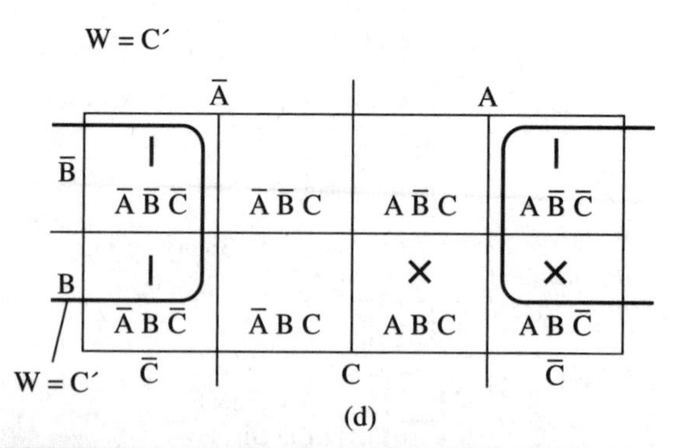

B11.9.1 The K-maps showing logic minimization for the digital die.

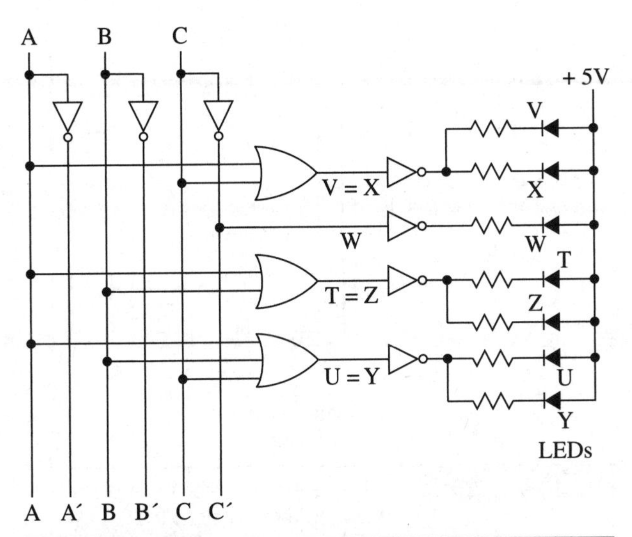

Figure B11.9.2 The simplified logic resulting from the K-map minimization of the digital die.

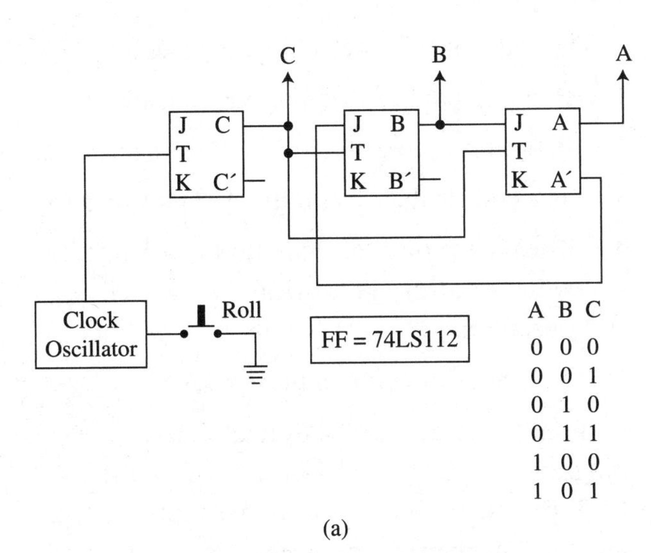

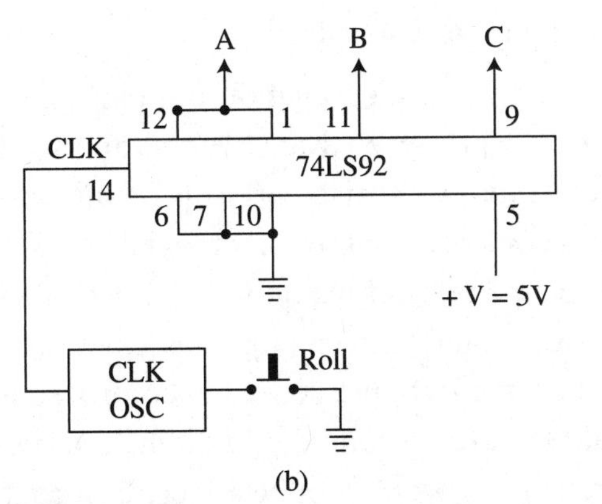

B11.9.3 A six state counter can be built with individual flip flops as shown in (a) or the count by six counter in the 74LS92 IC (b).

The A, B, and C signals come from one of the two circuits in Fig. B11.9.3. You can get that with the three FF circuit as shown in Fig. B11.9.*a*, or you could use the three FF segment of a 74LS92 counter as in Fig.11.9.3*b*.

Either will work. The counters are driven by a 555 timer oscillator at some high frequency 1 kHz or more. Pressing the ROLL push button sends clock pulses to the counter and allows it to count and recycle multiple times. When you let up on the button, the clock will stop, and the counter state will be some random value. The appropriate LEDs light.

Chapter 12 Programmable Logic Devices

None.

Chapter 13 Microcontrollers

13.1

Using the BASIC Stamp example earlier suggests one simple solution. The four wires in the test cable are connected to pins 11 through 14. The program sends a logic 1 down the first wire in the cable and then pauses for 2 s. The LED lights, and the program turns it off. The program continues by sending logic 1s down each wire in the cable and leaves the LED at the receiver on for 2 s, so an observer can determine if that wire is OK. The test then repeats.

```
DO
HIGH 11
PAUSE 2000
LOW 11
HIGH 12
PAUSE 2000
LOW 12
HIGH 13
PAUSE 2000
LOW 13
HIGH 14
PAUSE 2000
LOW 14
LOOP
```

There is probably a more sophisticated way to do this, but this way is simple and easy to implement. You could also add the push-button routine explained earlier to start the test.

13.2

The 7-segment truth table follows:

Decimal	7-Segment Binary abcdefg
1.	1111110
2.	0001100
3.	0110111
4.	0011111
5.	1001101
6.	1011011
7.	1111001
8.	0001110
9.	1111111
10.	1001111

These bit patterns would be stored in RAM, forming a lookup table (LUT). When a digit is to be displayed, its bit pattern would be called up from the LUT and sent to the MCU output where the displays are connected.

Assume a 2-digit 7-segment LED display. Instead of using the BCD to 7-segment ICs described in Fig. 13.11, you may be able to connect to the LEDs more directly if your MCU has enough output connections. The LEDs could be connected to these pins by way of some driver ICs. Then the program would output the 7-bit code to each display one after the other, but that would be too fast for the displays to retain an output. A holding register is needed to keep the numbers displayed. Fig. B13.2 shows how this might look.

The program in the MCU would call up a number to be displayed. The program looks up that number in the RAM LUT and sends the appropriate 7-bit code to the output pins. This code will be stored in an external register, such as the 74LS373 or 74HC373, an 8-bit register made up of D flip flops. One of the output pins would serve as a signal to load the register.

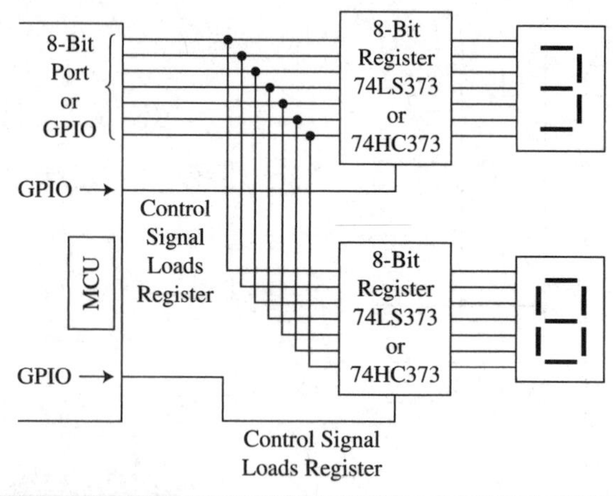

Figure B13.2 A multiplexed 2-digit 7-segment LED display for micro I/O.

The register IC mentioned also has output drivers that could connect directly to the 7-segment display.

The code for this goes something like this:

1. Program computes a number to display.
2. Program retrieves the 7-bit code from memory.
3. The code is then sent to the I/O pins or port.
4. The data is then latched into the register with a control signal on another MCU output pin.
5. Number shows up on the display.

If another digit is to be displayed, the program gets the related 7-bit code from the LUT, then sends it to the second external register. The register holds the code, and the second digit is displayed.

If there are not enough output pins on MCU, use just seven available pins and send the code to both registers. The program would send a control signal to load the first register. The second register would not be affected. The program would then call up the second digit and put it on the I/O pins. A second control signal would load the second digit holding register. The other register would not be affected. Repeat this procedure at a high clock rate, and it makes you believe it is one constant display.

That is only one possible method.

13.3

```
PORTB EQU $04
ORG $A000
LDAB #$A    Note: $A is the hex digit for
a decimal 10.
LDAA #00
REPEAT  STAA PORTB
INCA
DECB
BNE  REPEAT
STOP
```

Chapter 14 Component Selection

None.

Chapter 15 Troubleshooting and Debugging Solutions

A Linear Circuit

The reported problem in the linear circuit of Fig. 15.4 is that the output is not the desired or expected +3 V. The solution is shown in Fig. B15.4. If the supply voltages are OK at ±9 V and all the circuits seem to be working, what is the problem? The solution lies in calculating each and every op amp output and using these values to determine what the output really is.

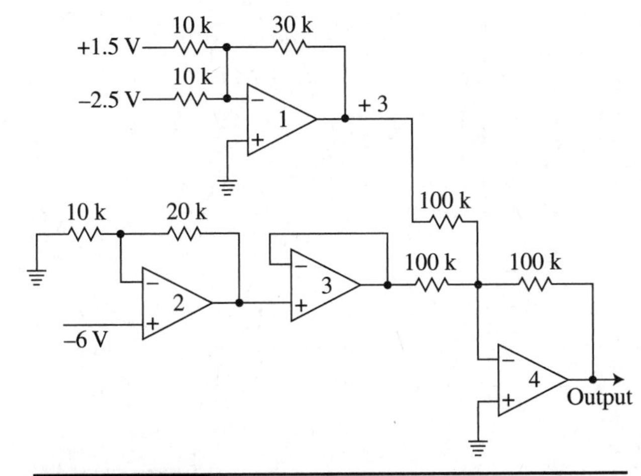

Figure B15.4 A hypothetical linear circuit for troubleshooting practice. Power supplies are ± 9 V.

Op amp 1 is a summer. The +1.5-V input is multiplied by a gain of 30 k/10 k = 3 to get 4.5 V. The second input is −2.5 V that is multiplied by a gain of 3 to get −7.5 V. These voltages are added in the circuit, giving 4.5 + (−7.5) = −3 V, but the circuit inverts that to +3 V.

The gain of the noninverting op amp 2 is (20 k/10 k) + 1 = 3. The input voltage of −6 V is amplified by 3, giving −18-V output. The op amp 3 follower should give the same or −18 V. Adding −18V and +3V should produce an output of −15V. The desired output is −3 V.

You should have spotted the problem when calculating the output from op amp 2. With 9-V supplies, the maximum possible output of each is 9 V. You calculated −18 V. If you measured the output from op amp 2, you would see that it was −9 V, not −18 V. Therefore, the problem is in op amp 2. The only possible problems are resistor values being incorrect (not likely) or too much input voltage. The latter is the real problem because the op amp output is saturated at its max at −9 V.

If the output from op amp 4 is supposed to be +3 V. You can work backward to figure out what the input to op amp 2 should be. It is −2 V.

A Digital Circuit Troubleshooting Example

The counter should count in a normal binary sequence such as

Decimal	Binary Out (CBA)
0	000
1	001
2	010
3	011
4	100
5	101
6	110
7	111

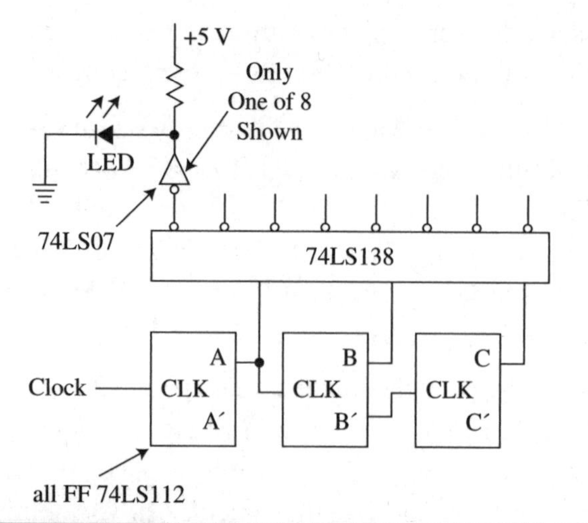

Figure B15.5 A hypothetical digital circuit to troubleshooting.

If this were the case, the decoder would turn the LEDs on in sequence. Instead the LEDs seem to turn on in some random order.

One way to troubleshoot this is to actually draw the inputs and outputs of the circuits based on the circuit connections. Start with the clock and draw the output of the A FF, the B FF, and the C FF. See Fig. B15.6. Note the binary states on the ABC lines. After each clock pulse, record the binary code. With the circuit shown, the count sequence seems to be

Decimal (number of clock pulses)	Binary Out (CBA)	Decimal Value
0	000	0
1	001	1
2	110	6
3	111	7
4	100	4
5	101	5
6	010	2
7	011	3

No wonder the LEDs are turning on out of normal binary sequence. The first thing to check is the wiring. Be sure you know how each circuit works. As it turns out, the B' output of the B FF is connected to the clock input of the C FF. The correct wiring should be the B output (not B') to the clock input of C.

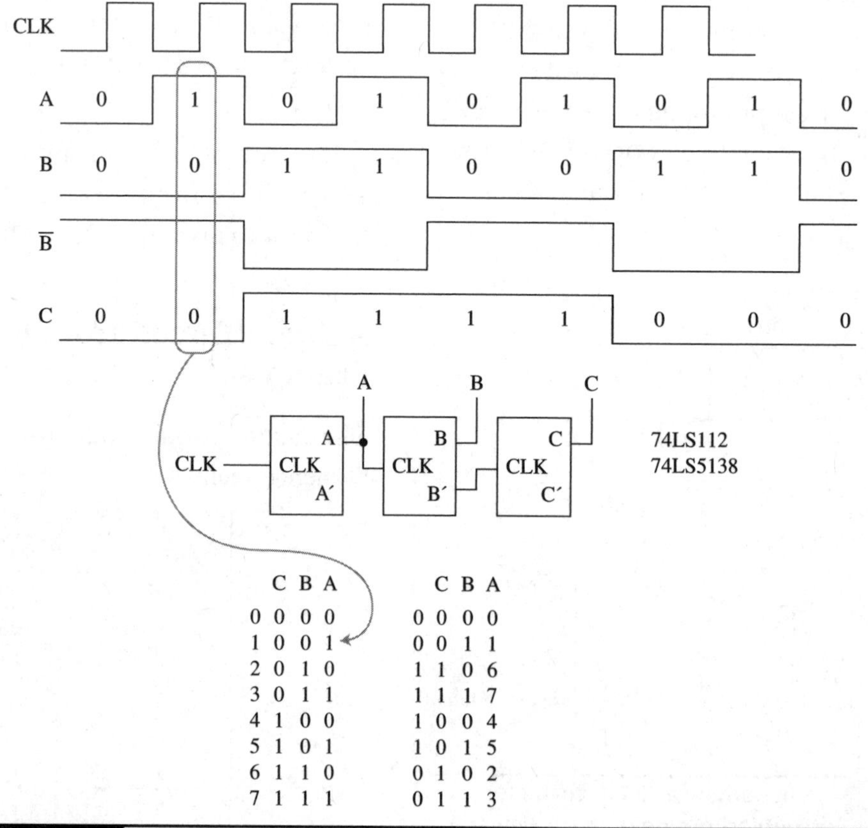

Figure B15.6 Drawing the output waveforms of each circuit illustrates which circuits are not working.

APPENDIX C

Transistor Amplifier Design

Discrete component transistor amplifiers are not used as much today as they were in the past, but every now and then you may need one. Here are two procedures to use if you ever need them. One is for a common emitter amplifier, and the other is for an emitter follower. This is definitely a "retro" subject, so skip this section if you want. However, the process lets you see the basic Ohm's and Kirchhoff's laws in action.

Common Emitter Amplifier

The amplifier to be designed is shown in Fig. C.1.

1. Identify the dc supply voltage (V_{CC}). This will determine the maximum output voltage variation. A good choice is 12 V. An alternative is 9 V. Don't go any lower than 5 or 6 V. A proven approach is to make the voltage at the collector with respect to ground one-half of V_{CC}. This usually results in the maximum undistorted output signal swing. An alternate starting point would be to state the desired output voltage swing then choose an appropriate dc supply voltage.
2. Choose a transistor. These basic amplifiers are for small signal levels, so no high-power special transistors are needed. Use something common and available like the 2N3904 or 2N2222 or one of the others given in the component recommendations. Almost any will work here. Let's select the 2N3904. You may need the data sheet if you do not already have it.
3. Choose a collector resistor value (R_C). This will also be your approximate output impedance. $Z_O = R_C$. Keep it low in comparison to the load but not so low that it draws considerable current. Something in the 500-Ω to 10-kΩ range is OK. Let's select 2.2 kΩ.
4. Identify the load resistance. It may be the input impedance of a following stage. Use 5 kΩ for this example.
5. Calculate the collector current I_C. The general design goal is to have the voltage at the collector one-half of the supply voltage. Therefore, the collector voltage to ground is $12/2 = 6$ V. For that reason, the voltage across R_C is $12 - 6 = 6$ V.
6. Knowing the value of R_C and the voltage across it, the collector current I_C can be found.

   $I_C = 6/R_C = 6/2200 = 0.0027$ $\quad I_C = 2.7$ mA
7. Determine I_E. The emitter current I_E is equal to I_C, so $I_E = 2.7$ mA as well.
8. Determine the value of the emitter resistor (R_E). To do this, we need to know the desired voltage gain (A). Let's choose a gain of 5. The gain (A) of the amplifier we are designing is approximately

   $$A = R_C/R_E$$

   Therefore,

   $$R_E = R_C/A = 2.2\text{ k}/5 = 440\ \Omega.$$

   Use the common resistor value of Ω.
9. The actual gain of the circuit will be lower than 5 when the load is connected.

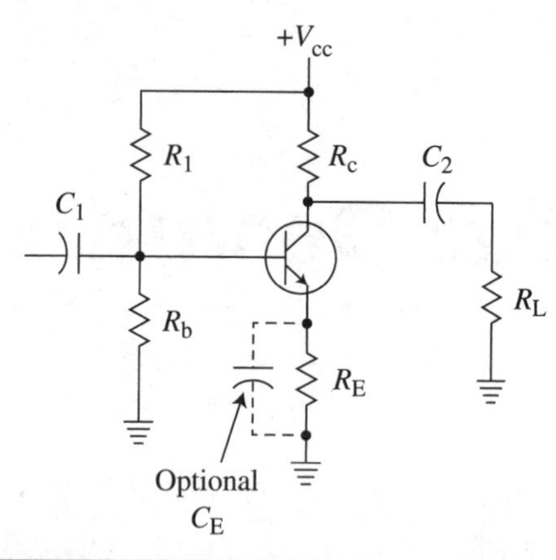

Figure C.1 A common emitter transistor amplifier showing bias and output load.

The actual gain is $A = (R_C||R_L)/R_E$. Note: $R_C||R_L$ means R_C in parallel with R_L or

$$R_C(R_L)/(R_C + R_L)$$

$$R_C||R_L = 2.2\text{ k}(5\text{ k})/(2.2\text{ k} + 5\text{ k})$$

$$= 11/7.2 = 1.57\text{ k}$$

10. Now calculate $V_E \cdot V_E = I_E(R_E) = 0.0027(470) =$ 1.28 V.

11. Next, calculate $V_B \cdot V_B$ is the voltage from the base to ground and equal to the voltage across R_E plus the transistor emitter-base voltage drop that is usually about 0.7 V.

$$V_B = V_{BE} + V_E = 0.7 + 1.28 = 1.98\text{ V}.$$

We will use 2 V.

12. Determine the base resistor (R_B) value. R_B will essentially be the input impedance to the amplifier. Make it as high as you can. One recommendation is to make it 5 to 20 times R_E. Let's assume 20 times or

$$R_B = 20R_E = 20(470) = 9400\ \Omega.$$

Choose 10 kΩ.

13. Now, determine the value of the bias resistor R_1. R_1 and R_B form a voltage divider with the supply voltage as the input. The load is the transistor base current. This is very small load, and we can ignore it. One direct way is to use the following expression that is derived from the basic voltage divider equation:

14. $R_1 = R_B(V_{CC} - V_B)/V_B$

$$R_1 = R_B(V_{CC} - V_B)/V_B = 10{,}000(12 - 2)/2$$

$$= 50{,}000\ \Omega$$

A standard 51-k resistor should work fine.

The purpose of the bias setting is to adjust the dc output at the collector to about one-half of the supply voltage. The calculations should get you close. Feel free to change R_1 to get your amplifier output to that level.

15. You can now calculate the input impedance to the amplifier. It is essentially R_B and R_1 in parallel, written as $R_B||R_1$, and that equivalent in parallel with the base input impedance. The base input impedance is $Z_B = (R_E + e'_r)\ h_{FE.}$ The term e'_r is the resistance between base and emitter. Its value is calculated with the expression $e'_r = 25\text{ mA}/I_C$. For this design, I_C is 2.7 mA so $e'_r = 25\text{ mA}/I_C = 25/2.7 = 9.26\ \Omega$. The value of $h_{FE.}$ is the transistor gain or β. Use the minimum value from the 2N3904 data sheet. Use 70. So

$$Z_B = (R_E + e'_r)h_{FE} = (470 + 9.26)70$$

$$= 479(70) = 33{,}548\ \Omega = 33.55\text{ k}.$$

$$R_B||R_1 = 51(10)/(51 + 10) = 8.36\text{ k}\Omega$$

Now Z_B is in parallel with that value to get Z_{IN}.

$$Z_{IN} = R_B||R_1||Z_B = 8.36(33.55)/(8.36 + 33.55)$$

$$= 280.47/41.91 = 6.69\text{ k}\Omega$$

16. Referring back to Fig. C.1, you can see that this amplifier is ac coupled. The input signal is through C_1 that blocks the dc on the base. The output is through C_2 to the load, and it blocks the dc voltage at the collector. These capacitors determine the low-frequency response of the amplifier. Here is how to calculate the capacitor values.

$$C_1 = 1/2\pi f_{LOW}R_B$$

$$C_2 = 1/2\pi f_{LOW}R_L$$

How low in frequency will the amplifier be required to accommodate? For audio, you could go as low as 20 Hz, but a more common response would be about 300 Hz. For this example, use $f_{LOW} = 300$ Hz.

$C_1 = 1/6.28(300)(10{,}000) = 0.053\ \mu F$

$C_2 = 1/6.28(300)(5000) = 0.106\ \mu F$

Standard capacitor values are 0.05 μF and 0.1 μF and so should work, but the lowest frequency would be higher, or 318 Hz. This is not usually a critical difference. If it is a factor, you can make the capacitor larger.

17. If you need more gain, you can put a bypass capacitor across R_E. The reactance of that capacitor X_C has to be less than one-tenth of R_E at the lowest frequency of operation. In this design, that is less than 470/10 = 47 Ω. Assume the lowest frequency is the 300 Hz used earlier.

Since $X_C = 2\pi fC$, we can rearrange the formula to solve for C.

$C = 1/2\pi fX_C = 1/6.28(300)(47) = 1.13\ \mu F$.

Use the next larger size, usually 2.2 or 3.3 μF.

With the emitter capacitor in place, the gain is going to be much higher. It will be dependent upon the transistor internal emitter resistance e_r' divided into the collector resistance.

$A = r_C/e_r'$

$e_r' = 25\ mA/I_C = 25/2.7 = 9.26\ \Omega$

The value of r_C is

$$RC||RL$$

and that value as determined earlier is 1.53-k. Theoretically then the amplifier gain will be

$$A = r_C/e_r' = 1530/9.26 = 165$$

Testing

It is always a good idea to breadboard your circuit and see how close it is to your design. Build the circuit, then do this after you apply 12 V:

a. Measure the dc voltage at the collector to ground. It should be about one-half—12 or 6 V or close to it.

b. Measure the dc voltage at the base to ground. It should be about 2 V.

c. Apply a 1 kHz sine wave to the input. Keep the input voltage low; otherwise, you will drive the transistor into saturation, and the output will be a rectangular wave rather than a sine wave. While observing the output at the collector on your oscilloscope, increase the input voltage until the output just begins to clip or flatten. At the point when clipping occurs, measure the peak-to-peak output voltage. Also measure the peak-to-peak input voltage. Calculate gain as

$$A = V_O/V_{IN}$$

Do all of this with and without the 5-k load.

Note: One problem you may have is keeping the input voltage low enough to avoid clipping. One way to do this is to put a voltage divider on the output of your function generator or other signal source. Refer to *Fig. C.2* for a couple of suggestions.

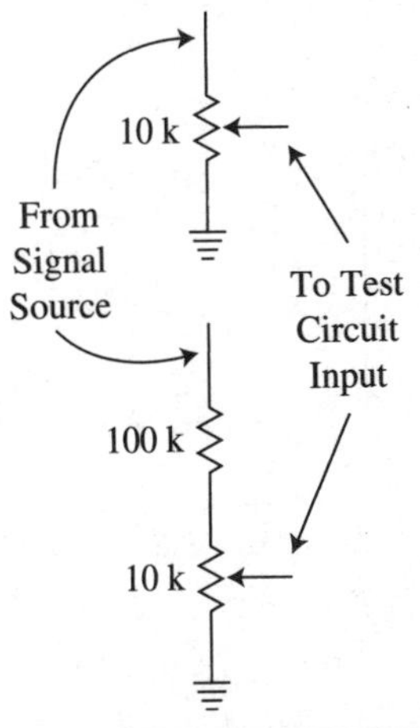

Figure C.2 Using a voltage divider and pot to reduce the amplitude of the input signal to the amplifier to prevent distortion and clipping.

d. Now put an emitter bypass capacitor across R_E. Then measure the gain with and without the load resistor.

Designing an Emitter Follower

A follower is a transistor amplifier that has a high input impedance, low output impedance, and a voltage gain of one. It doesn't appear that it would be useful, but it is. Its primary use is to act as a buffer between other circuits to eliminate or at least minimize the loading effect of one circuit on another. It can provide a low output impedance for another amplifier while its high input impedance will have little loading effect on the driving circuit.

Figure C.3*a* shows an emitter follower. It can be biased with a voltage divider like the transistor amplifier. The output at the emitter is set to roughly one-half the supply voltage. The voltage divider biasing resistors lower the overall input impedance. Not a good thing. The circuit is best used by connecting the base directly to the output of any previous amplifier as Fig. C.3*b* shows. The dc voltage at the collector of the amplifier biases the emitter follower. This collector voltage is typically one-half the V_{CC}.

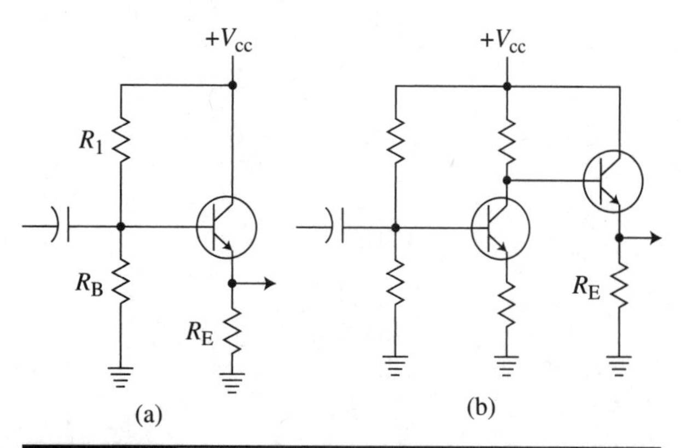

Figure C.3 An emitter follower (a) and an emitter follower biased by the transistor amplifier output (b).

That makes the voltage across the emitter resistor R_E equal to the dc input less the emitter-base voltage drop of 0.7 V.

The following is a simplified design procedure for this circuit:

1. Select a transistor. Let's continue to use the 2N3904. We need to know the transistor gain designated as h_{FE} in the data sheet. We also use the symbol β to represent gain. Typically, $\beta = h_{FE}$. Let's use the minimum of 70 shown in most data sheets.
2. Establish the supply voltage. It should be the same as the V_{CC} for the driving stage. Assume 12 V. Its collector voltage is about one-half that or 6 V.
3. Determine V_E. If you are making the direct connection between the base and the previous collector, V_E should be about half of V_{CC} or a little less. This establishes the maximum output voltage swing limits.
4. $V_E = V_{CC}/2 - V_{BE} = 12/2 - 0.7 = 6 - 0.7 = 5.3$ V.
5. Choose a collector current. It is usually something between 1 and 10 mA. Let's choose 2 mA. Remember $I_E = I_C$.
6. Calculate the emitter resistor R_E.

   $R_E = V_E/I_E = 5.3/0.002 = 2650\ \Omega$. This is not a critical value, so you can use standard values between 2200 and 2700 Ω. Choose 2200 or 2.2 k.
7. Calculate the input impedance (Z_{IN}).

   $$Z_{IN} = \beta R_E$$

   Using the value of β as 70 and R_E as 2.2 k,

   $$Z_{IN} = 70(2.2\text{ k}) = 154\text{ k}\Omega.$$
8. Calculate the output impedance Z_O. You may or may not need to know it, but an approximation is

   $$Z_O = Z_{OUT}/(\beta + 1)$$

   Z_{OUT} here is the output impedance of the driving stage. That is the same as the value of the collector resistor of the previous stage. Assume it is 5 kΩ.

$$Z_O = Z_{OUT}/(\beta + 1) = 5000/(70 + 1) = 70\ \Omega$$

That completes the design. You can also bias the follower with a voltage divider as indicated earlier and the procedure used with the single transistor amplifier should work. It is not recommended because the input impedance is lower and the circuit more complex. Go for the direct connection if you can.

APPENDIX D

How to Use Karnaugh Maps

Supplement to Chapter 11 and Related Solutions

Karnaugh maps, or K-maps, are a graphical technique for minimizing digital logic circuits. They are no longer widely used, but they are technically interesting. Most digital circuits are implemented with microcontrollers or programmable logic these days, so there is little need to simplify a complex logic circuit. Simulation and logic synthesis software has also taken away the need for a minimization step. However, if you are still designing with basic TTL or CMOS logic gates, flip flops, and functional ICs, using K-maps can lead to some savings. If you are curious about K-maps, this appendix will give you an introduction and a few design examples. Only "real" digital designers use K-maps.

The Design Projects 11.1 and 11.2 in Chapter 11 are used to illustrate the technique. These are the solutions.

You can use Karnaugh maps to minimize the circuits. It is often possible to reduce the number of circuits used and chips required to save money and PC board space. Minimization may also reduce the propagation delay through the circuit if very high speeds are required. Some designs result in circuits that do not lead to a minimization that uses less circuits, yet it is worth a try because the savings could also be great.

Design Project 11.1

The assignment was to design a logic circuit that lights an LED whenever the input is the binary expression for a prime number. Identify the primes between 0 and 15. This calls for a 4-bit input because with 4-bits you can represent 2^4 = 16 states. Remember that a prime number is one that is only divisible by itself and one. Note: 0 and 1 are not primes. Now do the rest yourself. Don't look at the following results until you are finished.

Here is the solution.

1. Identify the prime numbers from 0 to 15. The prime numbers from 1 to 50 are 2, 3, 5, 7, 11, 13, 17, 19, 23, 29, 31, 37, 41, 43, and 47.
2. For values between 0 to 15, we need 16 states so we can use 4-bits (2^4 = 16). Name the inputs A, B, C, and D and the output P.
3. See the following truth table. Note that when the 4-bit binary code for a prime number occurs, the output P is a binary 1.
4. Write the boolean equations for the output using the SOP. Simplify.

In the left column is the decimal value of each standard binary count sequence, 0 to 15 (1111) given in the second column.

You can write the boolean expression for this circuit from the truth table. There will be six terms, one for each of the inputs identified by

Table 11.1 Truth Table for Prime Number Identifier Circuit

	Inputs	Output
Decimal	A B C D	P
0	0 0 0 0	
1	0 0 0 1	
2	0 0 1 0	1
3	0 0 1 1	1
4	0 1 0 0	
5	0 1 0 1	1
6	0 1 1 0	
7	0 1 1 1	1
8	1 0 0 0	
9	1 0 0 1	
10	1 0 1 0	
11	1 0 1 1	1
12	1 1 0 0	
13	1 1 0 1	1
14	1 1 1 0	
15	1 1 1 1	

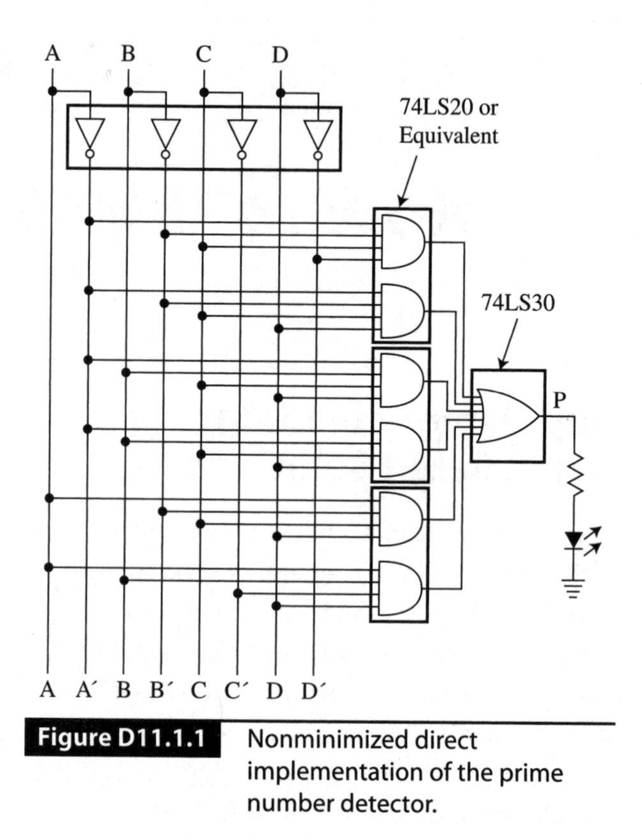

Figure D11.1.1 Nonminimized direct implementation of the prime number detector.

the binary 1 outputs. These will be in the sum-of-products format with multiple AND expressions ORed together. For each output where there is a binary 1 output, write an AND expression using the A, B, C, and D inputs.

Here are the boolean equations. For a binary 1, write the input name or letter. For a 0, write the input name or letter with an apostrophe (NOT).

$$P = A'B'CD' + A'B'CD + A'BC'D + A'BCD + AB'CD + ABC'D$$

Once you have the boolean equations, you can implement the circuit directly. The four inputs A, B ,C, D are each inverted to provide the complements A′, B′, C′, D′. Each state requires a 4-input AND gate. These product expressions are then ORed together in an OR gate. Lots of circuits are needed. With available logic ICs, it would take five chips. These are outlined in Fig. D11.1.1.

How to Use a Karnaugh Map

Figure D11.1.2 shows a Karnaugh map or K-map that is used to minimize the number of circuits needed. The map shows all possible 16 states of the three inputs A, B, C, and D. The decimal equivalent of each 4-bit expression is shown to help simplify transferring your boolean equation to the map. Take a moment to relate the K-map to the truth table.

AB \ CD	00	01	11	10
00	0 A′B′C′D′	1 A′B′C′D	3 A′B′CD	2 A′B′CD′
01	4 A′BC′D′	5 A′BC′D	7 A′BCD	6 A′BCD′
11	12 ABC′D′	13 ABC′D	15 ABCD	14 ABCD′
10	8 AB′C′D′	9 AB′C′D	11 AB′CD	10 AB′CD′

Figure D11.1.2 A 4-variable, 16-state K-map.

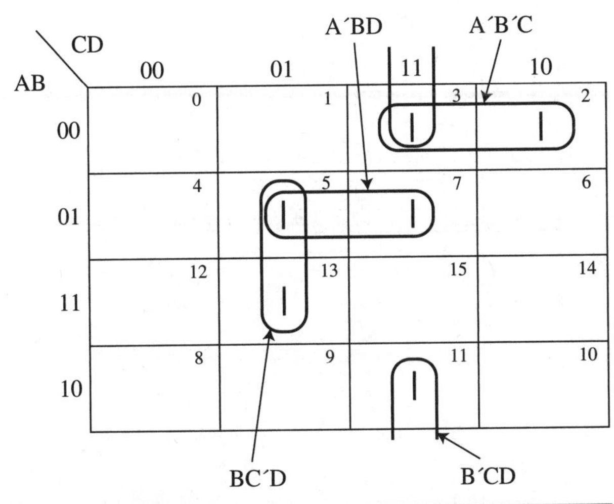

Figure D11.1.3 The prime number detector terms plotted on the K-map, showing minimization loops.

Figure D11.1.3 shows the boolean expressions plotted in the map. Those states defined by the equations are marked with a 1. The loops around the states with 1s in them lead to a simplification. You can make loops around two or four adjacent groups of 1s. A group of two will produce a three-input term, while a group of four will allow a group of two inputs. Note: squares 3 and 11 are linked with a loop as opposite edges (top and bottom or left and right) are assumed to be connected.

The boolean expression for a two-term loop is made by observing which input does not change in moving from one square to the adjacent within the loop. Try translating the loops to the simplified terms then compare to the following expression.

$$P = A'B'C + BC'D + A'BD + B'CD$$

With four inputs, the procedure is the same. A group of four can be formed with a loop. Again, observe which input variable does not change in moving from square to square in the loop. The correct expression will be the three variables that do not change.

Draw the logic circuit from the boolean expression.

$$P = A'B'C + B'CD + A'BD + B'CD$$

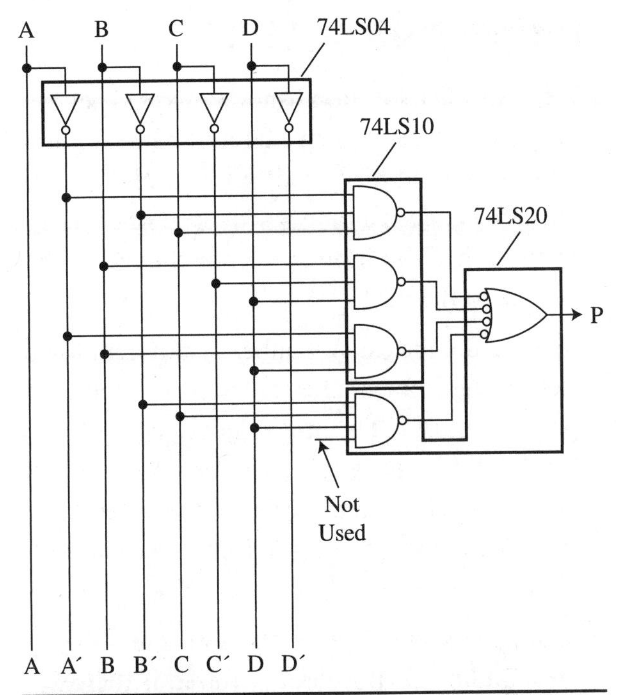

Figure D11.1.4 A minimized implementation of the prime number detector.

Convert the minimized boolean equation into logic gates connected to implement the logic. NAND gates are used. See Fig. D11.1.4. Only three instead of five chips are required, a major savings.

K-Map Confusion

K- maps are a nuisance. They are hard to understand and use, at least initially, but are learnable with persistence and experience. If the previous discussion confused you, made you stop reading, caused you temporary brain damage, or made you doubt your ability to design, there is a solution. Use some K-map software or design tools. There are multiple Web sites offering free Karnaugh map programs. Just enter your code or truth table and let the software do the map, boolean expression, and minimization for you.

Here are just a few sites that I found, but many others are out there. Search on Karnaugh map tools, Karnaugh map software, or Karnaugh map programs.

- Electronics-course.com
- www.charlie-coleman.com
- www.32x8.com
- Logicminimizer.com

Design Project 11.2

Here is another solution using K-maps. The assignment was to design a circuit that translates a 3-bit standard binary code into a 3-bit Gray code. Gray code is a binary sequence where only one bit in the 3-bit input sequence changes from one state to the next.

Determine the exact number of different logic states. Assign some names or letters to each input and output. The input signals will be a 3-bit standard binary code with bits A, B, and C. With three bits in, there are $2^3 = 8$ different states. The Gray code output signals we will call D, E, and F.

List all possible cases of the inputs and the corresponding outputs. For example, if there are three inputs, there will be $2^3 = 8$ possible input combinations. You define what you want the outputs to be for each input case. To get you started, the following truth table is shown.

Inputs	Outputs
Binary Code	Gray Code
A B C	D E F
0 0 0	0 0 0
0 0 1	0 0 1
0 1 0	0 1 1
0 1 1	0 1 0
1 0 0	1 1 0
1 0 1	1 1 1
1 1 0	1 0 1
1 1 1	1 0 0

Your job is to create the logic circuit for each of the outputs in the truth table (D, E, F).

Here is the solution. Write the boolean expression for each of the D, E, and F outputs. Look at each output vertical column and write the ABC expression where 1s appear.

$$D = AB'C' + AB'C + ABC' + ABC$$

$$E = A'BC' + A'BC + AB'C' + AB'C$$

$$F = A'B'C + A'BC' + AB'C + ABC'$$

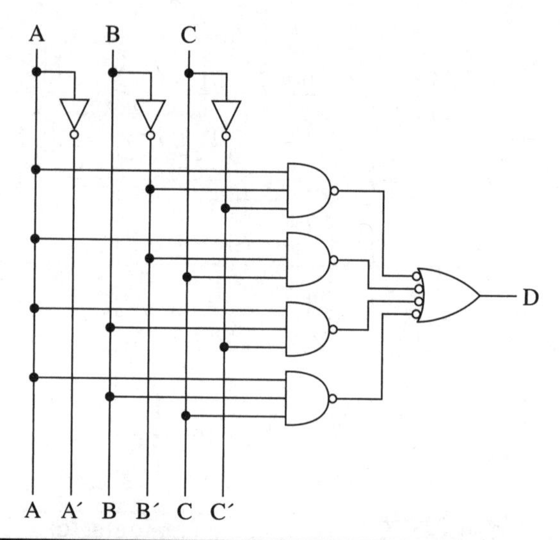

Figure D11.2.1 Direct implementation of the D output of the Gray code converter.

Figures D11.2.1, D11.2.2, and D11.2.3 show how these three equations would be implemented with standard NAND gates.

The K-maps are shown in D11.2.4, one for each output. Try to deduce the reduced expressions from the loops in the maps. A group of four translates to a single variable, in this case, A.

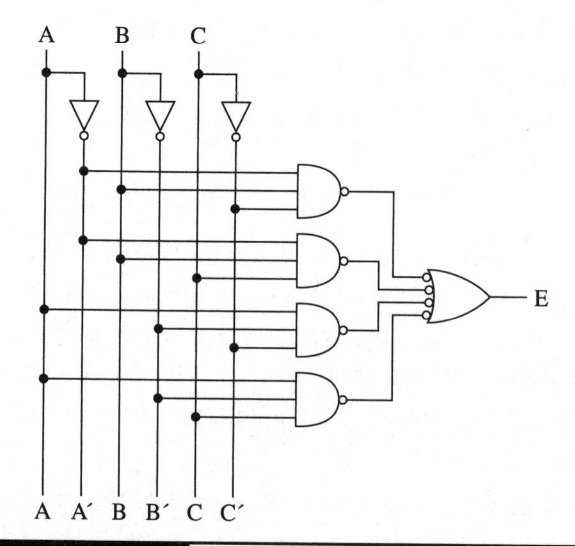

Figure D11.2.2 Direct implementation of the E output of the Gray code converter.

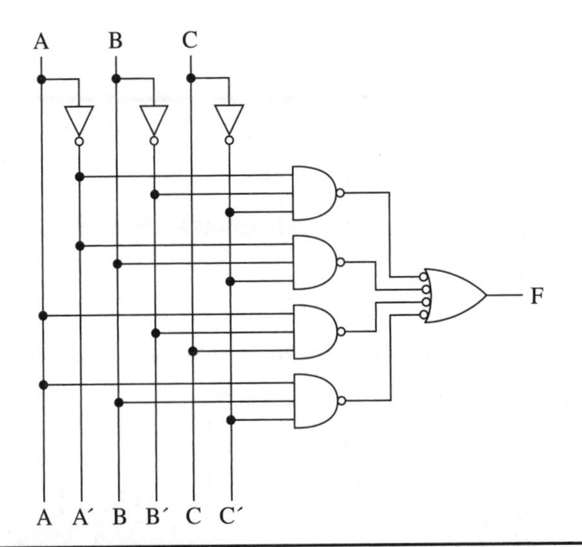

Figure D11.2.3 Direct implementation of the F output of the Gray code converter.

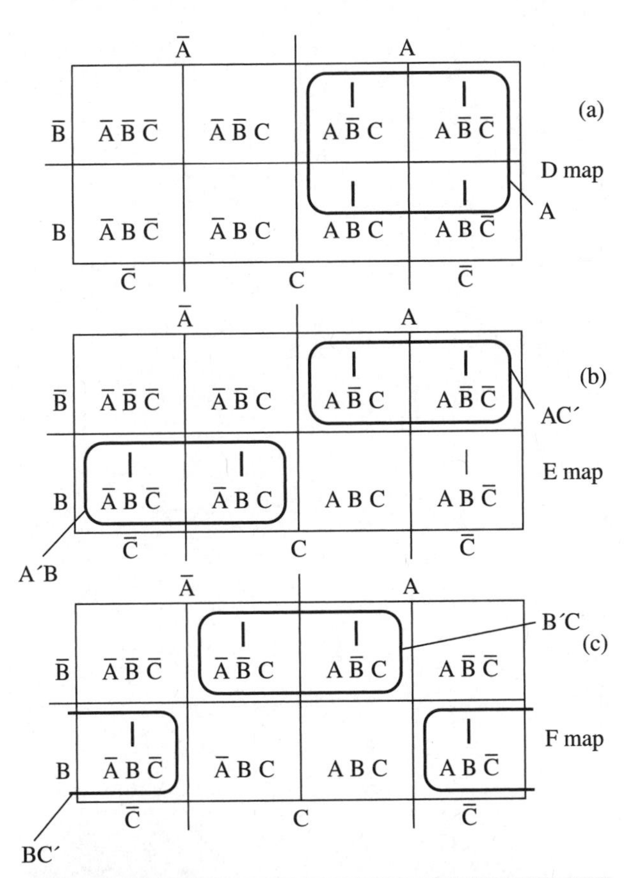

Figure D11.2.4 Karnaugh maps showing minimization of the D, E, and F circuits.

The reduced equations are

$$D = A$$

$$E = A'B + AC'$$

$$F = B'C + BC'$$

You should note that F is the DeMorgan's XOR function, so an XOR gate can be used.

See Fig. D11.3.5 for the final circuit—a significant reduction and savings.

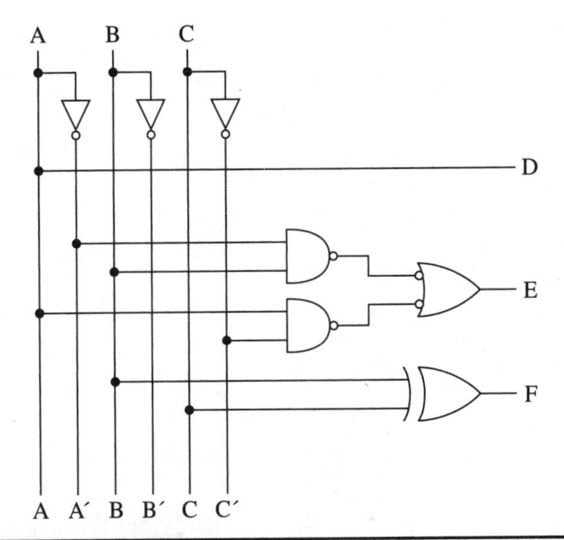

Figure D11.2.5 Circuit showing a major simplification of the Gray code converter.

Index

Note: Figures are indicated by *f*; Tables are indicated by *t*.

D

E

I

J

K

L

M

Q

R

W

X

Y